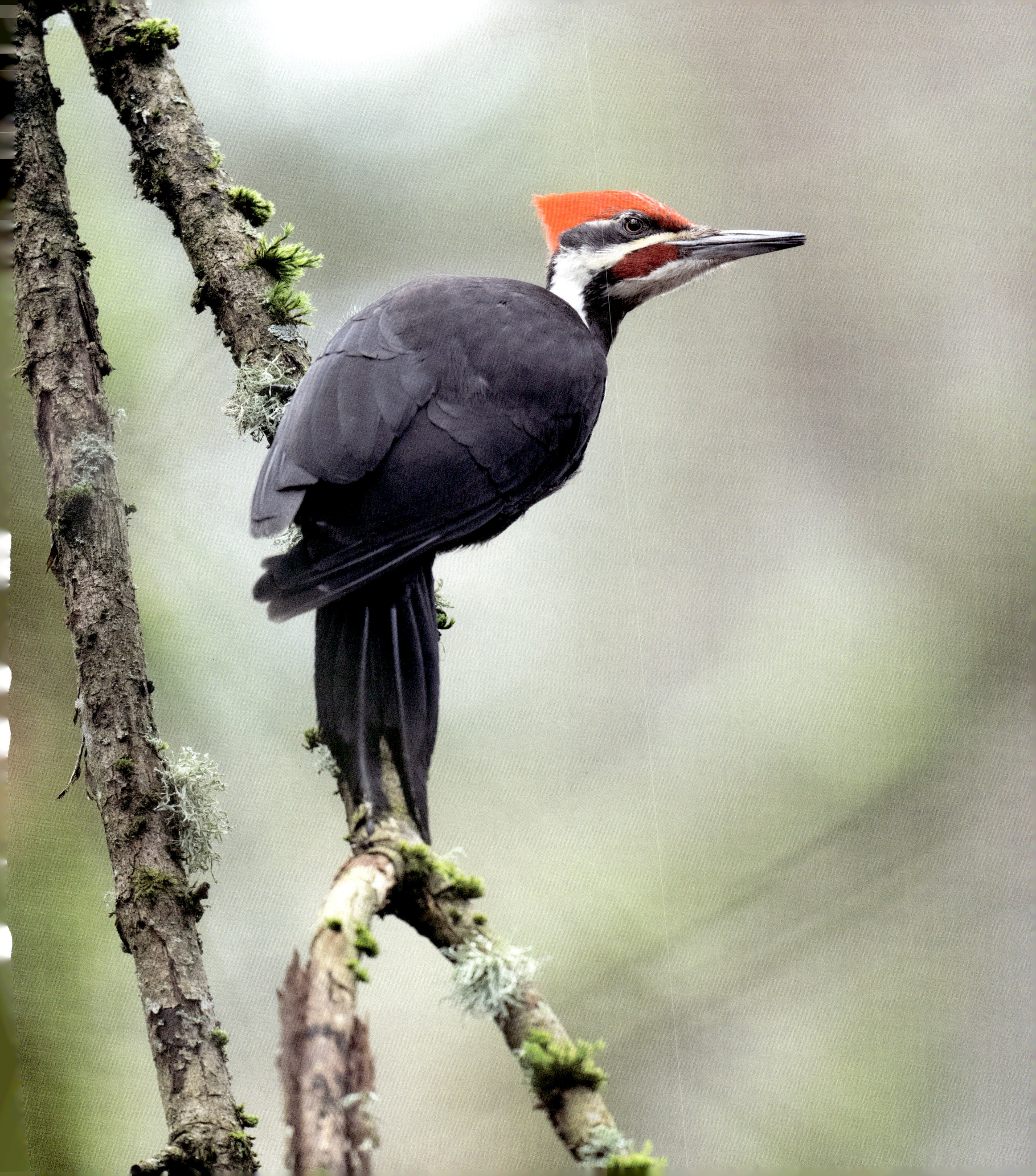

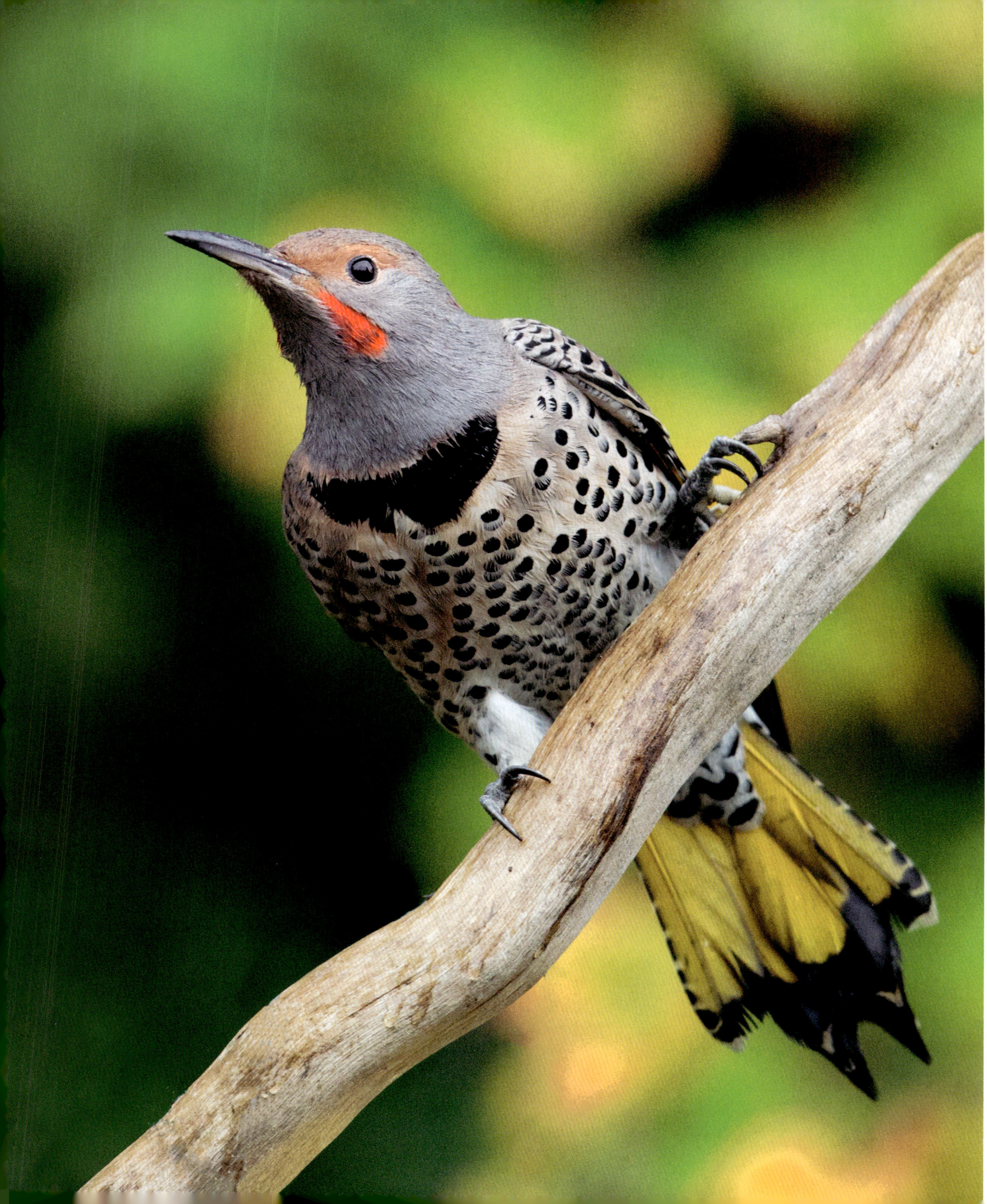

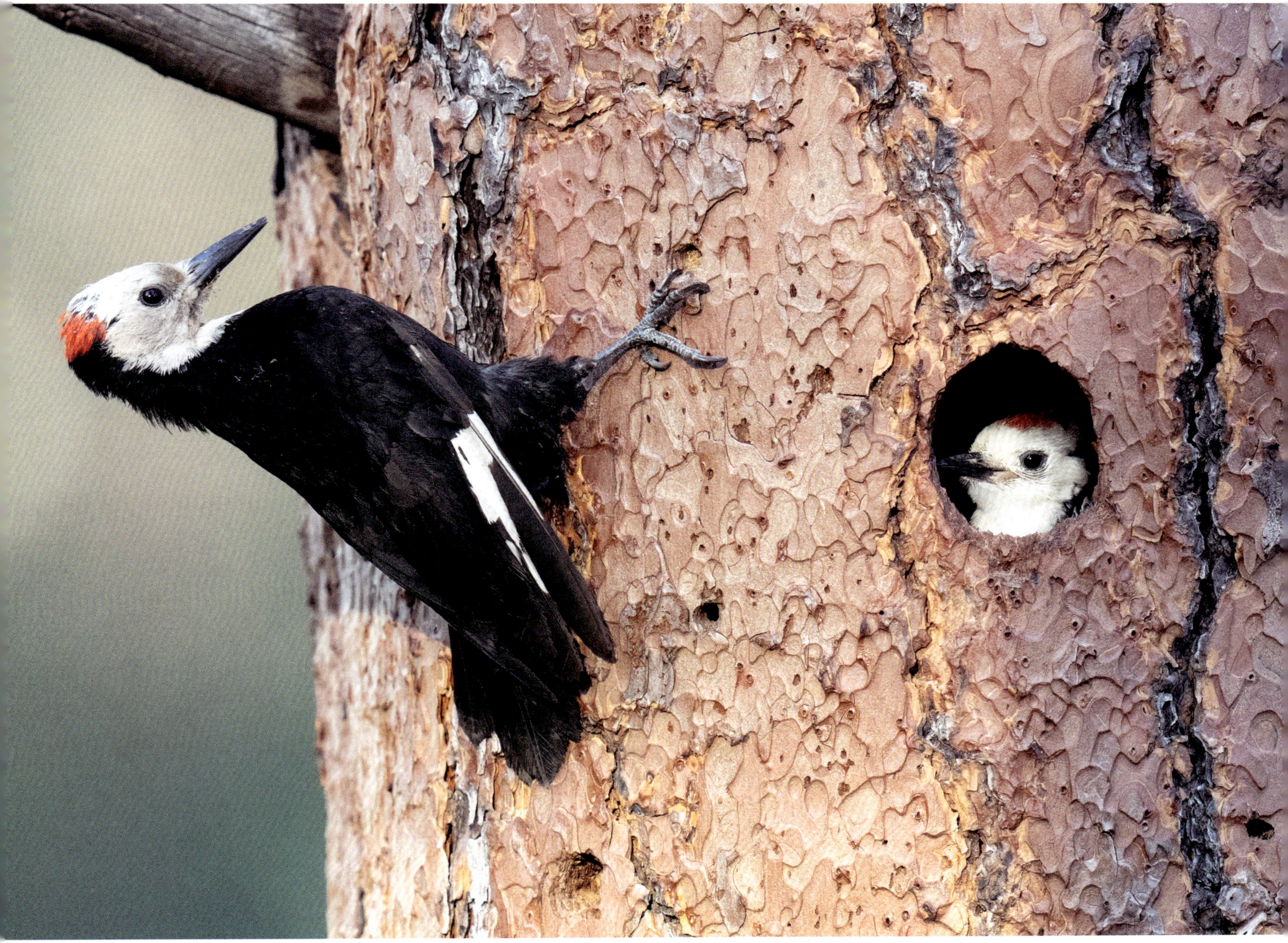

Page 1: Pileated Woodpeckers, like this male, have a broad range but are found in greater abundance in mature forests with plenty of moist, decaying wood—the ideal location for carpenter ants, their favored food.

Opposite: The Rocky Mountains separate breeding populations of Northern Flickers into two distinctive subspecies, with Yellow-shafted birds occurring east and north of this range and Red-shafted to the west. "Intergrade" flickers show a mix of traits of these two subspecies where ranges overlap. This Seattle visitor displays the red malar of the Red-shafted species and the yellow wing and tail feathers of the Yellow-shafted.

During the last week of attending to the nest, adults are more likely to land on the nest tree and maneuver toward the opening in view of the begging young, as this White-headed Woodpecker is doing, rather than flying directly to the entrance.

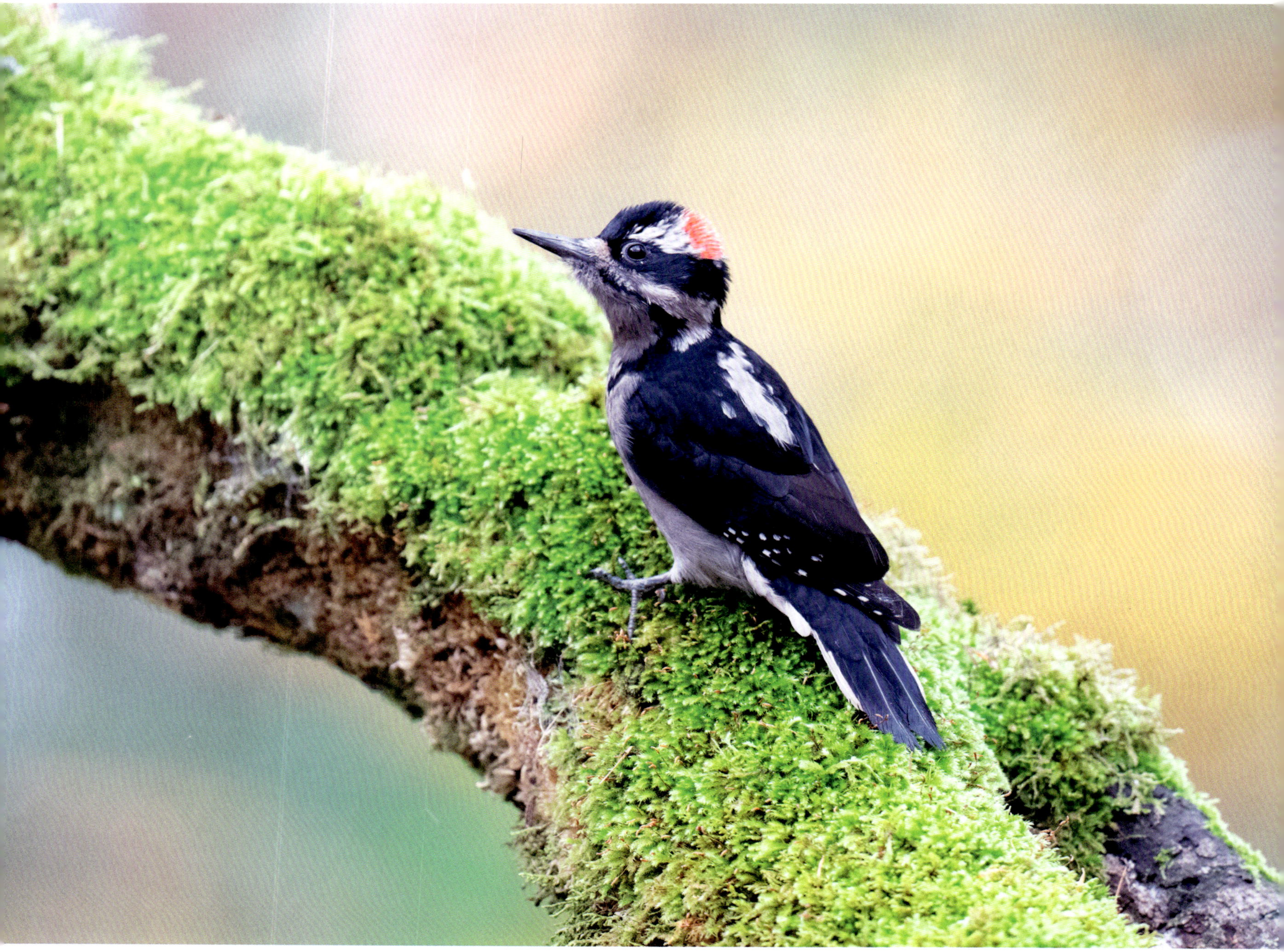

Woodpeckers show regional differences in color, markings, bill length, and overall size. Hairy Woodpeckers have fewer white spots, and males have more red, west of the Rockies with some birds from the Pacific Northwest displaying grayish rather than white feathers.

Opposite: An adult Red-headed Woodpecker is faced with a choice of whether to attempt to capture a dragonfly and risk losing the nut it holds in its bill or deliver the nut to the nestling peering from the cavity.

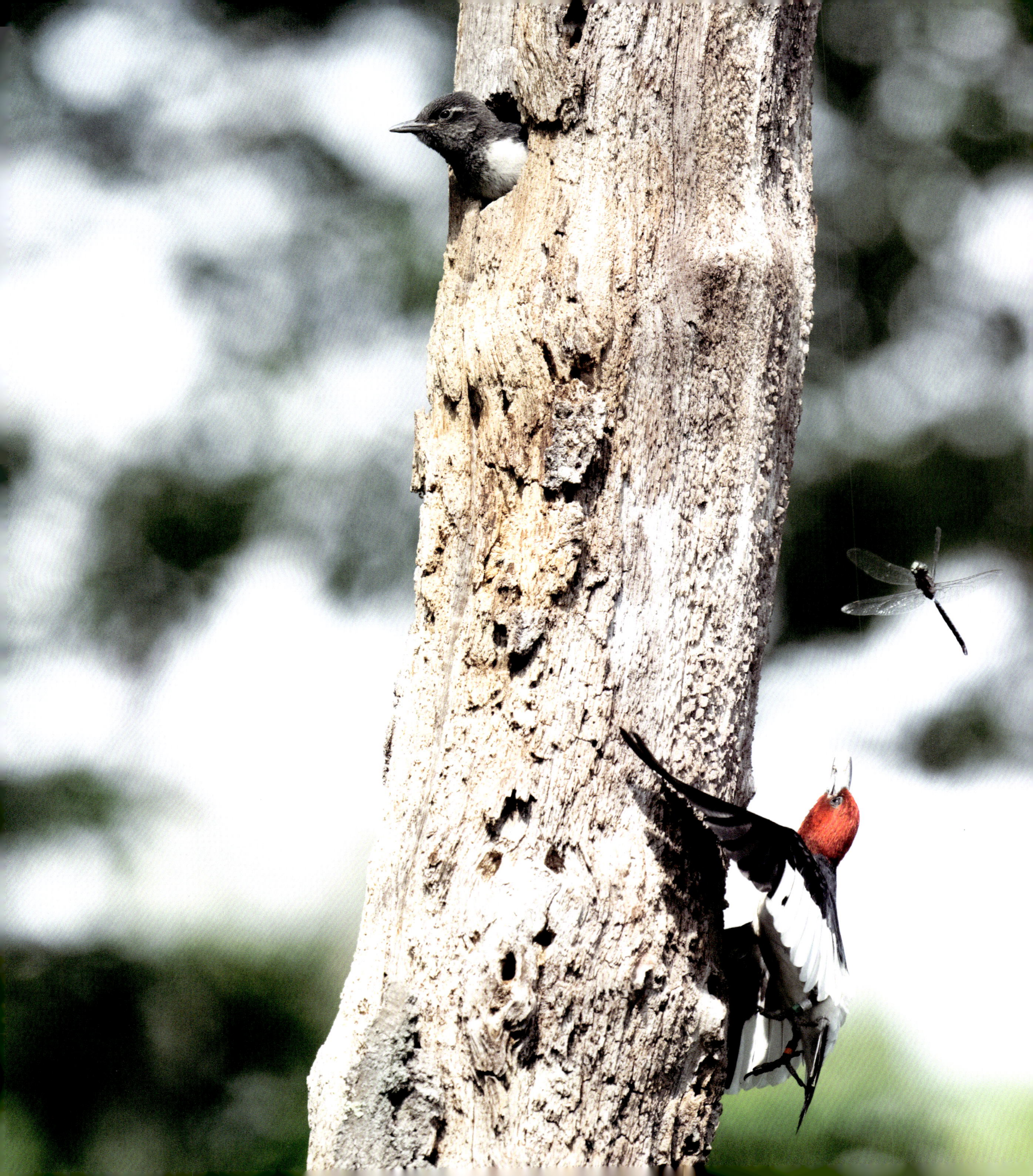

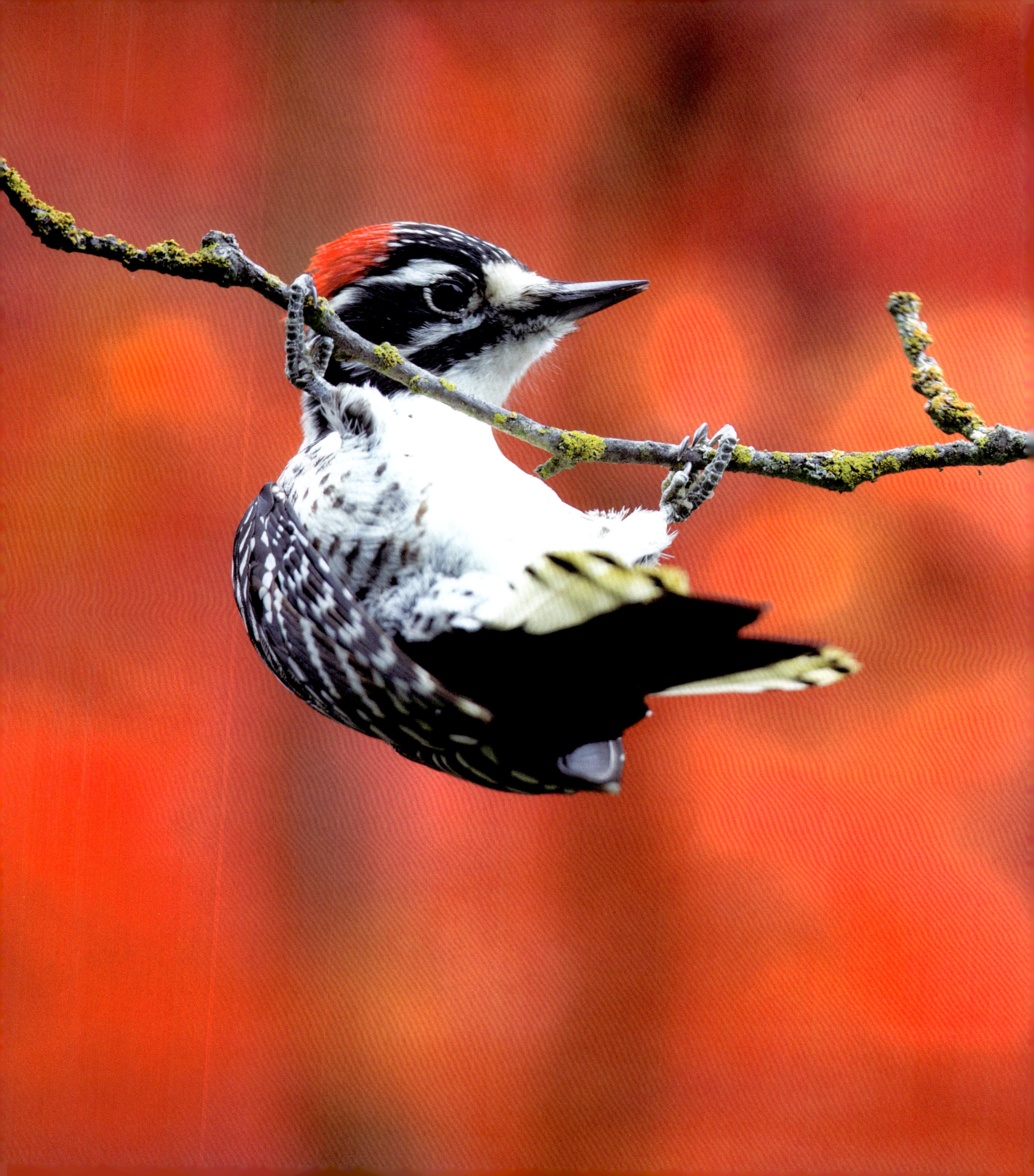

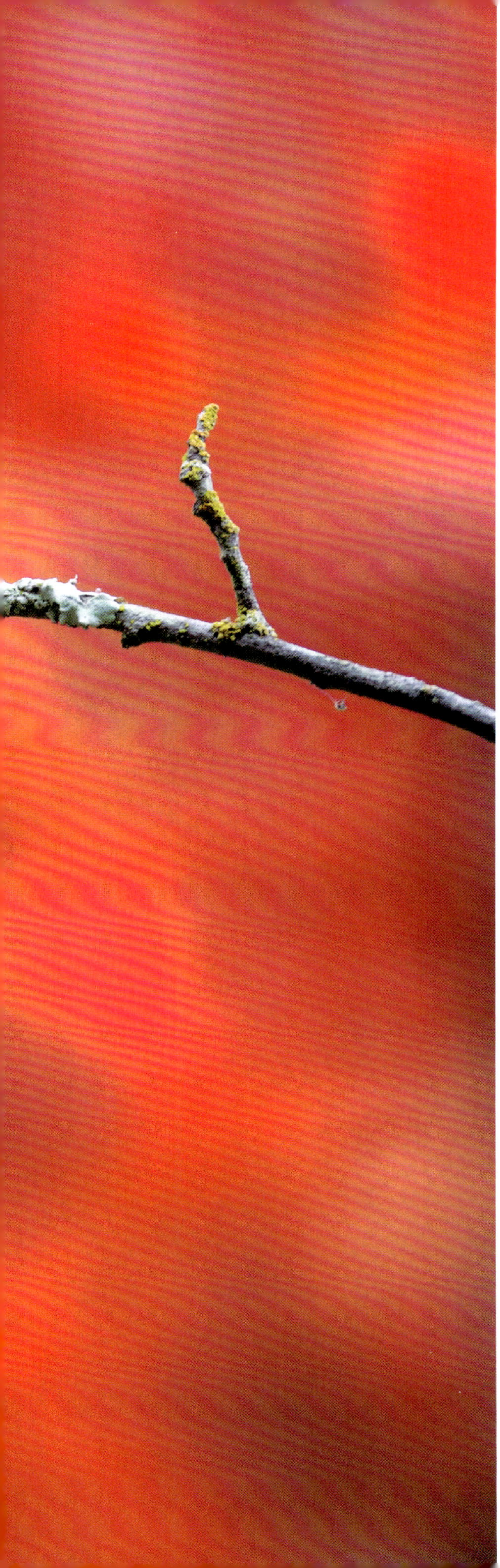

WOODPECKER

A YEAR IN THE LIFE OF
NORTH AMERICAN WOODPECKERS

PAUL BANNICK

BRAIDED RIVER

For Matthew J. Bannick,
a man of integrity who generously helps
improve the lives of others and
whom I am fortunate to call my brother

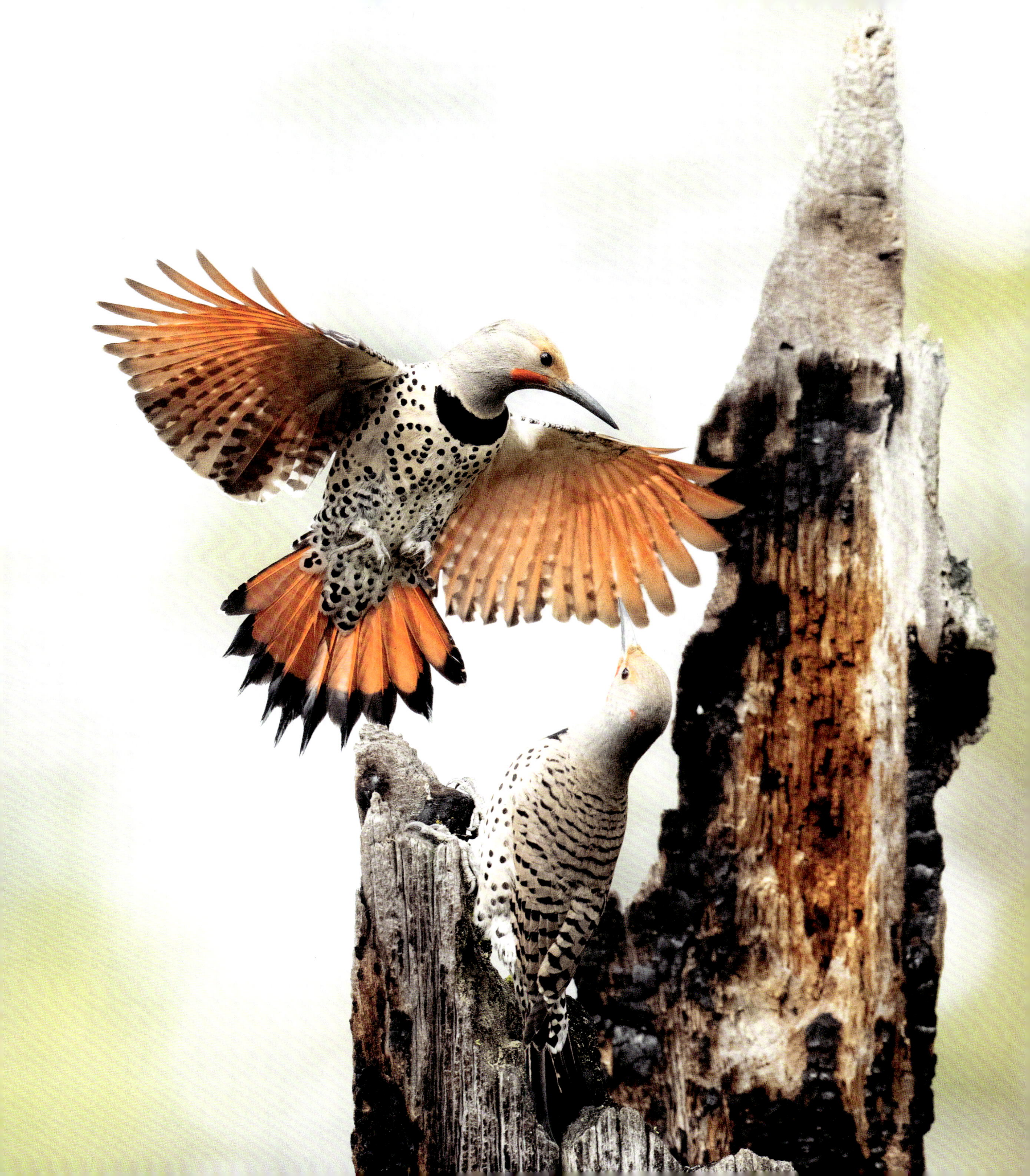

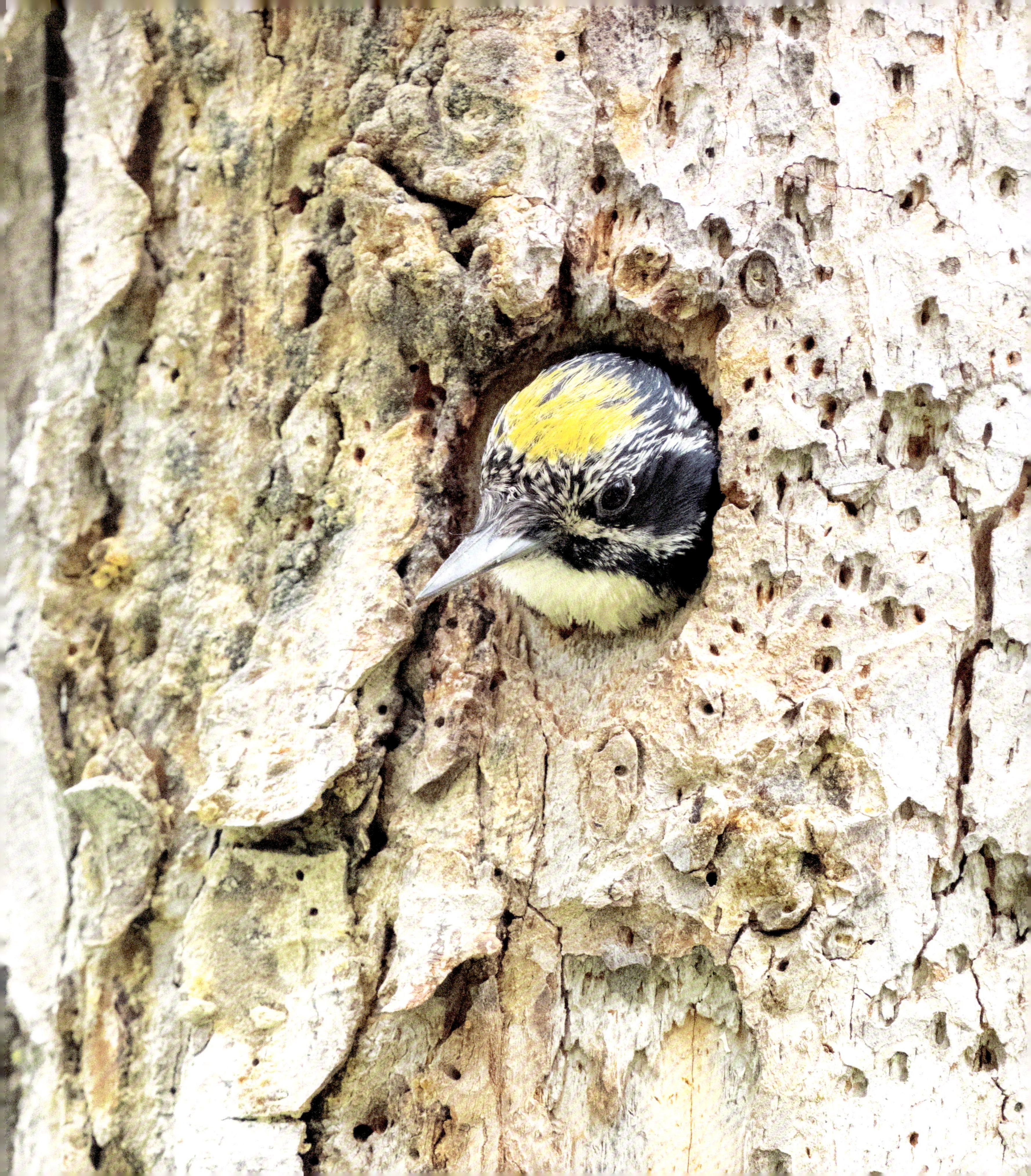

Contents

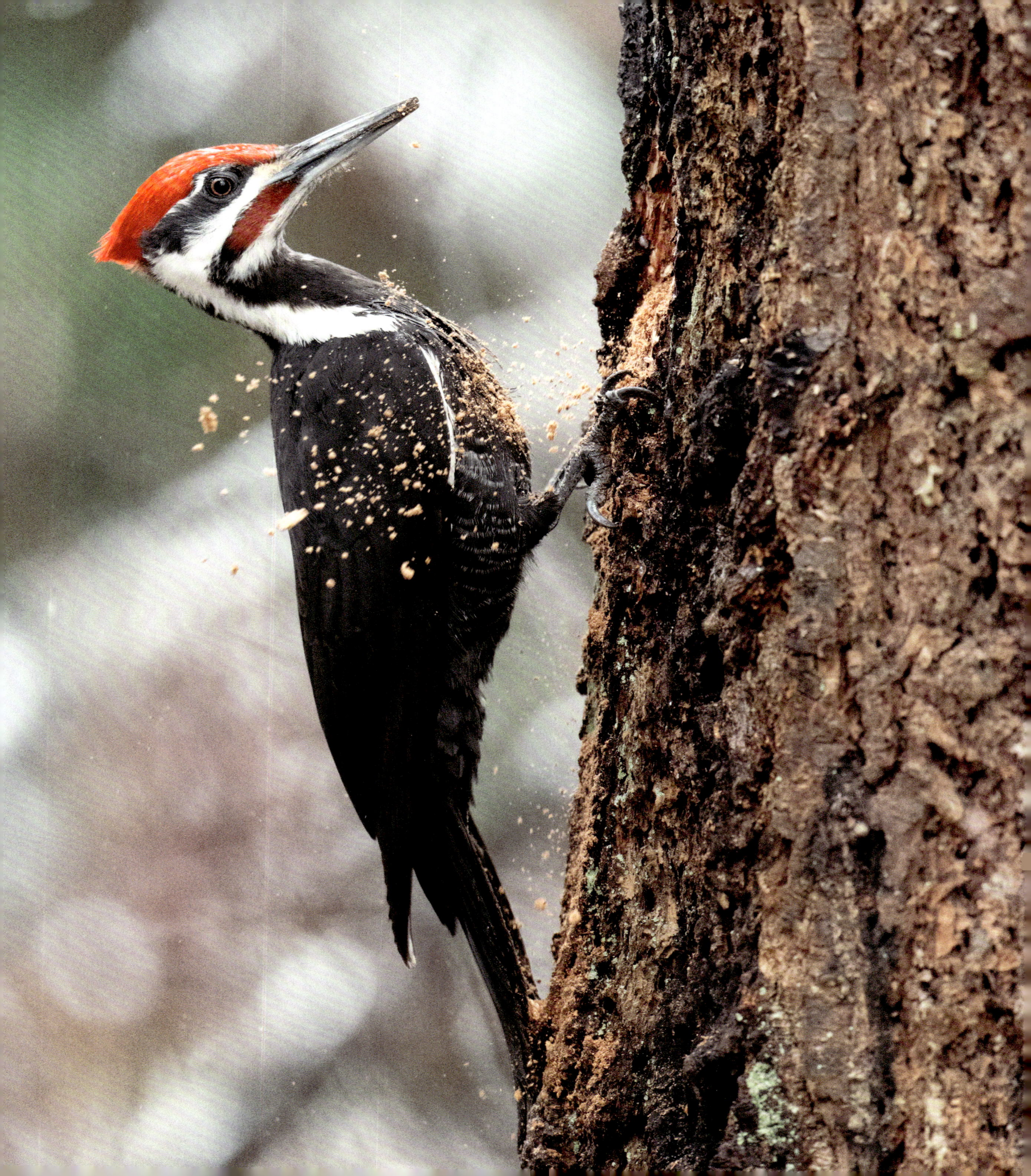

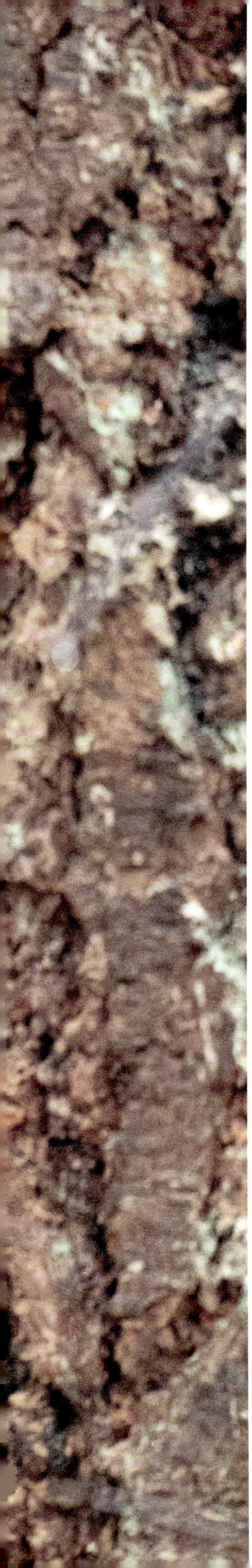

INTRODUCTION

The Heart of the Forest

In the early spring as night's darkest hour melts to dawn, you can hear the haunting calls of owls, soon followed by the coos of doves, and gradually joined and replaced by harmonious singing from secreted performers in every direction. The high-pitched melodic trills of robins below, the ascending flutelike calls of thrushes above, the accelerating whistling of warblers, and the cheerful song of wrens surround you. Each adds to a sense of peace as their exquisite songs blend as one choir, making it difficult to separate one singer from another until a primal beat from the heart of the forest adds rhythm to the mix.

Wood chips fly in response to blows from the dagger-shaped bill of a male Pileated Woodpecker. This species pursues ants by opening tree bark to expose carpenter ants, termites, and beetle larvae before lapping them up with its sticky tongue.

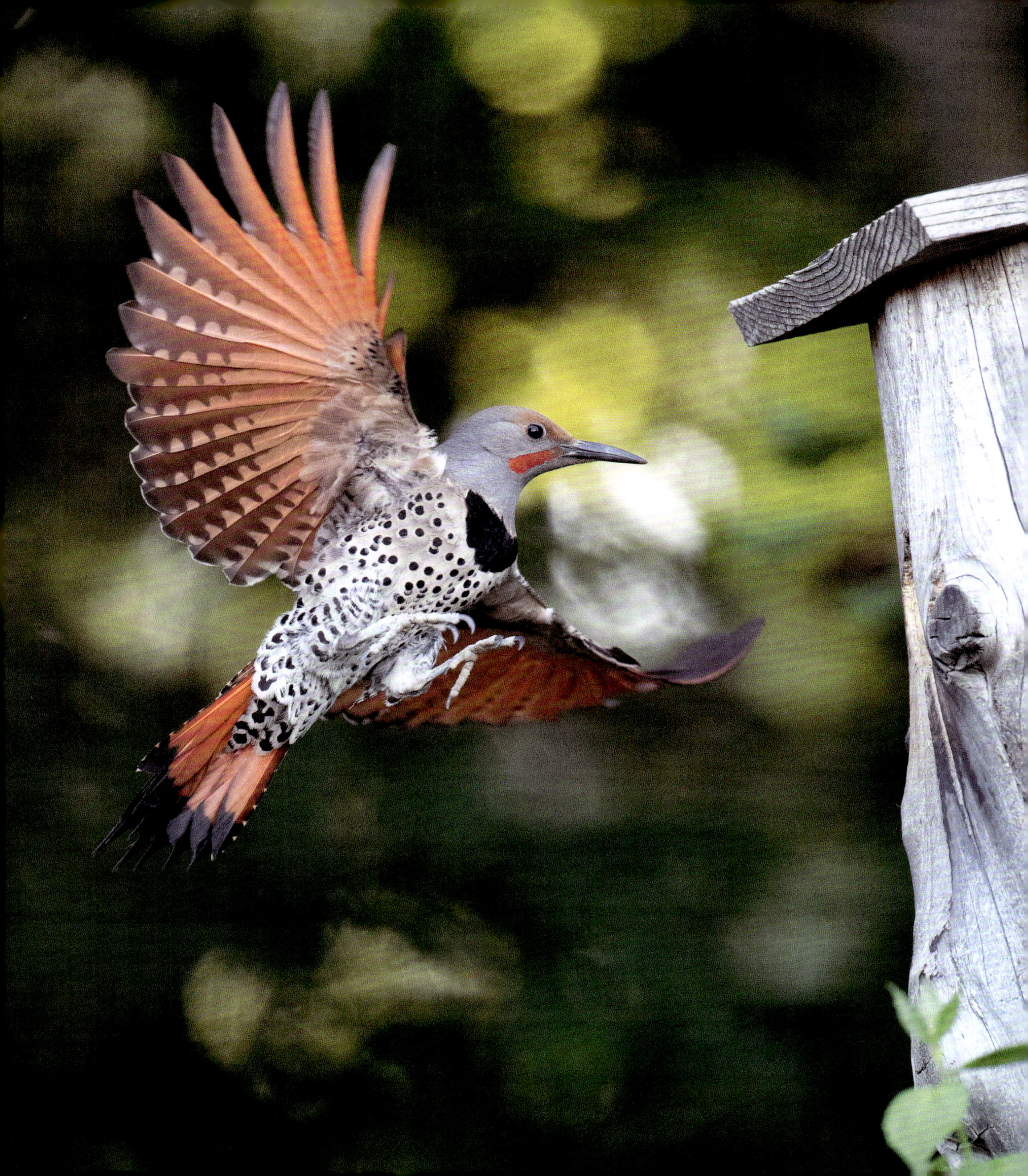

The woodpecker's pulsing cadence shifts the mood from transcendent to urgent. At moments of their choosing, these birds demand to be noticed. They raise crowns of gold or scarlet as they bob, sway, and scream, and find the most resonant objects to strike, drumming signature beats with their impressive bills.

Woodpeckers are deliberate. They drop or dart through the air and scramble up the vertical face of trees like the most skilled climbers, balancing the weight of stocky bodies between the anchor points of a stiff tail and the claws of large feet. These "tools" propel them up trees before they re-anchor. To close distances quickly while flying, they pull their wings against their body, lower their head, and free-fall as they weave between blossoms, twigs, limbs, and trunks like heavy arrows with bright feathered shafts. Intricately marked wings flash open like capes to break their fast fall and impress their rivals or mates.

Woodpeckers capture our attention with their charismatic and at times seemingly playful and whimsical ways, belying their mission as essential agents of transformation in the places they inhabit.

Every natural place is ephemeral and constantly transitioning. A favorite wildflower meadow may become a savanna as it gains isolated aspen or pine. Savannas become woodlands, which fill out as forests until fires or storms open them up again. Still, trees alone do not attest to the richness of habitat until you hear the woodpecker's drumbeat, the heartbeat of the forest. With each successive habitat change, woodpeckers enable the lives of dozens of species that can and will thrive in that specific place only with the woodpecker's handiwork. They do this first by carving cavities into trees. The wood chips hit the ground and retain water, which helps seeds germinate and amphibians survive. Finished cavities provide shelter not only for the woodpeckers themselves but also for dozens of other animals that cannot construct their own.

Woodpeckers evaluate timber with their eyes and ears like a master builder before carving signature holes of identical dimensions to ensure a mate's satisfaction. *Pajaro carpintero* is the Spanish word for woodpecker and literally translates to "carpenter bird." In China, children are taught that the woodpecker is the "tree doctor" for its role in eliminating forest pests. But this physician also senses rot in a tree's wood, including its heartwood, and figuratively breathes life into dying trees by surgically building nests that will later be used by dozens of other creatures—birds, mammals, insects—many of these serving as pollinators and seed spreaders, thus bringing even more life to forests. And some woodpeckers encourage these trees to share their last energy by drilling wells to draw the lifeblood to the surface. This sap feeds the woodpeckers and facilitates the migration of hummingbirds by providing food when it is most scarce. These and other services make woodpeckers keystone species—animals that positively impact the habitat in a way that is exponentially greater than their numbers might suggest and whose absence could lead to the collapse of an ecosystem.

Many woodpeckers are specialists, meaning they need habitat with trees of a particular species, size, and age within stands of the right density or canopy cover. Such precise requirements make them indicator species for many habitats: their presence and relative numbers attest to the health of an ecosystem. Woodpeckers are also umbrella species in that when humans manage habitats to meet these birds' needs, they inadvertently protect many other life-forms dependent upon the same elements of habitat.

Consequently, woodpeckers are perhaps the most important living catalyst in creating forest diversity, and many iconic forest organisms could not be part of the assemblage without their presence. Woodpeckers fill such a vital role in ecosystems that they radiated and adapted to most parts of the world and to various niches, and in turn many other animals and insects evolved to take advantage of their work.

A flash of color and subtle sounds from arriving males and departing female Northern Red-shafted Flickers every one to six hours are often the only signs that they have chosen a nest and are now incubating eggs.

A NOTE ABOUT GEOGRAPHIC SCOPE

A geographic term such as "North America" seems like it should be straightforward, and in most natural history books it is passively defined to include the continental United States and Canada. Yet this political designation ignores the fact that populations of woodpeckers native to the American Southwest do not stop at the border but continue south with their trans-border habitats, and that several other woodpecker species that nest in the US and Canada migrate to or live in Mexico and the Caribbean. The broadest political definition is offered by the United Nations and includes the US, Canada, Mexico, and the Caribbean islands. The most expansive geographic definition would include the contiguous lands from the US and Canada south to Panama's Darién Gap, including most of Central America.

This book focuses on the woodpeckers and habitats emanating from the continental United States and Canada, south through Mexico and the Caribbean islands, including the Bahamas, Cuba, Jamaica, Hispaniola, Puerto Rico, and Guadeloupe. I've excluded islands with distinctly South American flora and fauna, such as Trinidad, Tobago, and smaller islands off the Central and South American coasts. As such, my defined area of North America includes forty-one living species of woodpeckers plus two now believed to be extinct.

Thirteen of the twenty-two woodpecker species found in the United States live in or migrate to the desert, scrubland, and pine-oak habitats shared by the US Southwest and northern Mexico. Some species are gradually replaced with relatives of the same genera as you move south and toward the coasts. Specific Caribbean islands included here host migratory woodpeckers (the Yellow-bellied Sapsucker and Northern Flicker), see one occasional vagrant from the north (the Red-bellied Woodpecker), and share two resident species (the Northern Flicker and Hairy Woodpecker).

Biologists recognize eight biogeographic realms on the earth's land surface based on the distribution patterns of terrestrial organisms. All of Canada, most of the United States, and much of Mexico fall within the *Nearctic realm*. Coastal central and southern Florida through the Caribbean islands, as well as much of Mexico's coasts and areas south of the Sierra Madre, are considered part of the *Neotropical realm*.

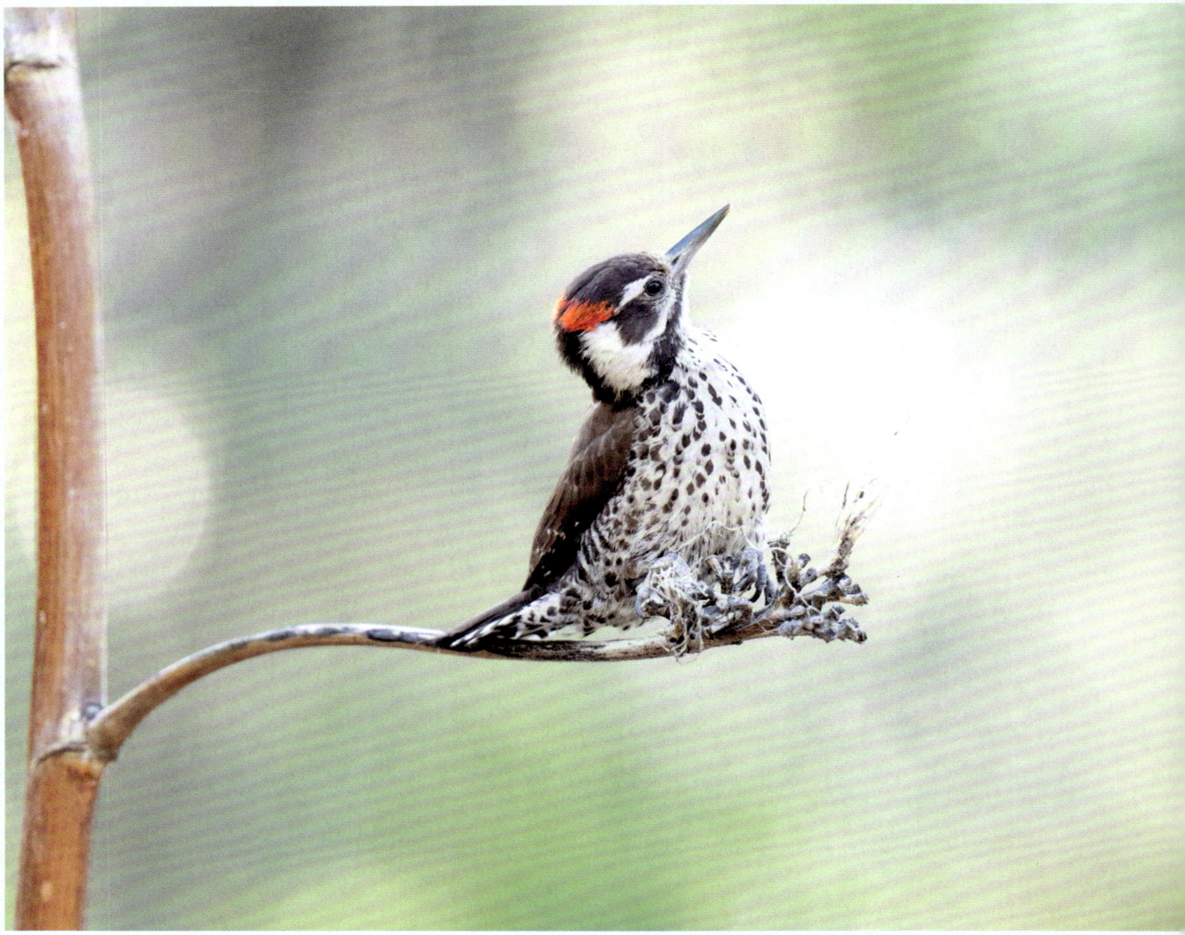

A male Arizona Woodpecker prepares to display to his potential mate in an effort to convince her to nest with him in this particular agave stalk. Arizona and Ladder-backed are the only two species to regularly nest in yucca and agave stalks.

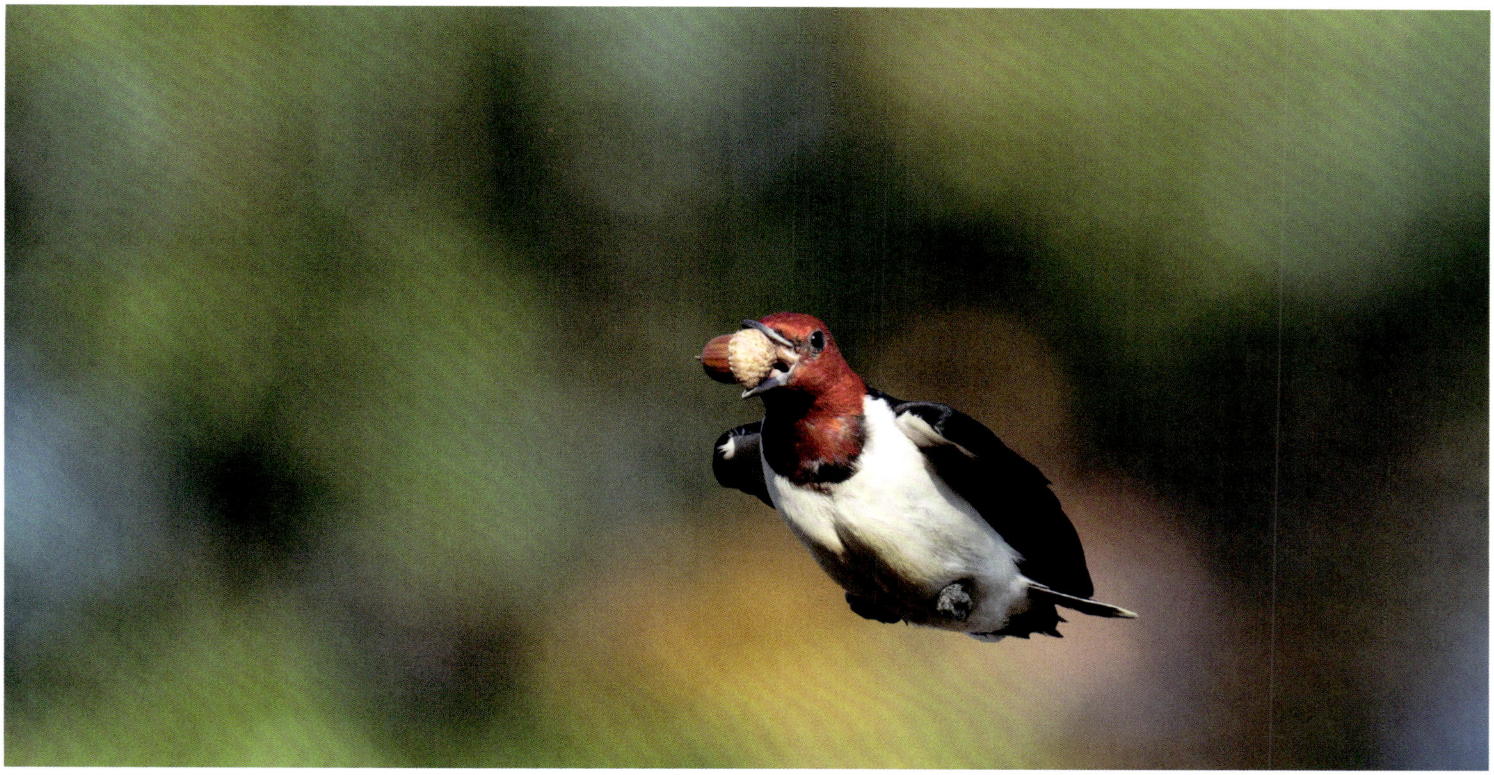

In addition to enriching the wild, these tree doctors help people by controlling termites, carpenter ants, grasshoppers, aphids, moths, and caterpillars—pests that can destroy or devalue homes, forest resources, and crops. They help protect trees by reducing populations of spruce budworms and wood-boring, bark, and mountain pine beetles. This work can prevent trees from dying, reducing the fuel load in our forests to promote natural fire regimes.

It is no wonder that such a consequential bird has been admired and portrayed in the myths, legends, and cultures of people throughout human history, not only Indigenous Americans but also ancient civilizations of Greece, Rome, Babylon, Persia, Africa, and Asia. Woodpeckers have been seen as supernatural beings and are considered prophetic from Europe to Southeast Asia to South America. Indigenous American cultures liken their drumming to the earth's heartbeat, and Norse mythology connects it with thunder, and by extension the god Thor. In China, the appearance of a woodpecker symbolizes good luck and represents new opportunities, creativity, optimism, and courage. Woodpeckers earned such cultural importance for their instrumental role pumping life into the forests they inhabit.

Woodpeckers hopped into my life in an unexpected place. As a young boy of eight or nine, I peered impatiently through the kitchen window into the backyard, waiting for the drizzle to stop so I might escape my tame suburban home to a wilder place where I could watch more than American Robins.

Suddenly, a presumed robin caught my attention. As it bounced closer, the polka-dotted breast, giant bill, bright red mustache, and veins of orange on the tail and wings revealed it was a male flicker! With what seemed like an explosion of red feathers, he was gone. From that moment on, I read everything I could about flickers and other woodpeckers, and I realized how patience, attention, and curiosity could make the common sublime. Instantly I became a woodpecker enthusiast.

An acorn-laden Red-headed Woodpecker flies from its harvest tree to a snag where it will process or store the nut.

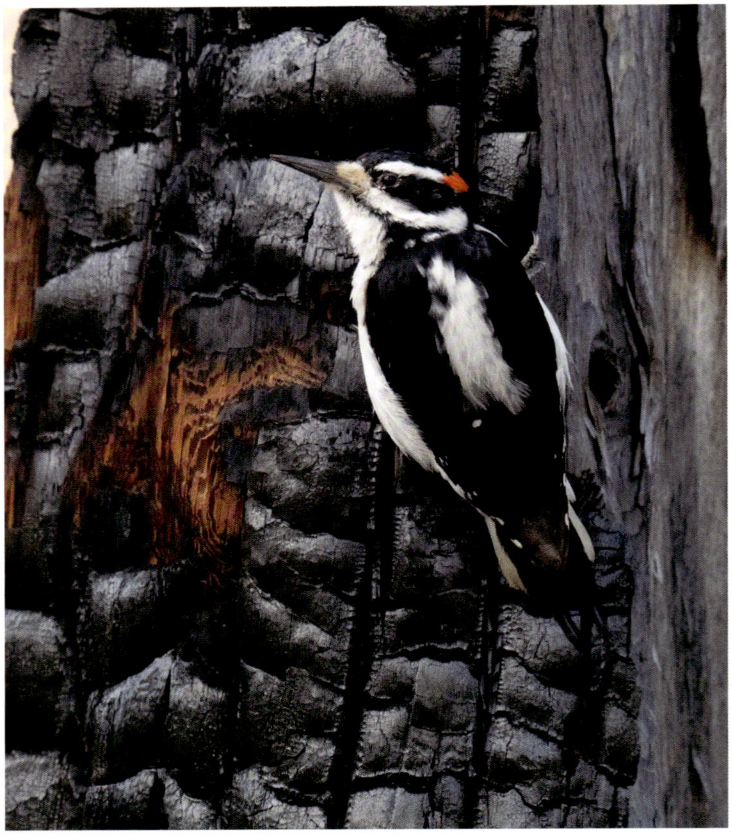
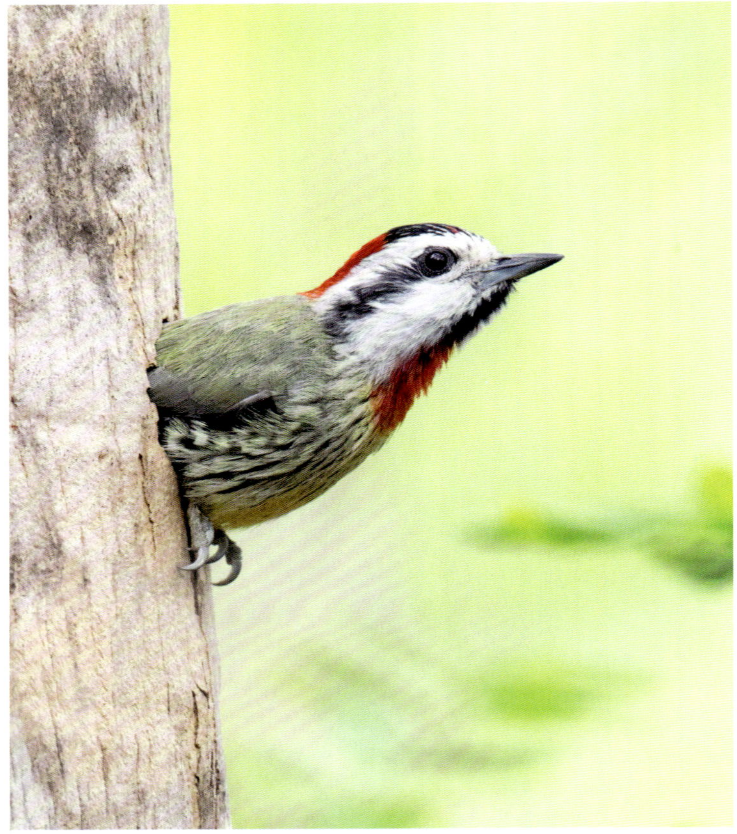
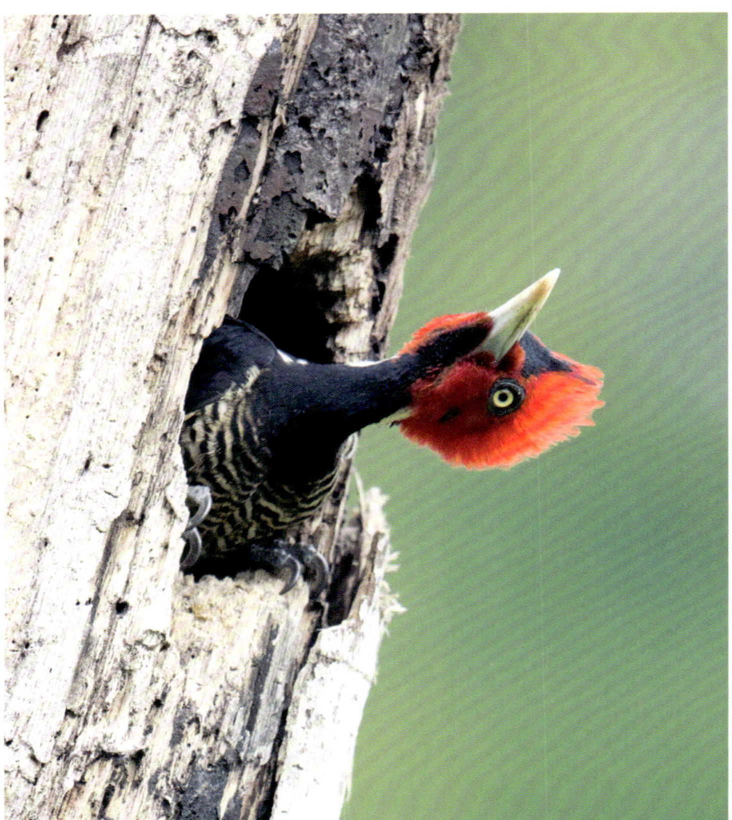
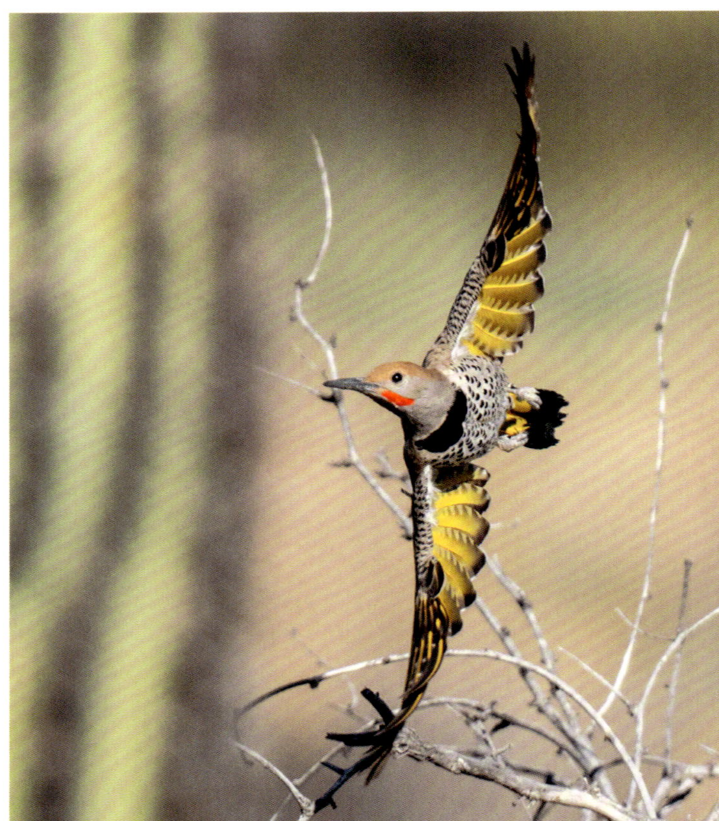

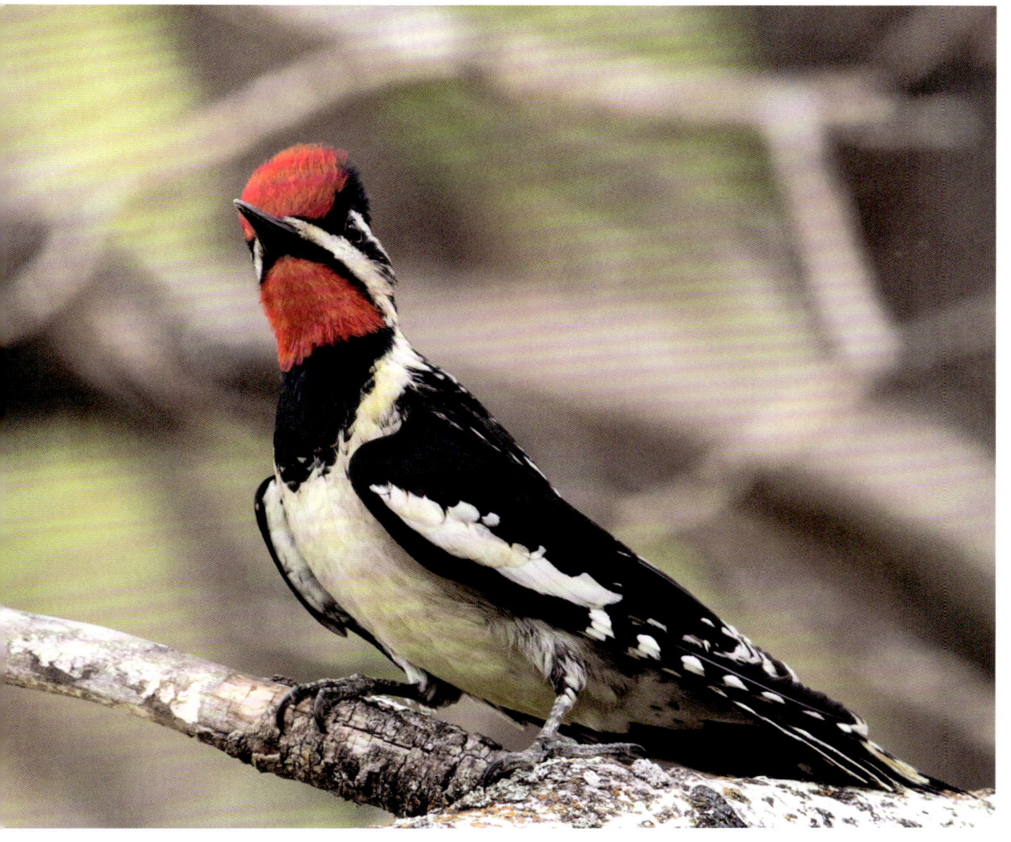

A male Yellow-bellied Sapsucker raises the feathers of his crown as he courts mates or confronts rivals.

Opposite, clockwise from top left: Black-and-white bodies and red caps on male Hairy Woodpeckers are the most common plumage combination in northern and mountain forests. Some tropical forest woodpeckers, such as the Cuban Green Woodpecker, sport green or olive feathers. Most flickers, such as this Gilded Flicker, show brightly colored feathers under their wings and tails during courtship, confrontations, and flight. The red crest and bill of the Pale-billed Woodpecker bring to mind their extinct relatives, the Ivory-billed and Imperial Woodpeckers.

Since that day, the call of the woodpecker has led me on many journeys from my backyard to distant continents and inspired me to rise before the sun so that their syncopated drumming might allow me to find them, follow them, and learn about their mysterious lives before they disappear back into the trees. My fascination has only grown over the years as I noticed that no matter where I was in North America, as long as there were large live or dead trees or cacti and surrounding natural landscape—whether the boreal forests of Alaska or the tropical forests of the Caribbean islands—there was at least one species of woodpecker to trigger my wonder and help me better see the richness around me and the woodpecker's role in enhancing it.

Woodpecker diversity brings wonder to us all. What other family of bird or animal might we see a few or several related species of on a single walk? So many related birds of different sizes, patterns, colors, markings, crests, and calls coexisting in what could superficially be seen as a single landscape defies our perceptions of what constitutes a habitat and its boundaries.

Whenever I observed multiple species in a particular location, I learned how each pursued different prey, on different species of trees, at different heights. Their choice of where and when to nest also varied. The needs of each individual species helped me see a variety of habitats where I had previously seen only one, and assisted me in more clearly understanding the factors that enable their survival.

Since woodpeckers live in every treed habitat in North America, their life stories are in many ways the stories of our wildlands. Every woodpecker alters its own surroundings in ways that create homes for friends and foes, improve the health of natural systems, and even benefit human beings. To witness this ourselves, all we need to do is heed the sounds, follow the flashes of color, and observe the vitality that follows in their wake.

These pages contain an exploration of the lives of the forty-one species of woodpeckers in North America, from the Arctic to the Caribbean, through four annual life phases, covering how they attract mates and create nests, how they find food for and raise their young, how the fledglings leave the nest and gain their independence, and how the birds survive winter. Along the way, I discuss the species-specific habitat components that woodpeckers need to successfully navigate each of these life phases.

In telling the stories of these woodpeckers, I focus my lens, narrative, and anecdotes on the species and behaviors that are most representative, while also highlighting interesting exceptions and contrasts. I look most carefully at species that represent the most threatened or sensitive habitats. I hope that by sharing intimate images and stories from the more secreted moments of their life histories, I can motivate people to better appreciate their critical role in enhancing our natural world and perhaps inspire the long-term protection of these birds and the habitats that they rely upon.

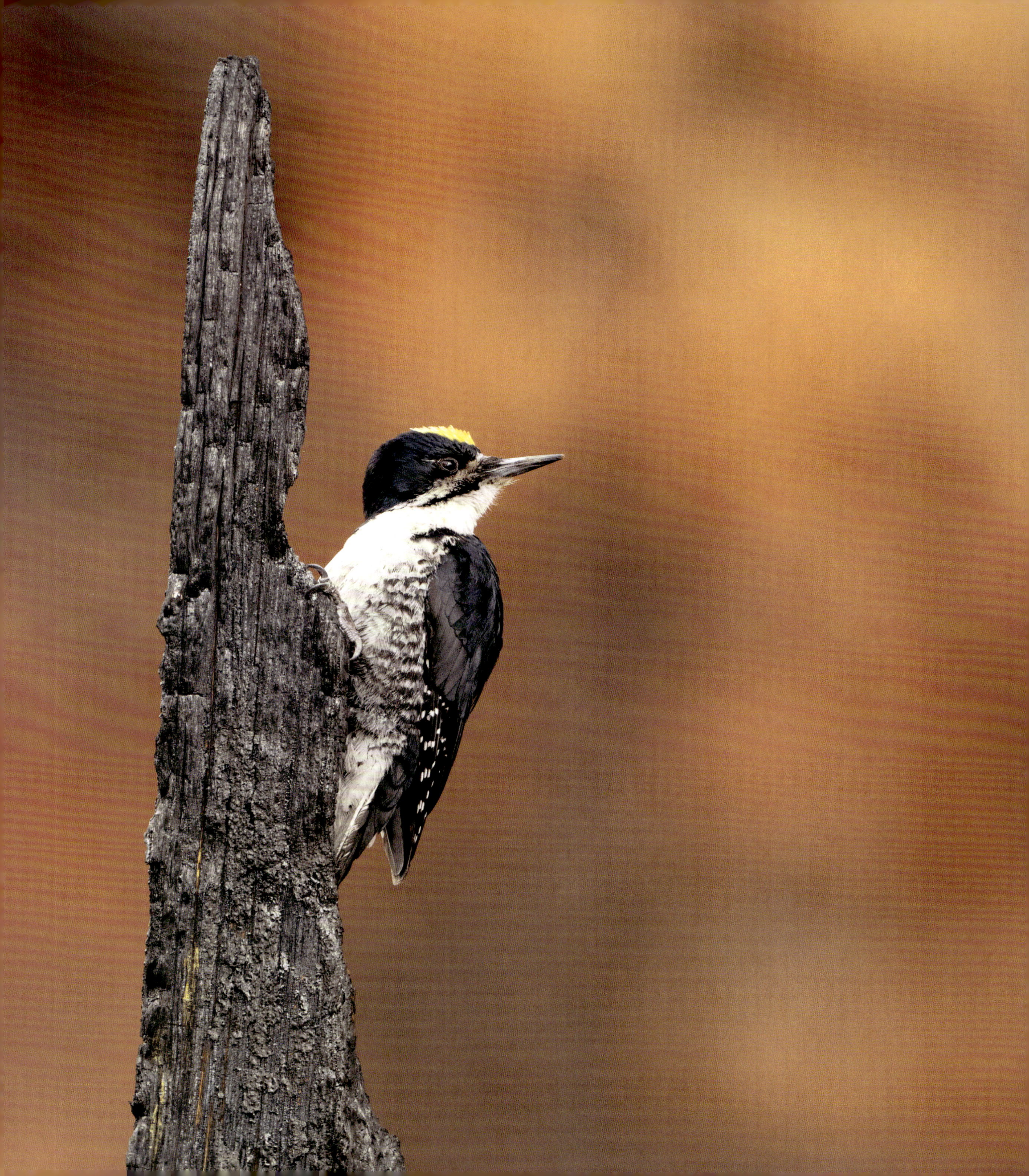

1

A Bird of Distinction

Woodpeckers are striking and easily recognized birds, especially when executing their most physically demanding behaviors, such as climbing vertical faces, pounding on hard surfaces, excavating cavities in trees, or employing their long tongue or bill to capture prey.

Most woodpeckers are well camouflaged overall, with isolated bright red, yellow, or orange on the crown, nape, throat, belly, or underwings, depending on the species. Woodpeckers that spend most of their time under the canopy of trees tend to have black-and-white base coloration to hide them in shadows, whereas those that forage on the ground have brownish base colors to blend with the earth. Woodpeckers living in tropical forests are sometimes green, which conceals them in year-round foliage, and other species, particularly members of genus *Melanerpes*, show off bright color on much of their bodies.

Species that are not closely related, such as the Downy and Hairy Woodpeckers, sometimes develop superficially similar plumage in response to the same habitat in a process called "plumage convergence." As a result, individuals of both of these species, for example, feature more black in the western mountains, more bright white feathers in dry eastern woodlands, and the white appears grayish in wet parts of the Pacific Northwest.

All woodpeckers are diurnal (active by day), cavity-creating, and cavity-nesting birds that live primarily in arboreal habitats. Most are nonmigratory, monogamous birds that dig and peck in wood or probe or glean from vegetation to feed upon ants, bugs, and insect larvae. Some species catch insects in the air, and others eat fruit and berries; still others feed on sap and store acorns or nuts that they consume during winter.

As forest dwellers, woodpeckers have adapted to every treed habitat in the world, except for Madagascar, Australia, New Zealand, and the neighboring Pacific islands. They have also adapted to some deserts with large cacti and even to treeless areas in South America, where the Andean Flicker nests in burrows in earth banks. The reliance of most woodpeckers upon forested habitats has

Two irruptive species, the Black-backed (shown here) and American Three-toed Woodpeckers concentrate in large numbers in response to insect outbreaks that occur following natural disturbances such as fire. Males of both species feature black feathers and yellow crowns.

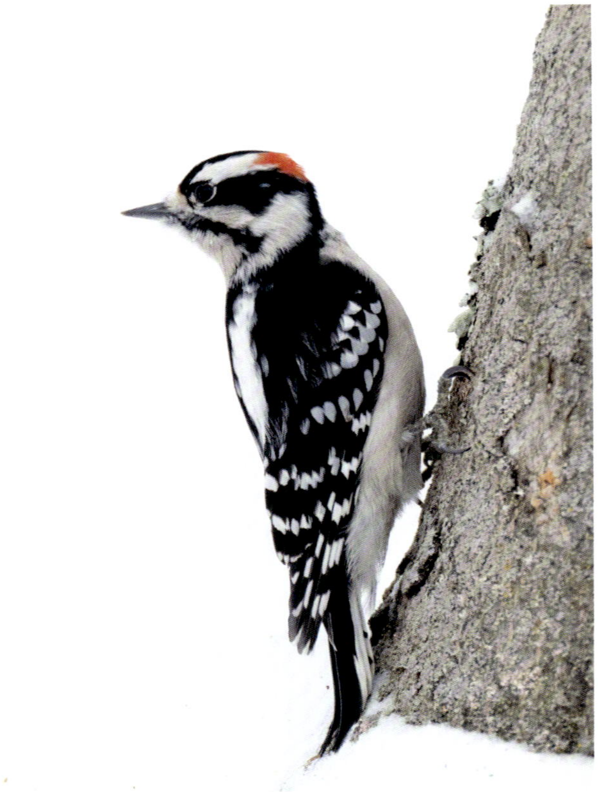

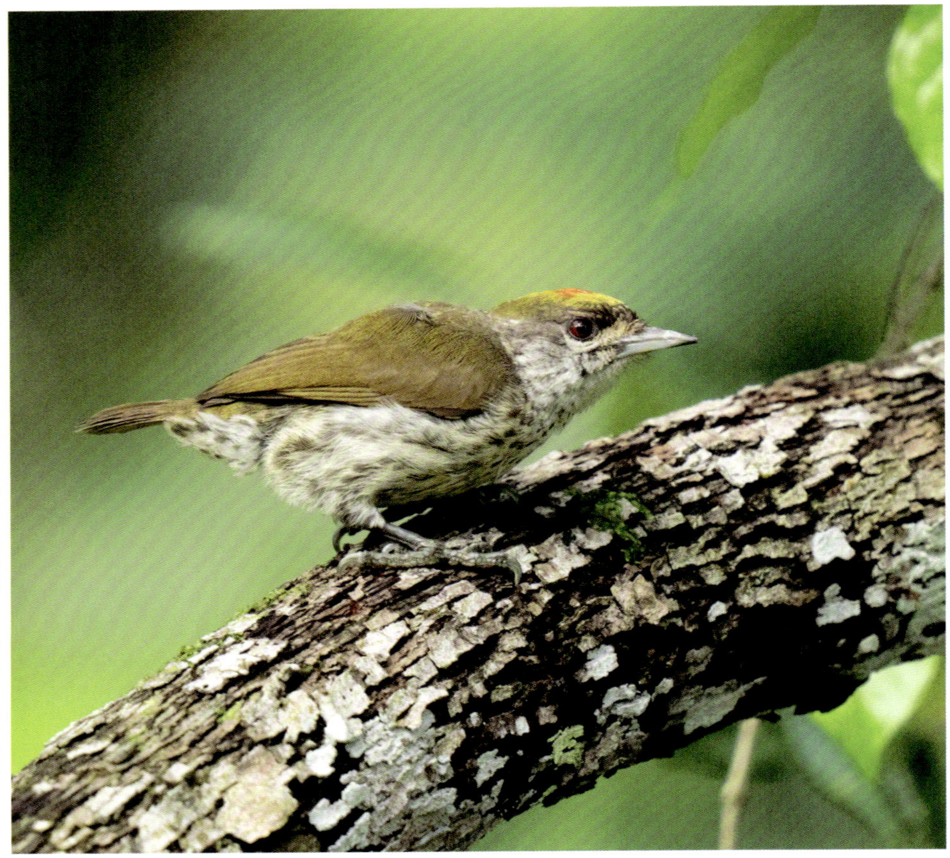

resulted in many species facing conservation challenges, including two now believed to be extinct. Human settlements often cut the largest trees for buildings and remove dead and diseased trees. These actions degrade habitat while eliminating prime nesting and feeding sites.

Woodpeckers became distinct from other birds roughly 50 to 53 million years ago and evolved into a variety of sizes to take advantage of multiple niches. In North America and the Caribbean islands, the shortest is the Antillean Piculet at roughly 5.5 to 6.2 inches long, from top of head to tip of tail. The lightest is the Downy Woodpecker, which weighs less than an ounce, while the Pileated Woodpecker is the largest and heaviest overall (up to 19.3 inches and 12.3 ounces). The now extinct Imperial Woodpecker was previously the largest in the world (up to 23.6 inches and 1.3 pounds).

Taxonomy is the practice of classifying all living things into smaller and smaller categories, from domain (the most general) through kingdom, phylum, class, order, and family to the most specific—genus (plural *genera*) and species. All woodpeckers belong to family Picidae, which includes three subfamilies: Jynginae (wrynecks), Sasia and Picumnus (most piculets), and Picinae (all species in this book, including woodpeckers, flickers, sapsuckers, and the Antillean Piculet). As of this writing, there are 235 woodpecker (Picidae) species recognized across 34 genera worldwide. The geographic scope of this book covers 41 species across 11 genera.

As fascinating and important as woodpeckers are, not many species have been studied to any great length, and even with those species, many facets of their lives remain mysterious, including the basic question of how long they live. Most woodpeckers likely live on average two to three years, although several North American species have been known to live at least six to fifteen years.

Left: Weighing less than an ounce allows the lightest woodpecker, the Downy, to find food in locations inaccessible to heavier birds.

Right: A stubby tail contributes to the Antillean Piculet's status as the shortest woodpecker.

Opposite: The absence of a fourth toe allows *Picoides* species like these courting Black-backed Woodpeckers to exert more force when excavating.

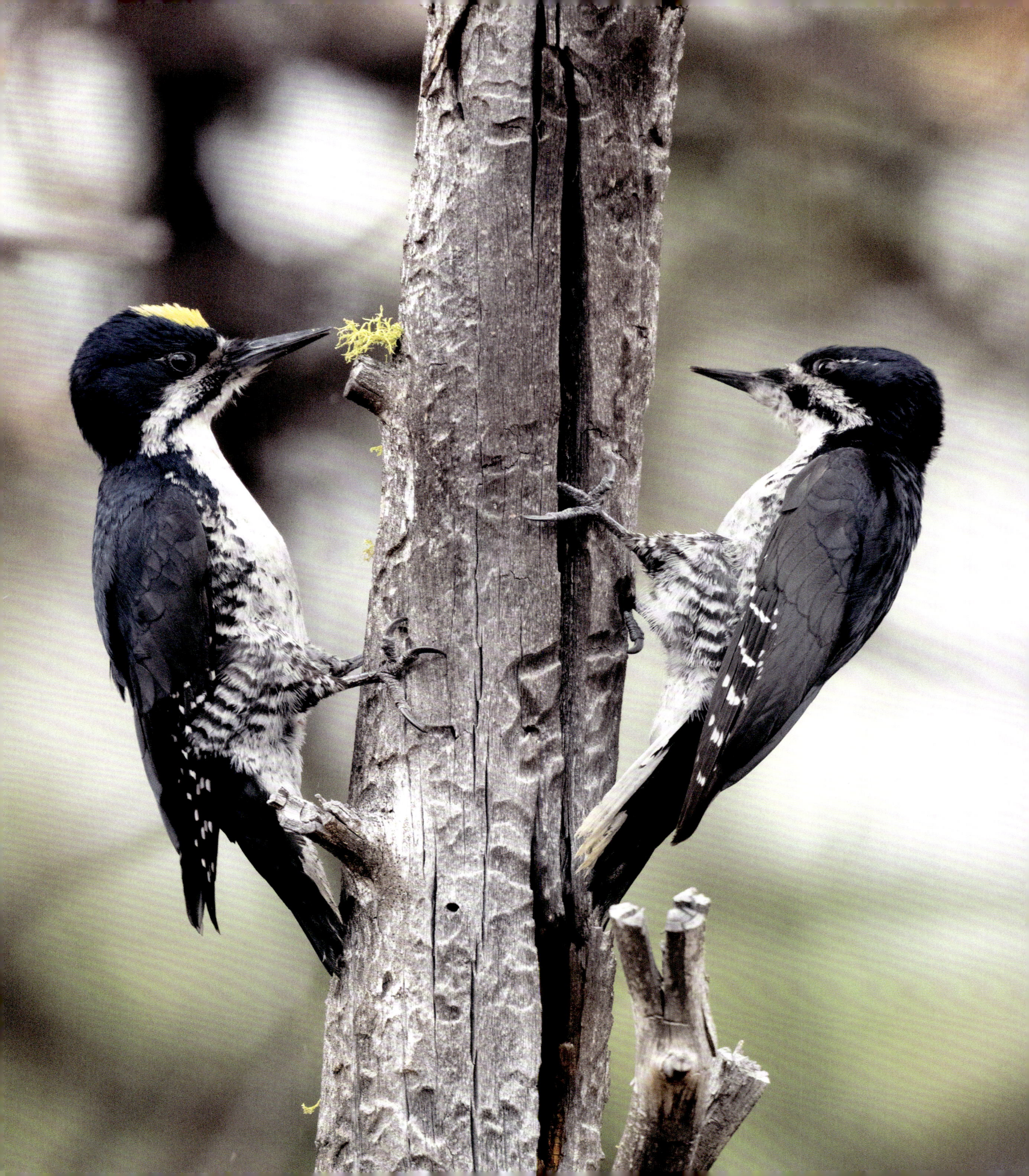

WOODPECKERS AND PEOPLE

Woodpeckers make an obvious impact on their surroundings, and people have strong opinions about their behavior. Fortunately, there are many ways to welcome them, or deter them if necessary.

Woodpeckers are keystone species that improve and create habitat through the nest cavities they create, the sap they make available to migrating hummingbirds and other animals, the seeds they disperse, the plants they pollinate, and even the wood chips they add to the soil. Since woodpeckers have adapted to every habitat with trees or large cacti, there is always at least one species, and often more, to improve each treed area, including our neighborhoods.

Pest Control

Woodpeckers remove pests that otherwise can wreak havoc on forests, farms, and homes. Arthropods are the favored food of most species.

Woodpeckers feed on destructive forest pests such as pine, wood-boring, and bark beetles. Left unchecked, these species kill large swaths of forests and leave young and eggs that can infect and kill even more. With their keen hearing and ability to rip bark from trees and expose insect tunnels, woodpeckers are deadly to these hidden wood-eating pests. The species targeting these pests work infected areas every day year-round until the unwanted pests are extinguished. By reducing beetle populations, woodpeckers not only save precious timber but also reduce the amount of climate-warming carbon emitted into the atmosphere.

Beetle-killed trees produce many potential nest snags that, along with the ample food supply, allow woodpeckers to breed in smaller territories and higher concentrations, produce young, and more quickly remove the problematic bugs. Woodpeckers have been reported at densities of more than fifty times their normal numbers during beetle outbreaks! Their excavations also expose beetle eggs and larvae to other predators, desiccation, and extreme temperatures.

Climate change is making beetle outbreaks more common as warmer winters allow more breeding cycles, and past management practices such as clear-cutting and fire suppression resulted in more vulnerable single-age stands, allowing beetles to kill more trees. Woodpeckers reduce the number of beetles and increase the time between outbreaks, helping reduce forest damage from the insects themselves and fires that start more easily and spread faster in dead and dying trees.

Farms and Food

Woodpeckers can protect some human food by targeting some of our most costly pests. They love grasshoppers, moths, beetles, and insect larvae such as tent caterpillars. Several, including Red-headed, Lewis's, Acorn, Gray-breasted, and Hispaniolan Woodpeckers, find the large open fields of farms, ranches, and agricultural areas ideal for fly-catching insects. Northern Flickers and Downy Woodpeckers are among those that feast upon the European corn borer (corn worm), a particularly problematic pest of corn and other grains. Woodpeckers also consume large numbers of weevils.

Garden and Home

Northern Flickers, Yellow-bellied and Red-breasted Sapsuckers, as well as Acorn, Black-cheeked, Cuban Green, Downy, Gila, Golden-cheeked, Golden-fronted, Hairy, Ladder-backed, Nuttall's, Pileated, and Red-bellied are among the woodpeckers that can adapt to our yards, where they target bugs in trees, flowers, lawns, snags, and wooden structures. Many will seek aphids, caterpillars, moths, mosquitoes, and other flying insects. Downy, Ladder-backed, and Nuttall's Woodpeckers are small enough to remove various pests from the tiniest twig or under slivers of bark and are particularly adept at removing wasps, flies, aphids, moths, and weevils from the galls (bulbous plant growths in response to a parasite such as an insect) on our garden plants and prevent them from spreading.

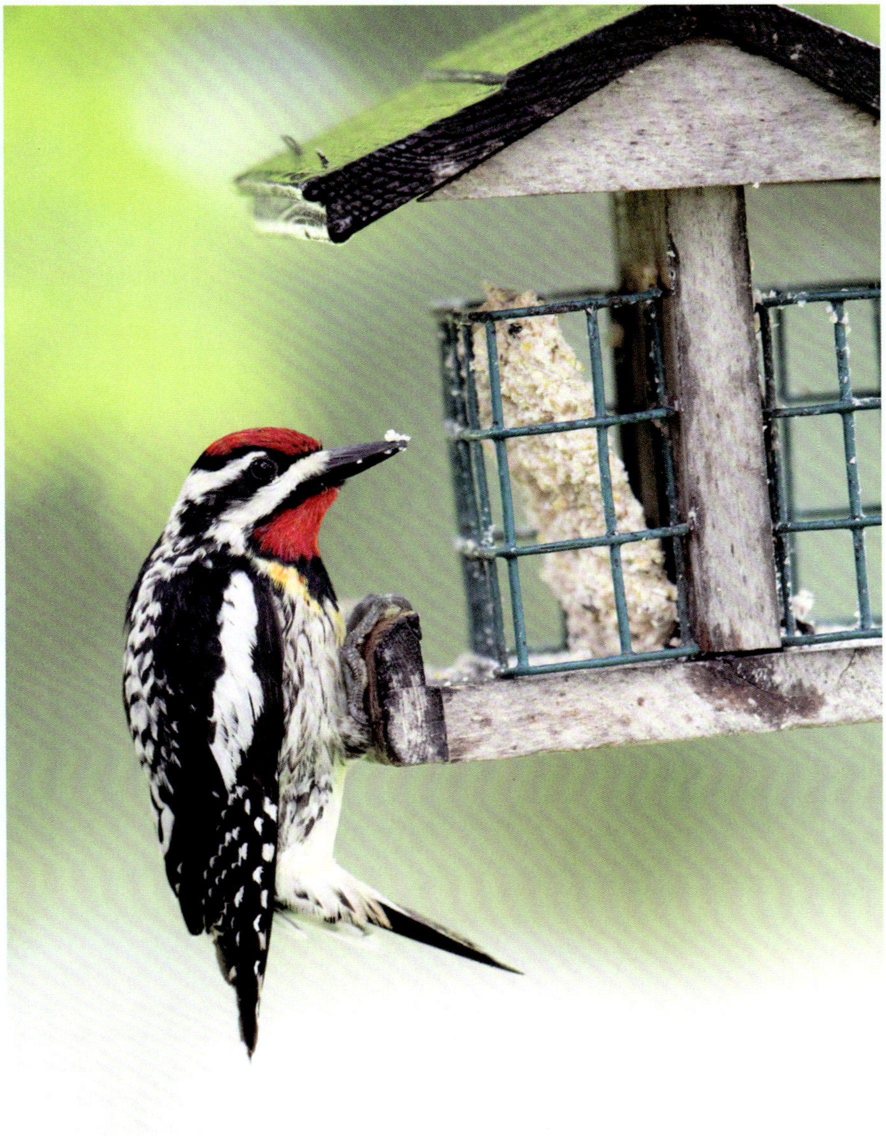

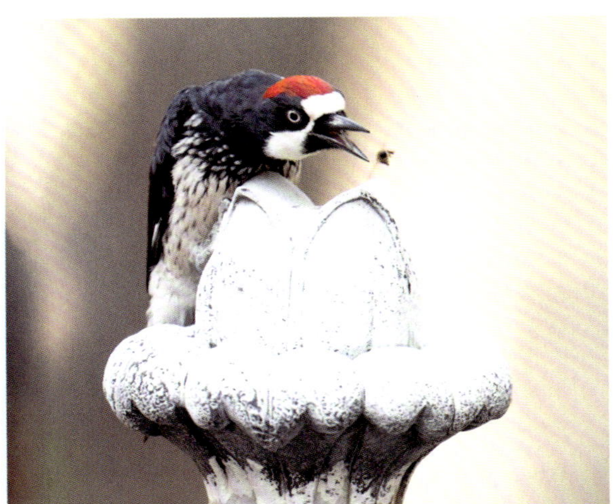

Above: Several species of woodpeckers, including the Yellow-bellied Sapsucker, will take advantage of feeding stations, especially those featuring suet.

Right: Insects, like the wasp being snapped up by this Acorn Woodpecker, are consumed in large quantities by woodpeckers in backyards, as well as in wilder woodlands.

Northern Flickers are attracted to suburban and even urban lawns and gardens, where they scour the ground, using their large decurved bill to probe the soil and their long tongue to capture and consume prodigious numbers of ground ants, beetles, their larvae, and crane fly larvae.

Woodpeckers can also alert people to potential infestations on homes as they are always listening and looking for subtle clues from hidden insects. If a woodpecker is pecking into wood between June and February, it is likely hunting or looking for a roost cavity but not building a nest. Pecking encourages their quarry to move, allowing woodpeckers to pinpoint where to dig and probe with their bill and tongue for carpenter ants and termites. Pileated Woodpeckers specialize in both insects, providing an early warning of potential infestation.

Water, native plants, suet, and sunflower seeds attract woodpeckers, although feeding stations can spread pathogens if they are not disinfected regularly. Use poles with trays to catch stray food, use baffles on poles, and keep feeders away from trees and homes to deter rodents. Consider leaving large dead trees standing or adding nest boxes to provide nest or roost sites.

In the early spring, woodpeckers sometimes run afoul of homeowners by choosing the loudest, most resonant object for advertising to potential mates, especially if that object is the frame of a window screen, a gutter, a chimney, or the side of a house. Some woodpeckers will eat nuts and fruit on home trees, including apples, pears, and cherries. Sapsuckers may create wells in trees, although they prefer trees that are already showing signs of stress and are expressing sap. These are often less adapted nonnative trees or those unlikely to thrive in their current locations.

In all cases, woodpeckers can be disuaded by using owl decoys, flashing objects, sounds, or the placement of a nest box of the appropriate size and dimensions at excavation sites to serve as roost or nest sites. You can find nest boxes through your local birding group or wild bird supply stores, or find plans at nestwatch.org.

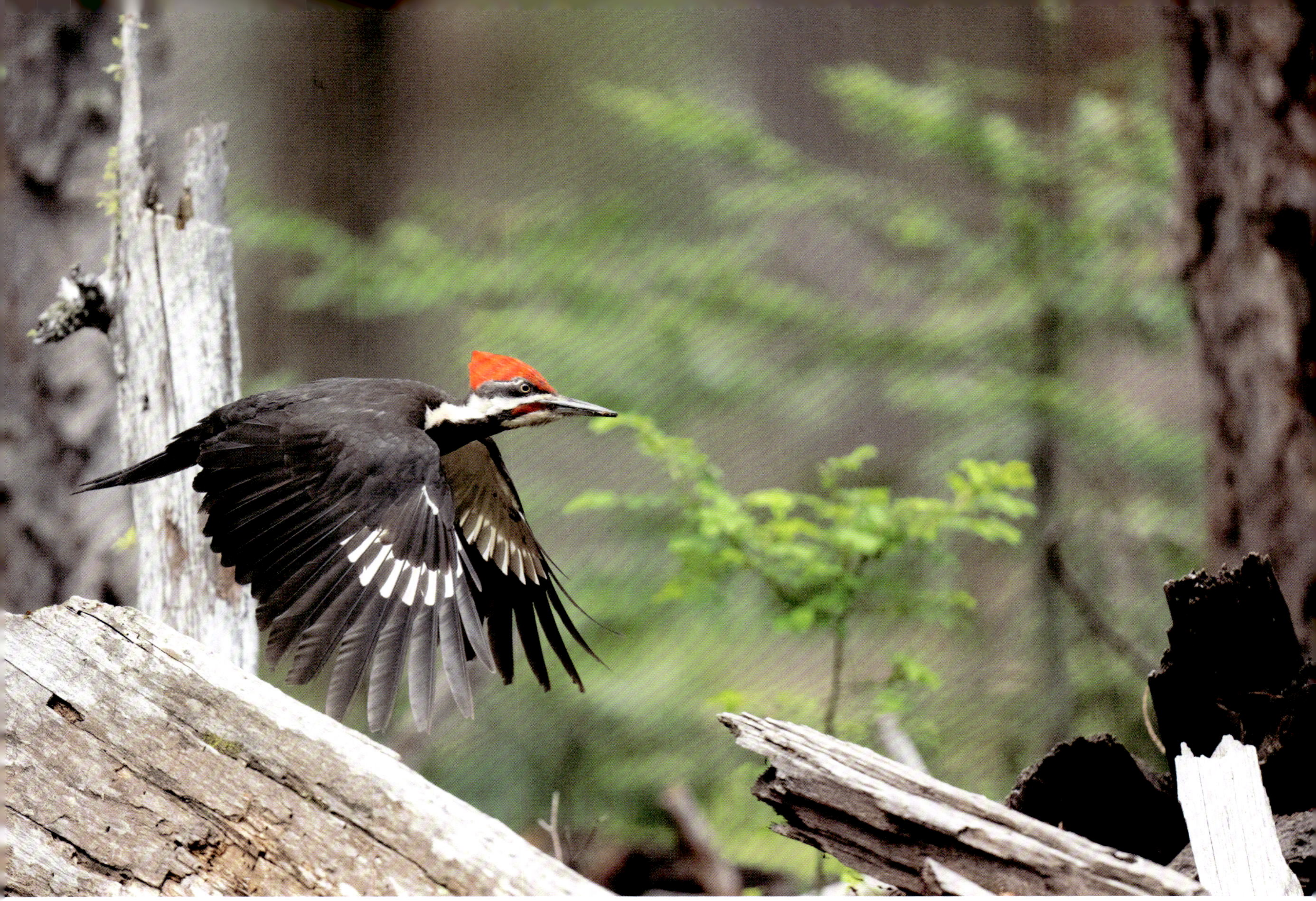

ANATOMICAL FEATURES

Woodpeckers are easily recognized thanks to their many adaptations for a highly specialized life of climbing vertical surfaces, excavating, probing, and signaling.

Feet

Most birds have four toes, and their feet are categorized by the direction in which the digits point. Passerines (perching birds) such as warblers and ravens have *anisodactyl* feet, with three toes facing forward and one pointing backward. The backward-facing toe is called the *hallux*.

Owls, cuckoos, most parrots, and most woodpeckers possess *zygodactyl* feet, with two toes facing forward and two facing backward, ideal for gripping branches. While the feet of flickers and piculets are more or less fixed in this position, most other woodpeckers have developed a modified version of this toe configuration, first called *ectropodactyl* by Waldron DeWitt Miller, in which they can swing their fourth (outermost) toe to the side or upward for better support when climbing.

Woodpeckers have evolved a few basic feet patterns based on their weight and the amount of energy they expend excavating while on vertical surfaces. The largest North American woodpeckers—the Pale-billed, Pileated, and Lineated—demand more of the hallux, which is fully engaged in climbing and can be splayed, along with the other toes, to provide more support. With the Pale-billed, the hallux is fully

As the heaviest and longest-bodied woodpecker and with an impressive red crest, the Pileated Woodpecker is hard to miss as it flies through the forest.

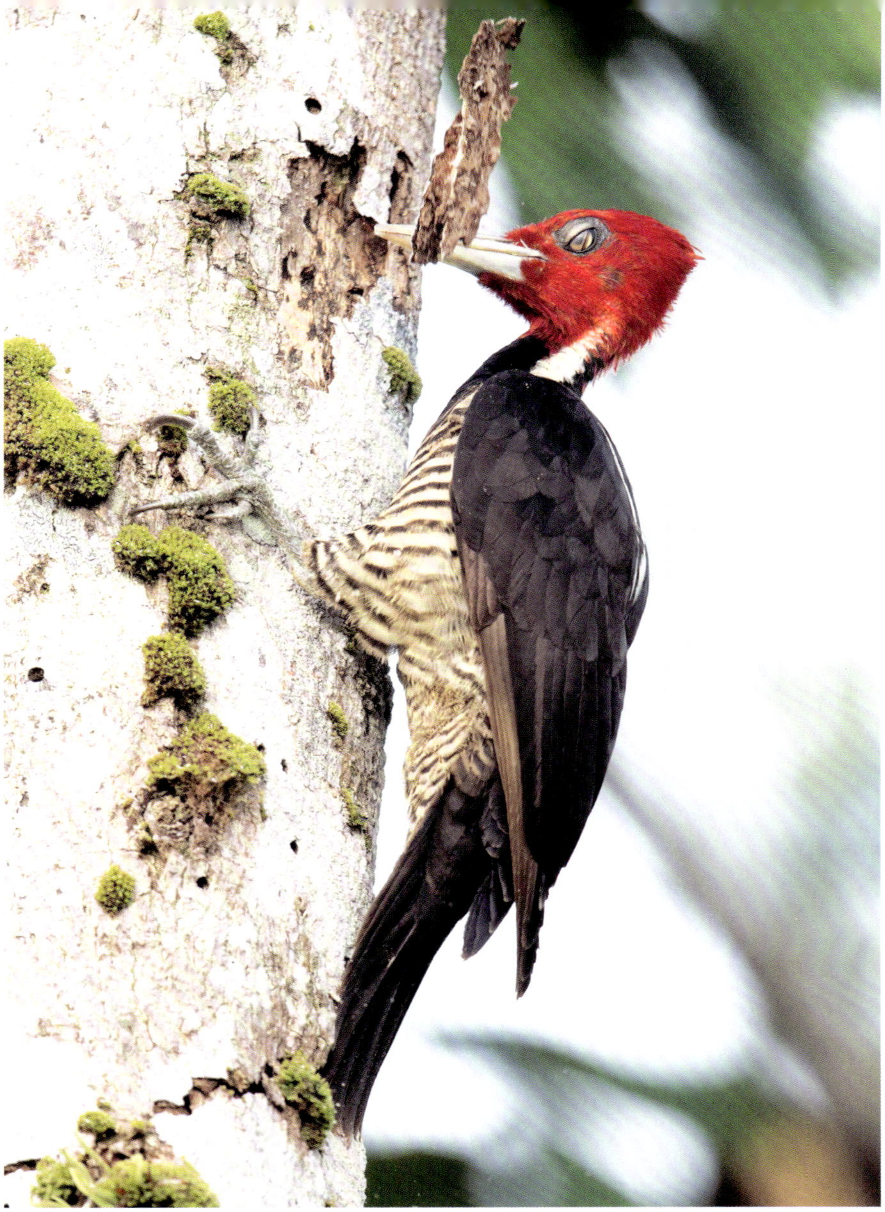

As the Pale-billed Woodpecker excavates on a vertical surface, it takes on an unusual posture with legs spread wide and the tarsus (the upper part of the leg) pressed against the tree, which reduces the load on the tail. The nictating membrane is lowered, protecting the eyes, and tufts of feathers protect the nasal passages.

engaged, the fourth digit is elongated, and the bird holds its unusually long legs out to the sides rather than underneath the body as in other woodpeckers. When on a vertical face, it can direct all four toes upward and deploy the padded part of the tarsus (the part of the leg that attaches to the foot) against the tree. The resulting wide posture with the extra contact point, combined with more engagement of the stiff tail, helps these heavier birds resist gravity. The most specialized North American excavators—the Black-backed and American Three-toed Woodpeckers—evolved in the other direction and lost the hallux. This allows these excavation-dependent woodpeckers to lean back farther, engage more of their body, and exert more striking force upon a tree.

Tail

The tail of most woodpeckers is stiff and inwardly facing so it can be pressed against the tree and bear much of the bird's weight as it climbs, allowing its belly to stay off the tree. The melanin in a woodpecker's tail makes it darker in color and more resistant to wear. Heavier and more excavation-dependent woodpeckers tend to have longer, stiffer, pointier, and more inwardly facing tails that maximize the surface area pressed against the tree, effectively pulling the center of gravity toward the tree. This is in contrast to the Antillean Piculet, which has a much smaller, flatter, and softer tail that it does not deploy when climbing. Its central tail feathers instead provide leverage and stability.

Nostrils and Eyes

Openings in the woodpecker's head need to be shielded from the wood chips sent flying during excavations. The nostrils of most woodpeckers are covered by tufts of feathers. Only the most infrequent excavators, such as the Chestnut-colored Woodpecker, lack such tufts. In more frequent pounders, such as the American Three-toed and Black-backed Woodpeckers and sapsuckers, the nostrils are farther apart, reduced to slits that sit behind ridges.

The bird's eyes are protected in many ways, most importantly by a relatively thick nictitating membrane—a transparent, protective eyelid that covers the eye during threatening situations, such as contact with aggressive young at the nest, a predator or rival, or flying wood chips. During excavations, the eyes close at the instant before contact, providing extra insurance.

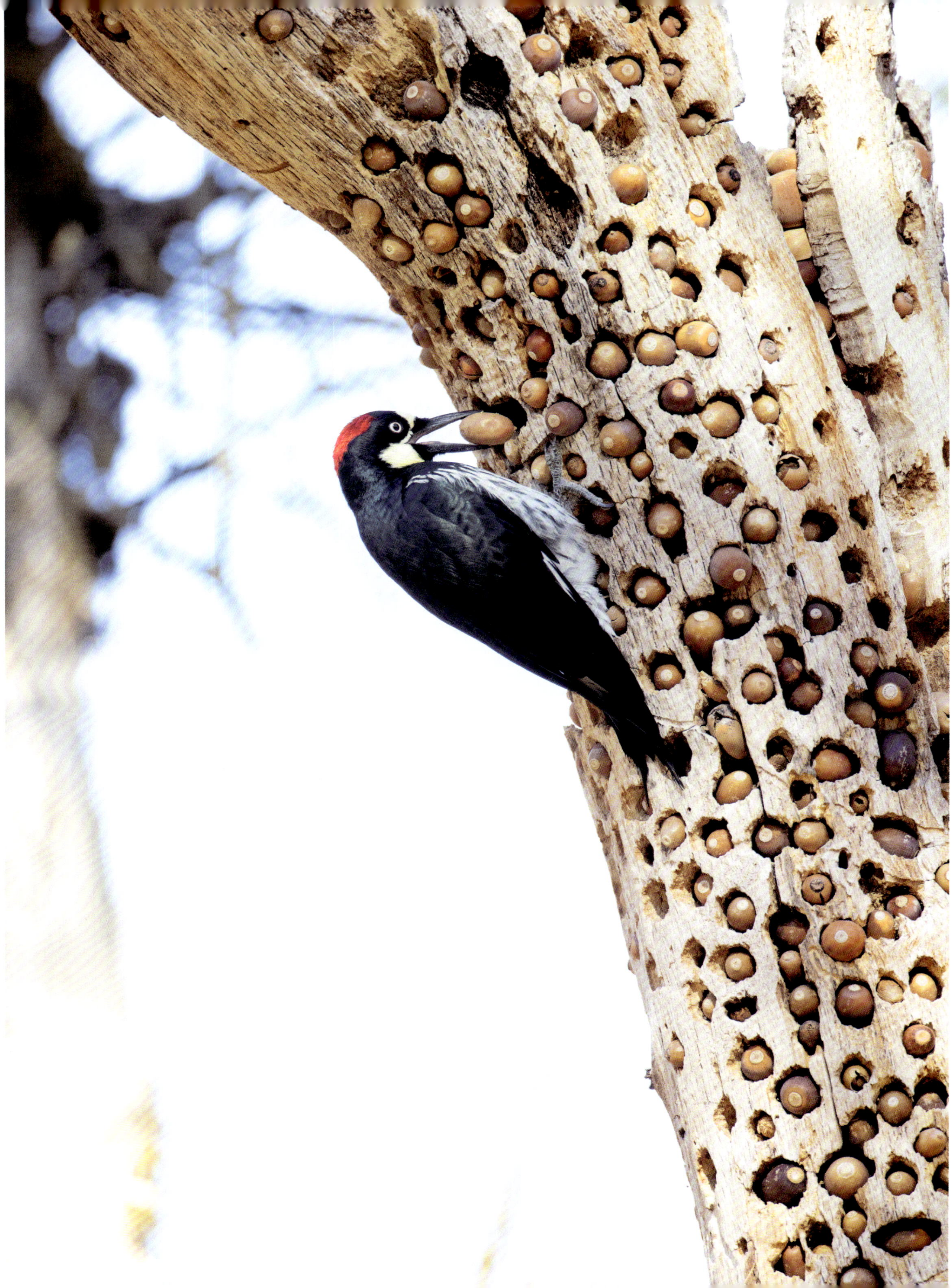

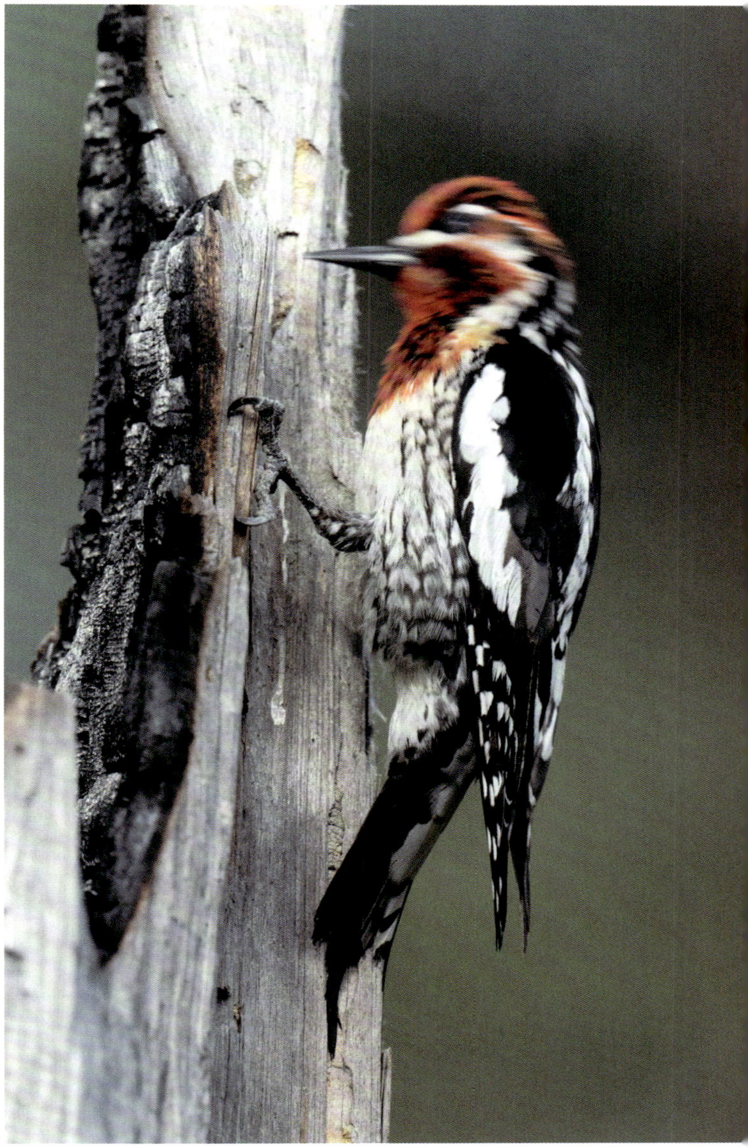

Bill

A woodpecker deploys its head and bill like a hammer fitted with a dagger-shaped, laterally compressed, chisel-tipped blade. Pecking wears away the fast-growing horny tip of the bill in such a way that it self-sharpens. The bills of species that rely most upon excavation, such as Black-backed, American Three-toed, and Pileated Woodpeckers and sapsuckers, are broader at the base, sharper, straighter, and reinforced by longitudinal ridges. In contrast, species that excavate only in soft wood, such as Lewis's Woodpecker and flickers, possess longer, more decurved bills that are better for probing versus excavating.

Woodpeckers, like most birds, can move their upper mandible independent of their lower bill, which allows them a wider gape for securing and manipulating large food. This flexibility helps them disengage their bill from a tree during excavations but increases the risk that the upper mandible could be pushed too far upward. To prevent this, several species have bulging frontal bones that extend beyond the back of the upper mandible, creating a more distinct rounded profile rather than a sloping one. This feature can be prominent or absent depending upon how much they excavate.

Woodpeckers, such as this hybrid Red-breasted/Red-naped Sapsucker drum or advertise by striking their bills against resonant surfaces in a species-specific rhythm. In contrast, when they excavate, they involve their entire body.

Opposite: A woodpecker's upper mandible can be moved independently of the lower bill, which allows this Acorn Woodpecker to manipulate a large acorn in a granary.

Skull

Woodpeckers slam their heads against trees several thousand times a day while excavating or while drumming up to twenty-two times per second. When an animal's head strikes a hard surface, the sudden deceleration force can damage the brain. For humans, a force of one hundred g's can cause a concussion, and repeating such incidents can cause permanent damage. The deceleration impact force of a Pileated Woodpecker striking a tree has been measured up to twelve hundred g's. How do they avoid causing severe brain damage or even killing themselves?

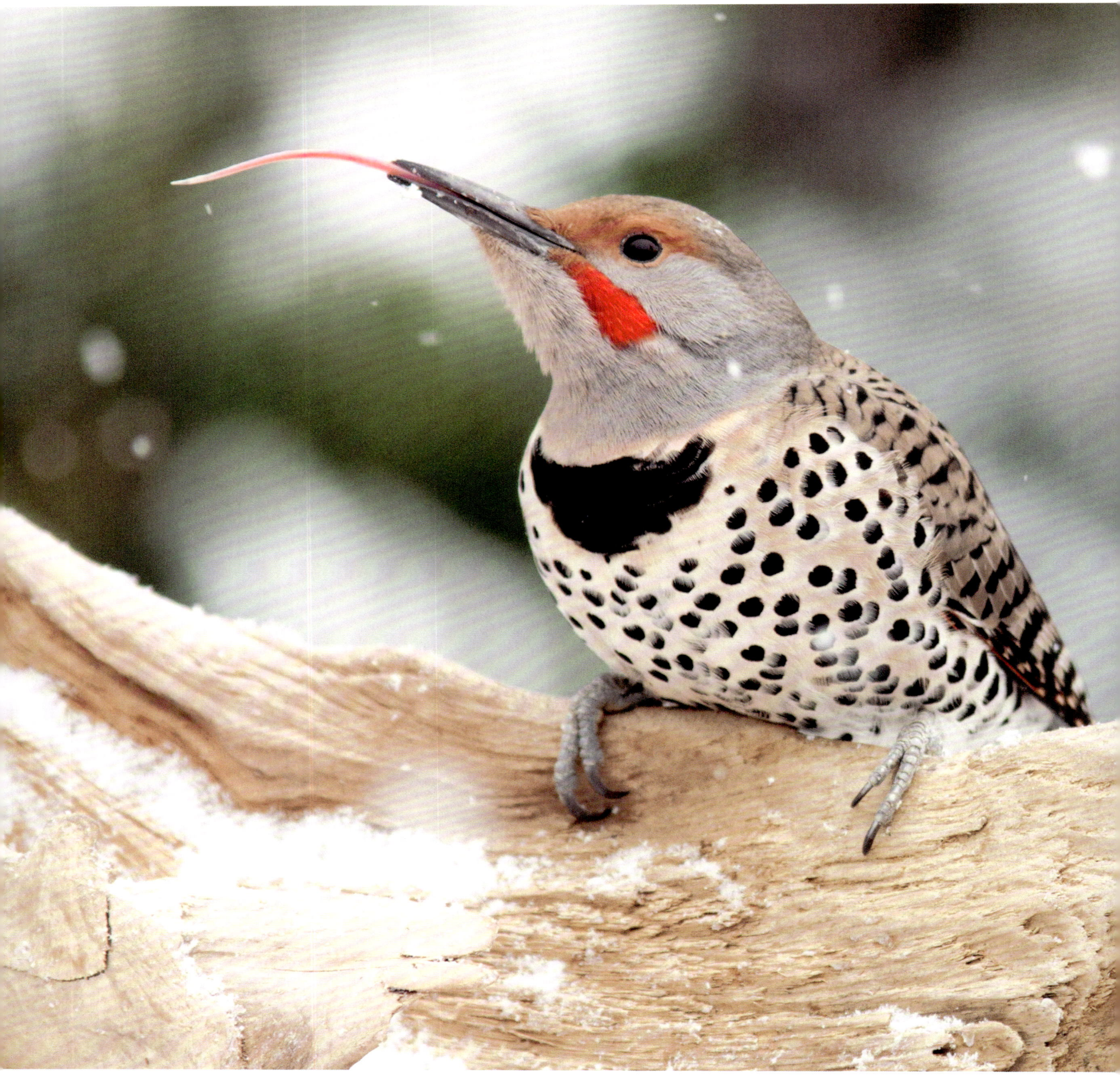

Woodpeckers are uniquely built, with many adaptations that likely add to their ability to withstand serious injury, including a chisel-shaped bill that reduces resistance; dense, spongy cranial bones; minimal cerebral fluid; and reinforced neck and rib structure. The shape and placement of the woodpecker bill in relation to the orientation of the brain likely limits impact. Surprisingly, the orientation and size of the brain are the biggest factors in woodpecker brains receiving 60 percent less pressure than what would be needed to cause a concussion in a human, even with the most violent pecks.

Birds have relatively small brains, with the woodpecker's being roughly seven hundred times smaller than a human brain, giving it a lower mass-to-surface ratio and spreading any impact over a larger area, thus reducing the risks. Woodpeckers would have to strike a tree at twice their current speed, or a tree four times as stiff, to sustain a concussion. It is possible that the protection afforded by the smaller brain helps explain why woodpeckers have not evolved to be larger.

We do not fully understand the reasons for all of the unique morphological adaptations of the woodpecker's skull, but they may include mechanisms to prevent and repair trauma to the brain from repeated strikes and help prevent long-term brain damage.

Northern Red-shafted Flickers possess the longest tongues of all woodpecker species. Woodpeckers' long tongues wrap around the brain, which may provide some protection to their brains when excavating and also allows the tongue to extend farther in pursuit of prey.

Tongue

A woodpecker's tongue can probe, spear, or stick to prey before retracting into the skull until needed again. Regardless of species, all tongues are cylindrical, slender, maneuverable, and barbed, but they vary in length and especially the number and type of barbs, particularly at the tip, depending upon the type of food they pursue. The tip of the tongue is touch sensitive, allowing the woodpecker to feel for prey, but also stiff with the spinelike barbs angled backward to better retract food. The length of the tongue also reflects the way a woodpecker feeds. Flickers can extend their tongue two inches or more beyond the tip of their bill, an adaptation to pursuing ants deep inside their tunnels.

Many species have enlarged salivary glands that coat all or part of the tongue with a sticky substance as it leaves the bill. This feature is particularly important for woodpeckers that pursue ants, including *Colaptes* species such as flickers, Golden-olive, and Gray-crowned Woodpeckers, as well as *Dryocopus* species such as Pileated and Lineated Woodpeckers. In these species, saliva often covers most or all of the tongue.

In other species, the saliva covers only parts of the tongue so the birds retain the flexibility and sensitivity to feel for prey without getting it stuck. Woodpeckers that specialize in subsurface excavation, such as the Black-backed and American Three-toed, have a shorter but more barbed tongue to snare beetle larvae as they retract it from tight spaces. Sapsuckers that glean from wells three-quarters of an inch or less have a short, brushlike tongue that laps up sap. The Acorn Woodpecker's tongue is similar to that of sapsuckers and makes it well adapted to licking sap.

Together, these anatomical features arm and shield the woodpecker's body for its physically demanding lifestyle.

MOVING THROUGH FORESTS

Woodpeckers propel themselves through their forest homes in two unique ways that help distinguish them from other birds.

Climbing

Woodpeckers make their way up vertical surfaces utilizing a three-part process called "hitching." First, they lower their breast close to the tree with their bill pointed in the direction they intend to move. They then propel their body upward with both feet and tail, pushing downward against the surface as they leap through the air with feet reaching upward to secure a new higher position. The bird moves downward either by rotating the body parallel to the ground and shuffling down sideways or by lifting the tail and feet, falling, and finding a new lower grip. This second method is far riskier as the bird cannot see obstacles that might catch the tail. Young birds often have a particularly fraught time during such maneuvers.

Flying

Most woodpeckers are easily identified by their undulating flight pattern as they flap-bound through the air. At launch and throughout much of their typical flight, woodpeckers alternate between holding their wings against their body, causing them to fall, and flapping, which brings them back upward. The undulating pattern is particularly obvious during longer flights but can sometimes be difficult to see in forested environments.

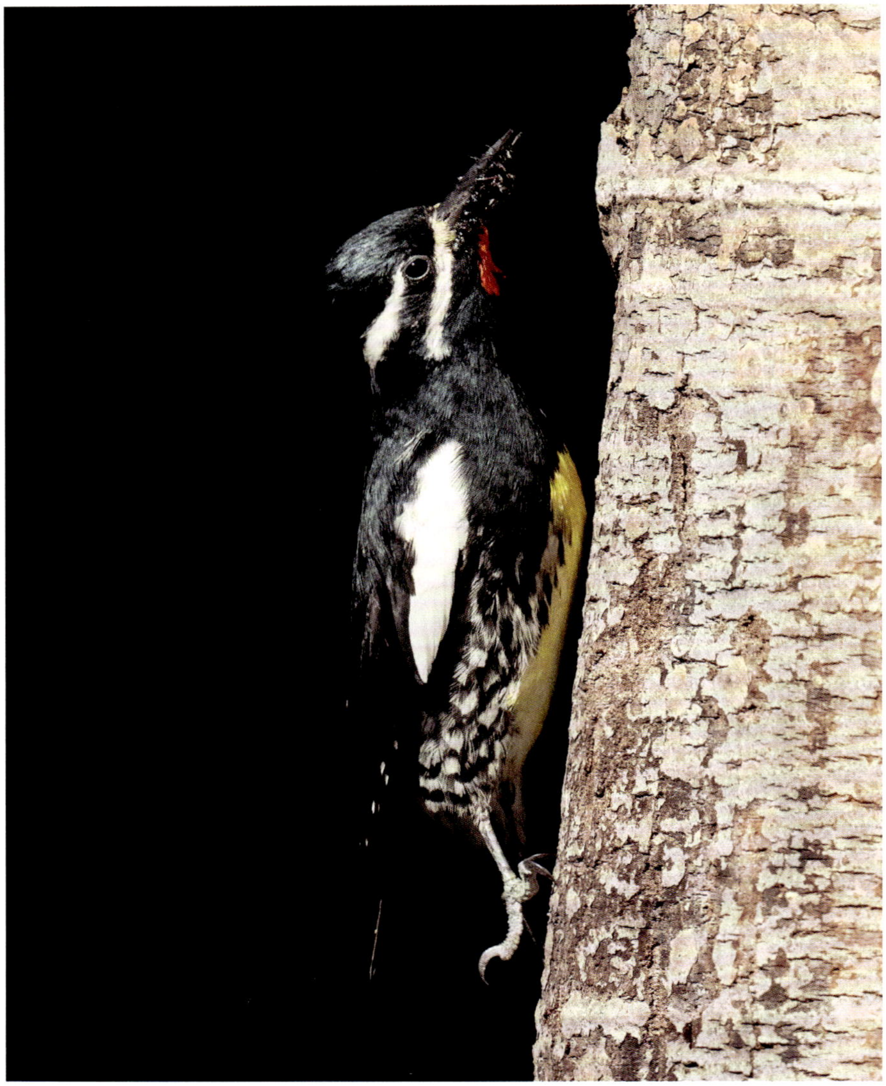

Some larger woodpeckers, such as those in genera *Campephilus* and *Dryocopus*, as well as Lewis's and Guadeloupe Woodpeckers, fly with stronger wingbeats and a less bouncy flight. When fly-catching, woodpeckers fly more vertically into the air, execute complex aerial maneuvers, hover, and return to the same perch. They hold the wings in for a longer moment to pick up speed or hold the wings out to the side to slow down in preparation for landing.

These unique physical adaptations and movement behaviors—along with their distinctive drumming and excavating behaviors—equip woodpeckers for their annual journey through four life phases.

Woodpeckers, like this Williamson's Sapsucker, propel themselves up trees by pushing with both feet at once.

Opposite: Its ability to swoop and hover allows this Red-headed Woodpecker, as well as Lewis's and Acorn Woodpeckers, to be the most specialized at capturing flying insects.

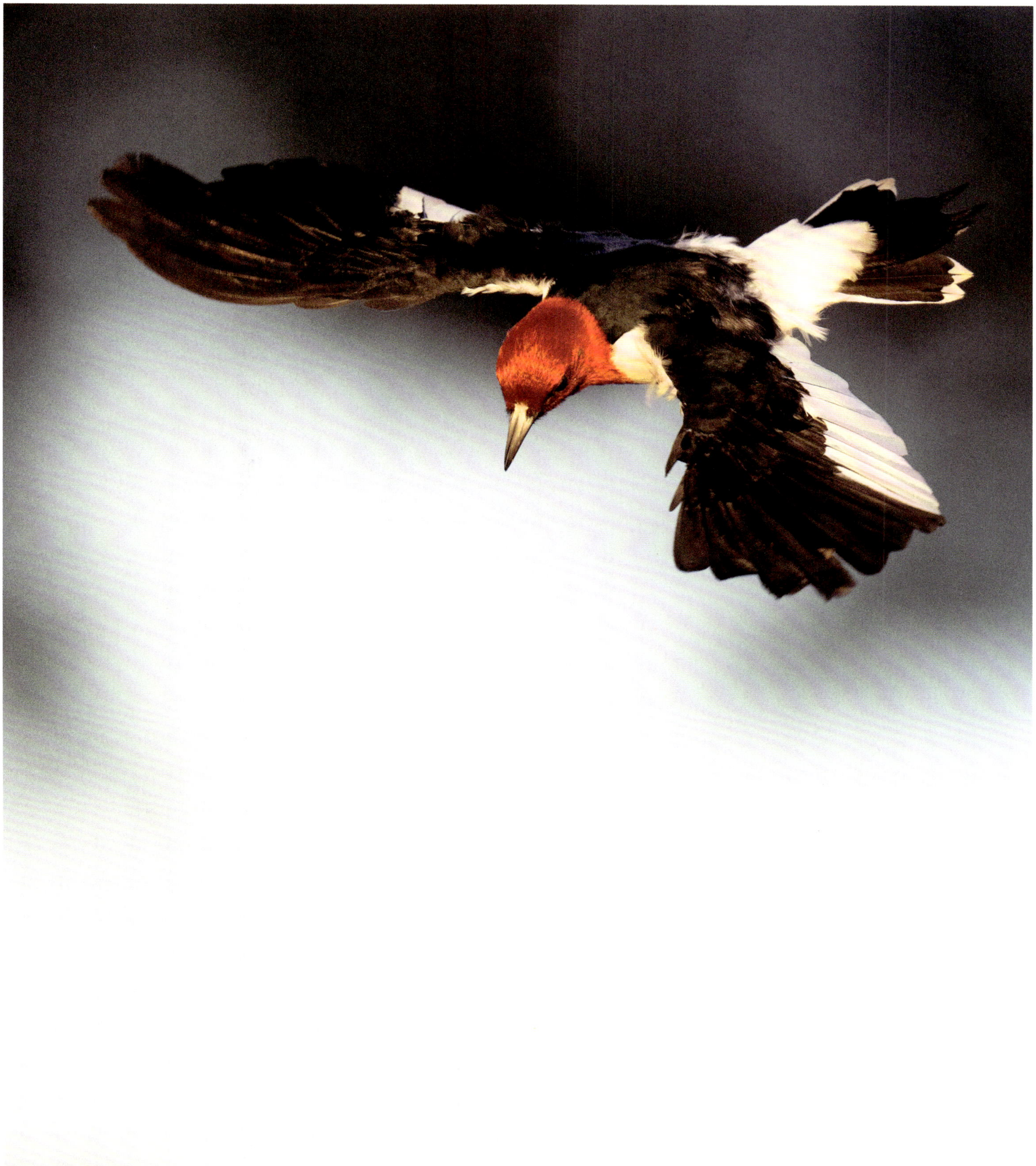

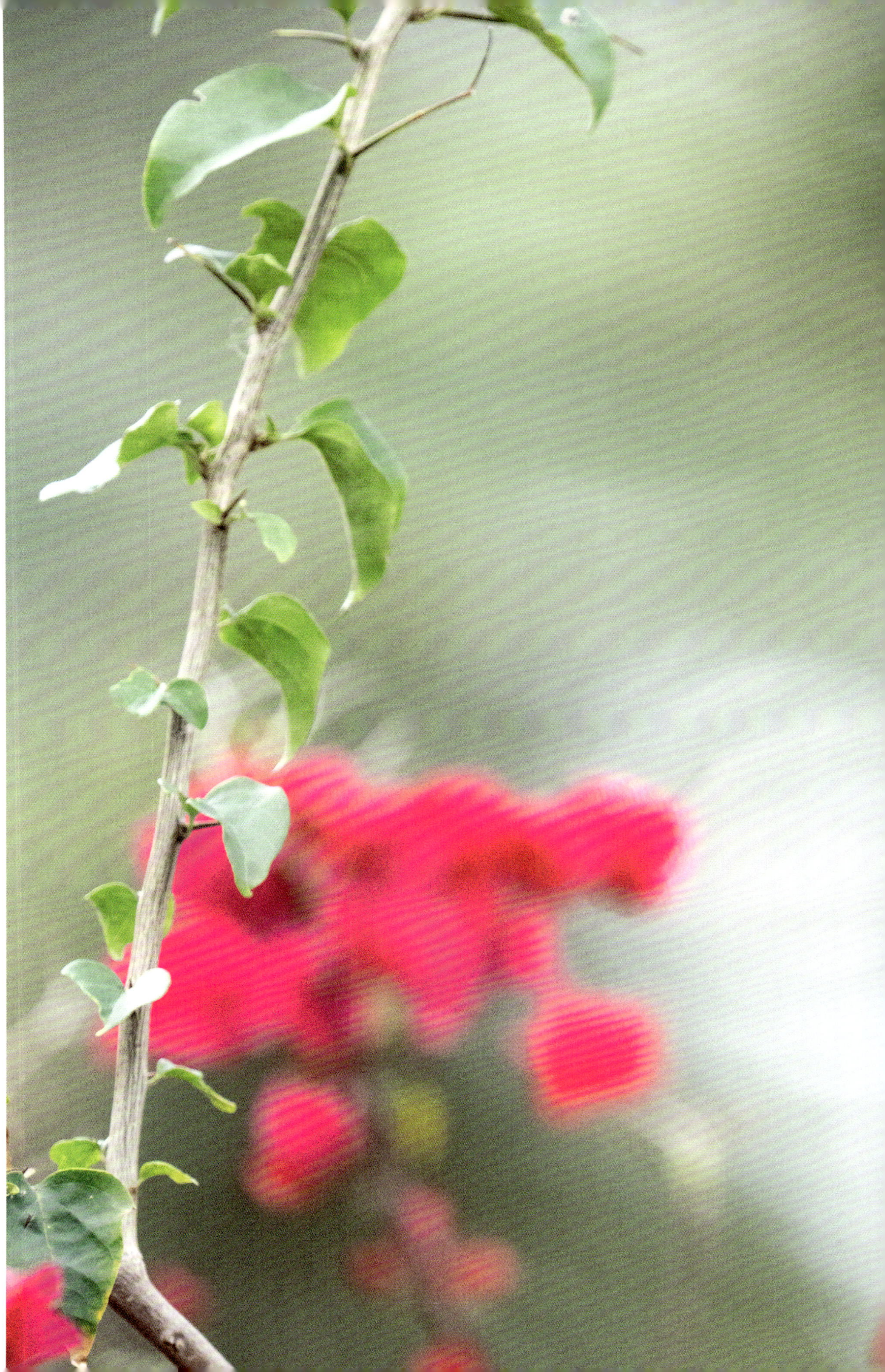

A Hispaniolan Woodpecker pushes off from a perch with its wings pressed against its body, allowing it to close distances quickly and discreetly.

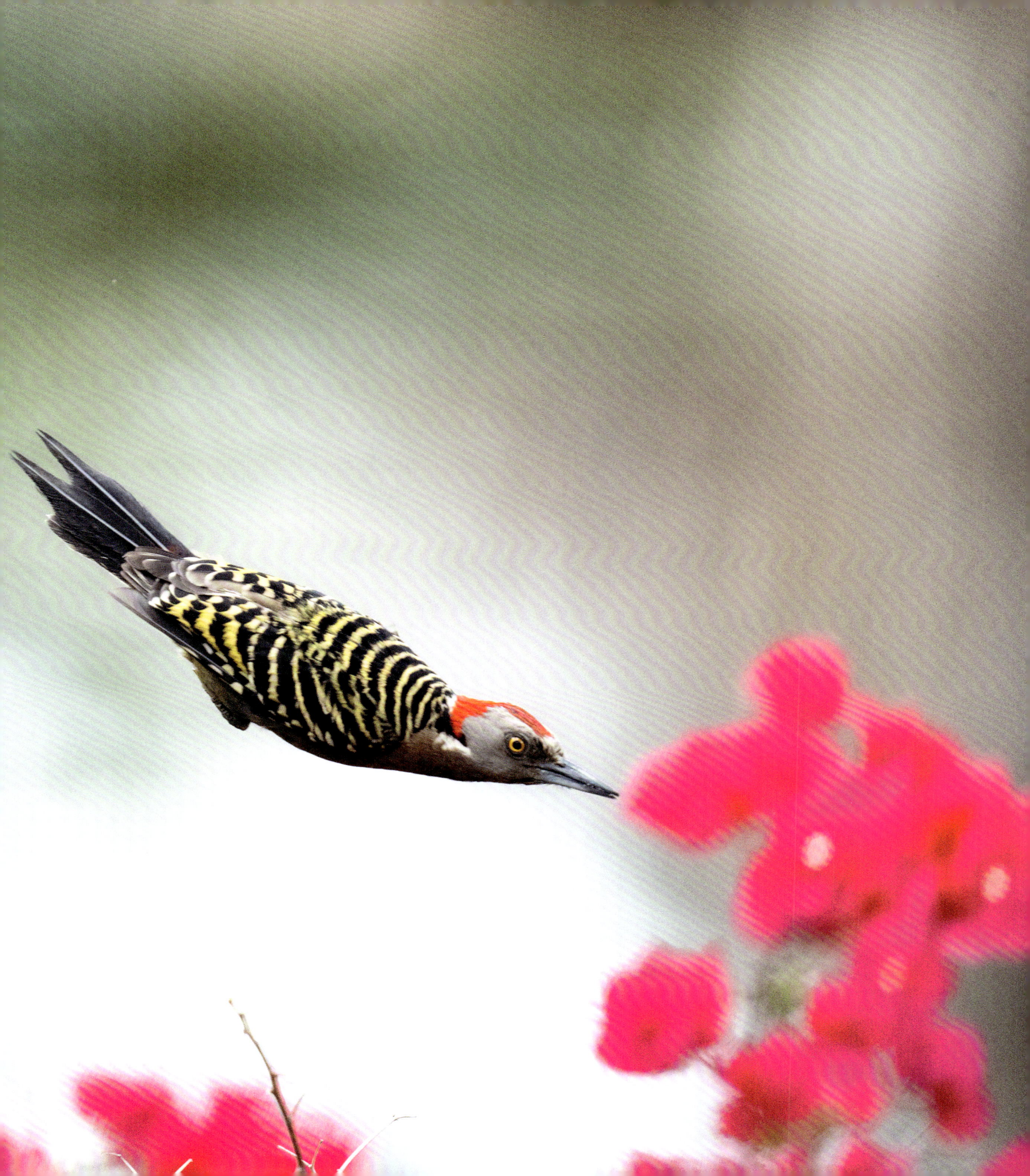

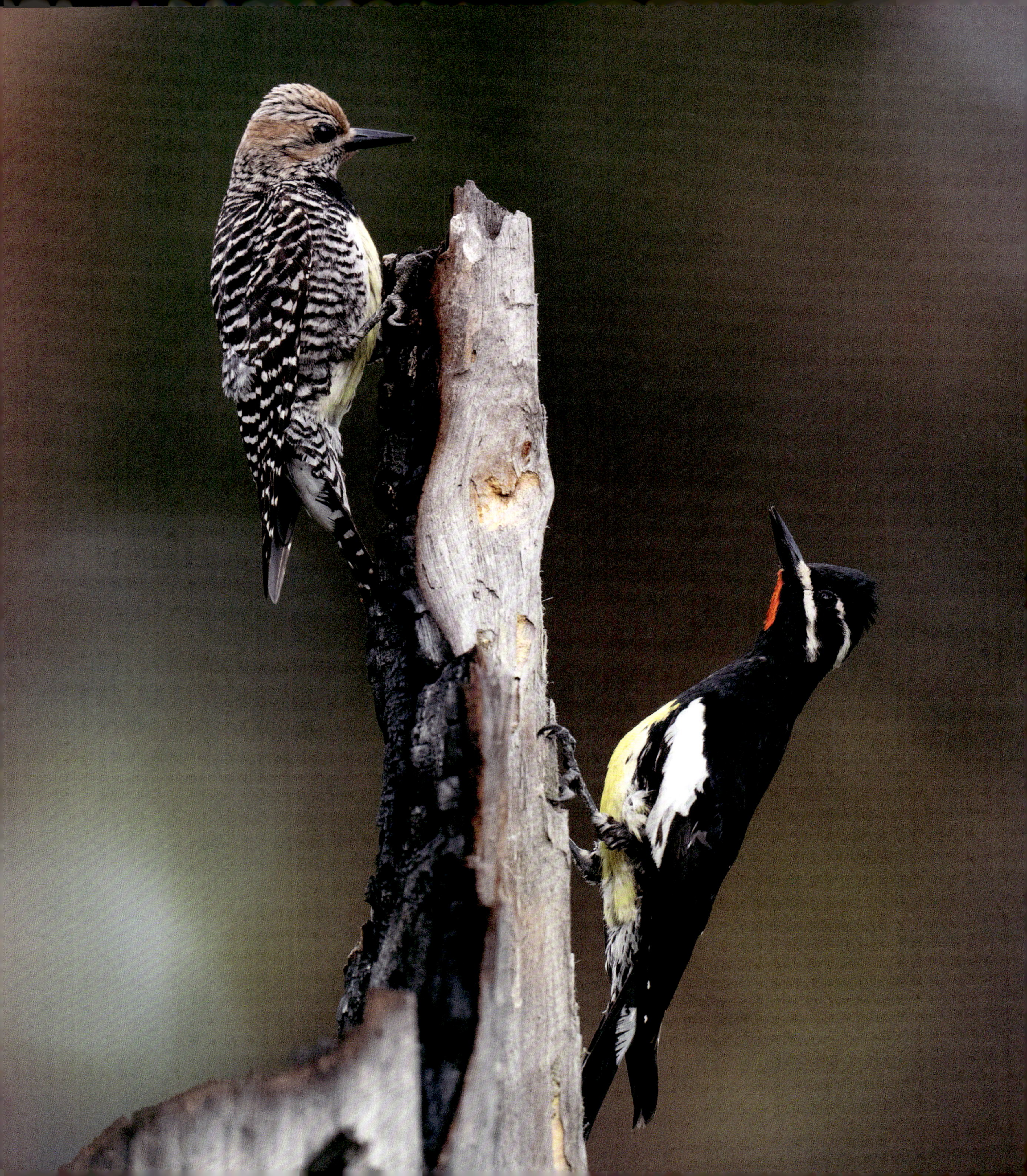

2

Territory, Courtship, and Nest Creation

As the first golden rays of light splintered through bundles of needles on giant red-armored ponderosa pines in Washington's Cascade Range, I began to hear a cacophonous mixture of shrieks, mewing, churrs, rattles, warbles, cackles, and chipping blended with loud, rhythmic drumming of various cadences and amplitudes, creating a tremendous sense of urgent vitality.

White-headed, Hairy, Black-backed, and Pileated Woodpeckers, along with Williamson's, Red-breasted, and Red-naped Sapsuckers, as well as Northern Flickers at times, competed with others of their own species and, in some cases, other species to stand out from the din, each at a subtly different stage in their efforts to breed. All amplified their testimony for potential mates or warnings to rivals by pounding their bills on the most resonant surfaces, from the face of a giant hollow aspen to the dead limb of a live fir to the metal roof of an outhouse.

Spring is analogous to dawn for woodpeckers, the revelatory instant when percussion and color erupt in response to the moment when weather and light tip beyond the threshold needed to lay winter to rest. Longer days trigger hormones that stimulate woodpeckers to move to breeding areas, establish or reestablish territories, secure mates, create or find nest sites, and prepare to breed. These behaviors vary by species, geography, ever-changing habitat, and the resulting availability of nest sites and prey, but are always aimed at contributing young to the next generation.

A male Williamson's Sapsucker raises his crest and stretches his body to reveal his colorful red chin and yellow belly in response to a female who answered his drumming and calling.

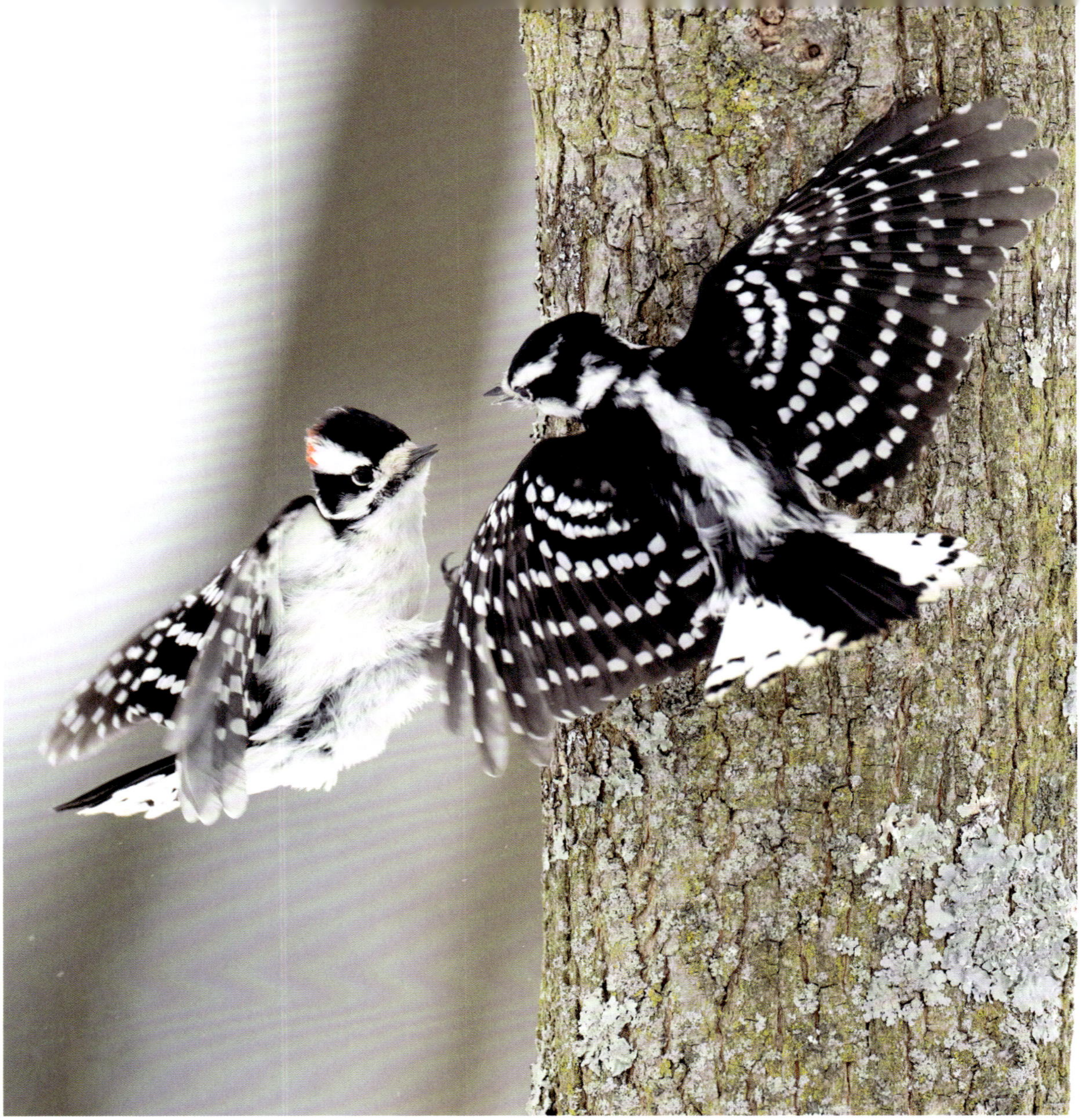

Sites with an abundance of the vital elements of food, nest sites, cover, and water draw together candidates from multiple species in a contest to determine which dominant birds will remain a few weeks longer, forcing others to move on to less-than-ideal areas where they may or may not succeed in breeding.

Most woodpeckers are resident on home ranges year-round and simply begin defending the most important resources in that location, whereas others migrate or fly to prospective or historical breeding territories to do the same. Spring is the season when woodpeckers most frequently confront others of their own and other woodpecker species as they compete for drumming posts—resonant and often prominent perches from which they advertise to potential mates or rivals—as well as nest sites and mates. Males arrive first, attempting to establish their relative dominance, secure and defend territories, and begin creating potential nests, followed by females, who choose from the most fit males through often elaborate courtship.

During this season, the seemingly indefatigable woodpeckers are at their most arresting as they seek attention from potential mates by flamboyantly advertising their vigor through calls, drumming, and courtship displays. It can seem as though you are witnessing a competition of percussion and dance with the trilling males waving their colored caps and capes, which all abruptly falls silent as the sun rises.

Some woodpecker species may enforce sex-exclusive feeding niches. Here, a male Downy lunges toward his mate, who responds by spreading her wings before retreating.

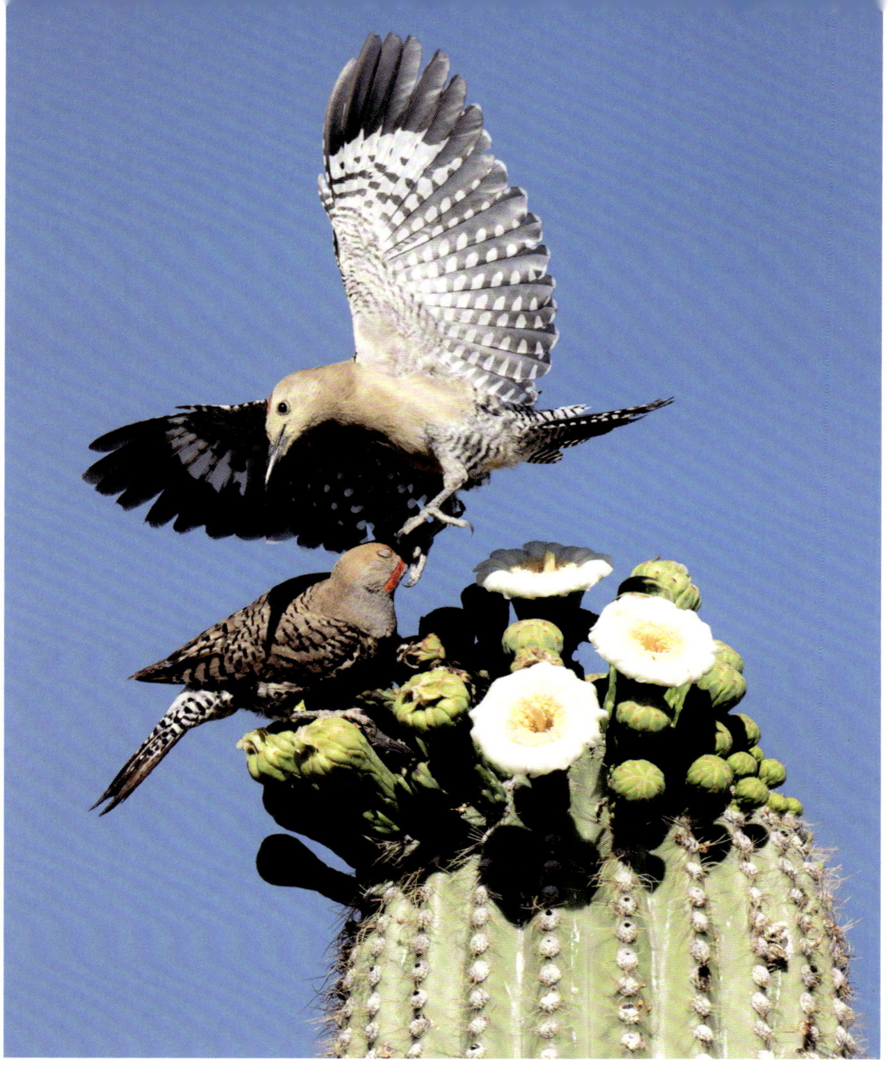

Aggressive physical contact is uncommon between woodpeckers but can occur when they compete for food. In this case, a Gila Woodpecker grabbed the head of a Gilded Flicker before they both tumbled to the ground.

HOME RANGE AND TERRITORY

Woodpeckers live in a home range, which comprises the areas they utilize to meet their needs throughout the year, such as foraging, breeding, and finding shelter. Migratory woodpeckers live in separate winter and spring/summer home ranges. A territory is a defended subset of a home range that is typically centered around potential nests and food sources.

It is difficult to quantify the size of a woodpecker's home range or territory since they utilize more vertical dimensions, the quality of which determines the quantity they defend to meet their needs. The size of the defended territory shifts throughout the year depending on if it is centered around nesting, guarding food stores, or temporary foraging opportunities.

All woodpeckers are territorial to some degree, varying by species, age, season, and the scarcity of the resource they are attempting to control. The most aggressive territorial defense peaks just before incubation and is directed at the individuals and species that present the most competition or threat, beginning with members of a particular woodpecker's own species and gender. If a woodpecker hopes to breed, it must establish a territory with control over, or at least access to, scarce resources, although "floaters," unpaired individuals living discreetly on the territory of another woodpecker or pair, are sometimes tolerated by species such as the Pileated Woodpecker. Larger woodpeckers are generally dominant over smaller species, although when the smaller species is more abundant or defending its own established food resource or nest site, it may defy the rules. Within a species, males are typically dominant over females. During the nesting season, woodpeckers are more tolerant of other woodpeckers and other cavity-nesting species when potential nests and food sources are plentiful.

Scarlet light illuminated the ridgelines of the surrounding mountains, yet the cactus-studded valley floor of the Sonoran Desert was still a muddy gray. Even in this darkness I could see the undulating shapes of woodpeckers flying forty feet to the tops of saguaro cacti and hear the whirring churrs and insistent screams from all directions. As the sun rose over the mountains, a saffron wave dropped first on the hillsides and then to the tops of the giant cacti themselves, where clusters of three-inch-wide milky-white blooms with citrus-yellow centers lured a variety of birds, including Gila Woodpeckers, Gilded Flickers, and Ladder-backed Woodpeckers, as well as tanagers, orioles, doves, and thrashers.

I could feel a mix of excitement and trepidation just before each woodpecker dipped its head deep into a bloom to consume the rich nectar, since during these delicious moments the submerged eyes would not see incoming threats, whether the daggerlike bill of another woodpecker or the talons of a nesting Ferruginous Pygmy-Owl.

THE WOODPECKER AND THE OWL

Woodpecker cavities are well protected against the elements and predators. As such, many other species that cannot excavate themselves readily use and compete for vacant woodpecker nests. Studies on Pileated Woodpeckers and Northern Flickers, for example, have found that their cavities are used by more than thirty species, including ducks, bluebirds, nuthatches, wrens, chickadees, bluebirds, kestrels, swifts, owls, and small mammals, plus insects.

Most forest owls prefer to nest in tree cavities, but it takes several decades for those cavities to form naturally via lightning, fire, or rot. And cavity-nesting owls prefer spaces with small entrances and large interiors on the vertical faces of trees—exactly the type of nest that woodpeckers create. Twenty North American owl species, including several species of screech-owls and pygmy-owls, as well as Saw-whet, Elf, Flammulated, and Northern Hawk Owls, will take advantage of an empty cavity found in their habitat. Woodpecker excavations may also indirectly benefit as many as fourteen other owl species that nest in larger cavities by introducing heartrot to trees, which causes them to decay and hollow out more quickly.

When required snags are present, there is still competition for available nest cavities of the appropriate size, and whereas squirrels, woodpeckers, trogons, and even chickadees might make adjustments to the size of existing cavities, owls must accept what is available, making it crucial that the right-size woodpeckers are continuing their work. Among the pygmy-owls, for instance, Northern Pygmy-Owls show a preference for Hairy Woodpecker–created cavities, and Cactus Ferruginous Pygmy-Owls for Gila Woodpecker cavities to limit predation. When larger owls such as Northern Hawk Owls utilize woodpecker cavities, only those of the Pileated Woodpecker will suffice.

Woodpeckers may benefit when owls consume rodents that might predate their nests; however, they may have their cavities usurped or preyed upon by their owl neighbors. Owls, too, may come under attack from woodpeckers that instinctively recognize and react to an owl's shape and attempt to drive it away through calls, displays, or even physical attacks.

The lack of adequate nest cavities can limit owl populations, making small and medium-size owls reliant upon healthy populations of woodpeckers within their habitat for their long-term survival.

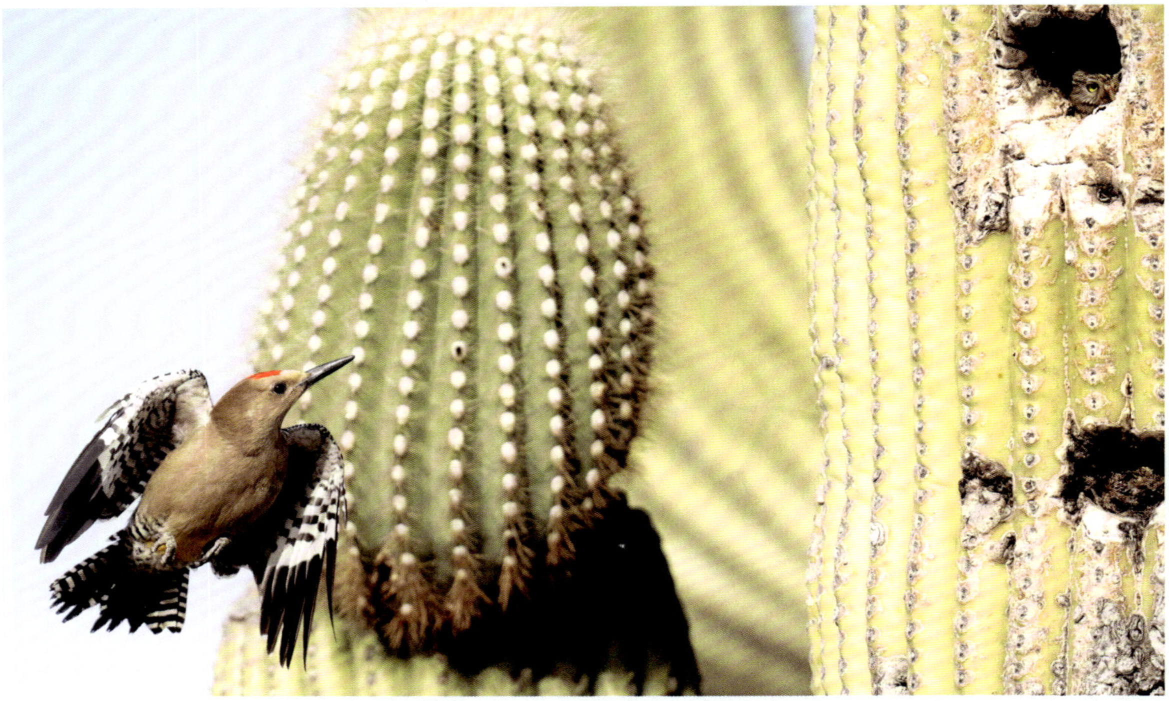

A Gila Woodpecker flies toward a Cactus Ferruginous Pygmy-Owl in an attempt to discourage it from nesting nearby. Many owls require woodpecker-created cavities for nesting, but they can threaten the woodpeckers that create the cavities, spurring potential conflict.

These bountiful blooms are open for less than twenty-four hours, so competition for the nectar can be profound. A woodpecker seemed to stake claim to each. I watched as a Gilded Flicker spread its glowing underwings and tail like a golden cape to slow its momentum before landing on one milky bloom, and almost instantly a much smaller Gila Woodpecker barreled into its belly and grabbed the flicker's head with its claws until the flicker retreated.

At another saguaro, the same flicker drove away a Gila Woodpecker with similar intensity. A much more civil encounter occurred when the flicker's mate fed at an adjacent bloom. Initially the pair greeted each other by bobbing their heads in unison and flashing the gold under their tails, but they quickly settled down to feed. All woodpeckers lunged at and displaced tanagers, orioles, and doves.

I would later find the nests of the Gila Woodpecker and Gilded Flicker and realize that each had exerted more energy to establish dominance at the saguaro closest to its nest regardless of the relative size or dominance of the competing species, a pattern I would later see with other woodpeckers in other habitats in relation to scarce resources.

Three hours and 25 degrees Fahrenheit later, the warm golden hues of the rising sun had been replaced by gray tones cut harshly by shadows amplifying the hostility of the surrounding thorny vegetation, and the birds had fallen silent and invisible. In the scorching heat of the Sonoran Desert, survival efforts are condensed into the first and last hours of daylight, creating a sense of rich color and life in places where potential food resources or nest sites must be established in competition with those species with similar preferences.

It is easier to see the boundaries of defended territories in the Sonoran Desert than, for example, a seemingly contiguous pine forest. The low-lying seasonal washes that host saguaros and their nest sites are often narrow cracks of green breaking up a plain of brown earth. Walking away from such an oasis is like stepping out of a noisy marketplace, as woodpeckers do not defend or fight over the large areas that do not include resources such as food, nests, water, or drumming sites.

HABITAT PARTITIONING AND NICHES

Woodpeckers may be aggressive toward competitors and even partners year-round as they struggle to secure mates, nest sites, and food. Since they are such specialized birds with seemingly similar roles in an ecosystem, it is remarkable how many species can coexist in a relatively small area. When resources are scarce, they reduce potential conflict by partitioning their habitat, pursuing different food and nest sites.

Interspecies Competition

A woodpecker's potential foraging area begins inches below the ground surface for woodpeckers such as flickers, which pursue subterranean ants, and rises hundreds of feet above the tallest treetop for fly-catching woodpeckers such as Lewis's.

Species may pursue insects on the surface and subsurface of stalks, trunks, branches, and twigs of varying diameter, degrees of decay, angle of lean within a different stratum, or relative height within a stand. Some may find their food in the open, others along the edges of stands, and some under the darkest canopies. Other species harvest buds, fruit, berries, nuts, acorns, pollen, the eggs and young of other birds, or frogs, lizards, and even snails. Some drill wells to eat sap, visit feeding stations, or even feed on dead animals.

Despite these specializations, the competition for nest snags, food resources, water, or even drumming sites can lead to direct conflict between species that normally avoid one another. In such situations, one species establishes a relative dominance to reduce the chance of risky physical altercations. Such dominance is usually context dependent and is influenced by hormones, available time, and the resource at stake.

Worldwide, the greatest diversity of woodpecker species generally occurs in tropical rain forests, peripheral fragments of which can be found in parts of the Caribbean islands and southeastern Mexico. In such forests in southeastern Mexico, you can find several species in a single stand, but similar tropical forests on many large Caribbean islands host only one resident woodpecker (see the "Caribbean Islands" sidebar in chapter 5, Surviving Winter) because woodpeckers are poor dispersers.

In the eastern states and provinces of the United States and Canada, species diversity increases from north to south and peaks in mature longleaf pine woodlands (see the "Longleaf Pine Habitat" sidebar in chapter 3, Life in the Nest), where you might find eight species. Woodpecker species diversity is lowest where the habitat is less treed and with less topographical diversity, such as in North America's interior scrublands, grasslands, and deserts lacking pine, oak, or large cacti.

A dozen or more woodpecker species can be found within a relatively small area in and adjacent to the western mountains from British Columbia, Canada, through several western US states and into the Sierra Madre of Mexico. In these places, habitats change with elevation gain, resulting in more diversity over a small area (see Merriam's life zone concept in the "Sierra Madre" sidebar in chapter 3, Life in the Nest). Topographic complexity in these areas results in a mix of wet and dry habitats and a diversity of tree species and canopy cover to which unique species of woodpeckers have adapted. Habitats transition and canopies close as elevation increases from drylands to oaks to pine to fir and finally to spruce.

In some such locations several species may advertise from or nest in a common area that provides scarce snags but seek food far from one another. In these cases each species excavates cavities proximate to their preferred foraging areas and in many cases advertise and nest on slightly different schedules, which limits conflict.

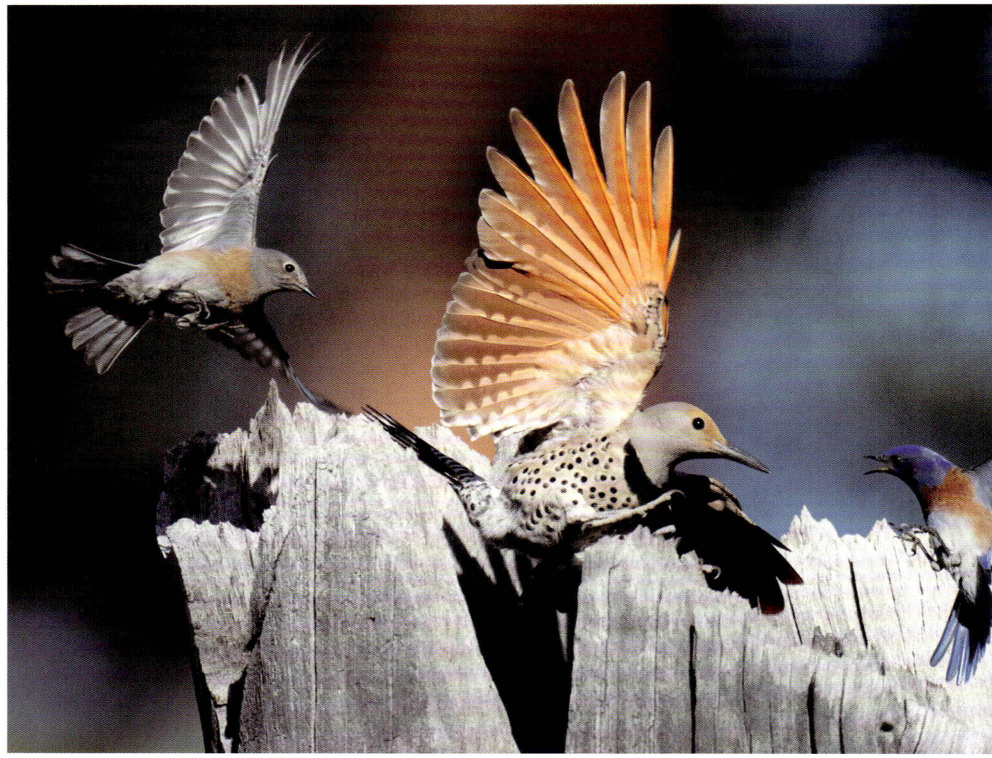

When periodic fire kills trees, the pine and pine-fir forests of the western United States provide a strong example of how this plays out when a number of species are able to partition dense nesting habitat through subtly different timing, nest placement, and foraging preferences.

Generally, Hairy, Black-backed, and Pileated Woodpeckers; Williamson's and Red-breasted Sapsuckers; and Northern Flickers begin their spring cycle in these forests first, followed by White-headed, Downy, and American Three-toed Woodpeckers and Red-naped Sapsuckers, with Lewis's Woodpeckers nesting last. Several of these species occupy similar niches in the East, with the Yellow-bellied Sapsucker and Red-headed Woodpecker occupying a similar relative position to the Red-naped Sapsucker and Lewis's Woodpecker.

Downy Woodpecker and Red-breasted and Red-naped Sapsucker nests open toward their preferred foraging areas in aspen and other decidu-

Rather than creating their own cavities, Northern Red-shafted Flickers often compete with secondary cavity nesters. Western Bluebirds, such as this pair, are aggressive enough to dissuade this female Flicker.

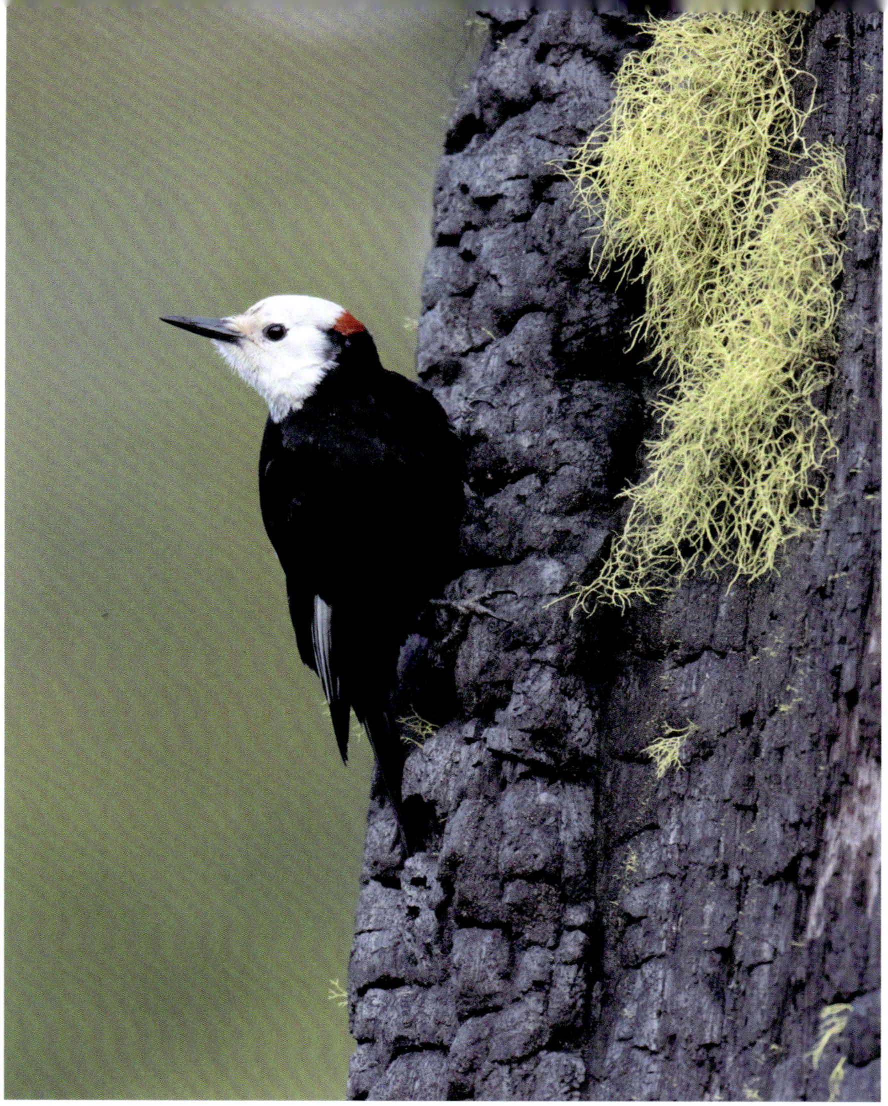

Forest fires create an abundance of food for several woodpecker species including the White-headed, one of many species that benefit when burned trees rot and soften, making them easier to excavate.

ous trees. The Downy prefers denser vegetation near water, whereas the sapsuckers spend more time managing their sapwells on trunks and even fly-catching. Red-breasted Sapsuckers forage in more mature forests than Red-naped Sapsuckers or Downy Woodpeckers.

Williamson's Sapsuckers are more conifer dependent than other sapsuckers and choose forests dominated by ponderosa pines, Douglas-fir, and sometimes western larch, where they forage for tree ants and carpenter ants and tend sapwells. Where they occur, Pileated Woodpeckers prefer cavities near moist, closed-canopied forests with large decaying snags, where they forage for carpenter ants, wood-boring insects, and their larvae. This largest of North American woodpeckers forces others to alter their feeding strategies, obliging smaller woodpeckers to focus lower in the tree column.

White-headed and Lewis's Woodpeckers and flickers all carve nests near open woodlands where they forage. The Hairy Woodpecker is more of a generalist but also thrives in these areas, where they nest before White-headed Woodpeckers. Lewis's Woodpeckers nest later to take advantage of flying insects and prefer sites within or adjacent to large openings from which they hunt. Northern Flickers spend a great deal of time foraging on the ground, including in some of the same open-canopied areas where the Lewis's Woodpecker might be fly-catching in the air or foraging on branches, greatly reducing the chance for conflict.

Hairy and White-headed Woodpeckers both like to forage in coniferous stands with a lot of snags and downed woody debris. Hairy Woodpeckers may be able to breed earlier because of their greater facility in extracting subsurface wood-boring insect larvae that are available year-round. White-headed Woodpeckers also pursue these foods but spend more time than Hairy Woodpeckers foraging for surface prey such as ants, termites, scale insects, and beetles, which are not available until temperatures are warmer.

Burned forests and insect infestations in mountains attract many species of woodpecker due to the bounty of prey. The most specialized for such situations are the Black-backed, American Three-toed, and Hairy Woodpecker. The Black-backed Woodpecker specializes in wood-boring beetle larvae on trunks and large branches. The American Three-toed Woodpecker focuses on denser areas of spruce and lodgepole pine where it seeks pine beetles and their larvae. The Hairy Woodpecker is more of a generalist, seizing a variety of beetles and their larvae. Burns are also a bounty for Arizona, Downy, Lewis's, Pileated, and White-headed Woodpeckers, which relish beetles and beetle larvae.

In the Sonoran Desert, the Ladder-backed Woodpecker nests earlier and in woody vegetation like agave, yucca, and mesquite in contrast

THE SONORAN DESERT

Iconic, multiarmed, spiny saguaro cacti rise several dozen feet above the thorny shrubs of the Sonoran Desert. This rich ecosystem lies west of the northern Sierra Madre Occidental pine-oak forests in extreme southeastern California, southern Arizona, and northwestern Mexico. The combination of rocky soils and less than five inches of annual rain eliminates most of the grasses and trees found farther north. The Sonoran Desert is the only subtropical ecosystem in the western United States. It is the hottest desert in both the United States and Mexico and is warm year-round, in contrast to neighboring deserts, which are cold in winter. It has two seasons of rainfall rather than the typical one, resulting in more plant species than any other desert in the world.

This vibrant ecosystem boasts a wide array of species, including two woodpeckers that are indicator species for this habitat: the Gila Woodpecker and the Gilded Flicker. They serve, along with the more broadly distributed Ladder-backed Woodpecker, as keystone species because they create cavities in the saguaros later used by Elf Owls, Western Screech-Owls, and the threatened Cactus Ferruginous Pygmy-Owl; Ash-throated, Brown-crested, and Great-crested Flycatchers; Purple Martins; Lucy's Warblers; and many other creatures of the desert.

Gila Woodpeckers and Gilded Flickers find their richest habitat in close proximity to riparian or *xeroriparian* (dry stream) areas, where dry washes carry a lot of water and soil during extreme monsoon rains. The result is an impenetrable vein of thorny trees and shrubs that include ironwood, palo verde, mesquite, and of course the saguaros that germinate in their shade. These thickets facilitate safe roosting, predator avoidance, and thermoregulation.

Gila Woodpeckers and Gilded Flickers are relatively common but have limited ranges, the core of which is in the Sonoran Desert. Both are threatened by conversion of habitat into farmland and housing and by wildfire fueled by nonnative grasses, which can kill large saguaros.

The shy Gilded Flicker often leaves populated areas, whereas Gila Woodpeckers are more impacted by the quantity of surrounding intact habitat. Gilas suffer losses due to ever-increasing numbers of European Starlings. The losses from development and starlings are expected to become severe.

Recent population declines in the United States motivated Partners in Flight, a network of more than 150 bird conservation organizations, to place the Gilded Flicker on its Yellow Watch List. In California, the Gilded Flicker is listed as "state endangered." Threats to both species are alarming, and their declines could directly impact the survival of the saguaro itself, as they are likely important pollinators.

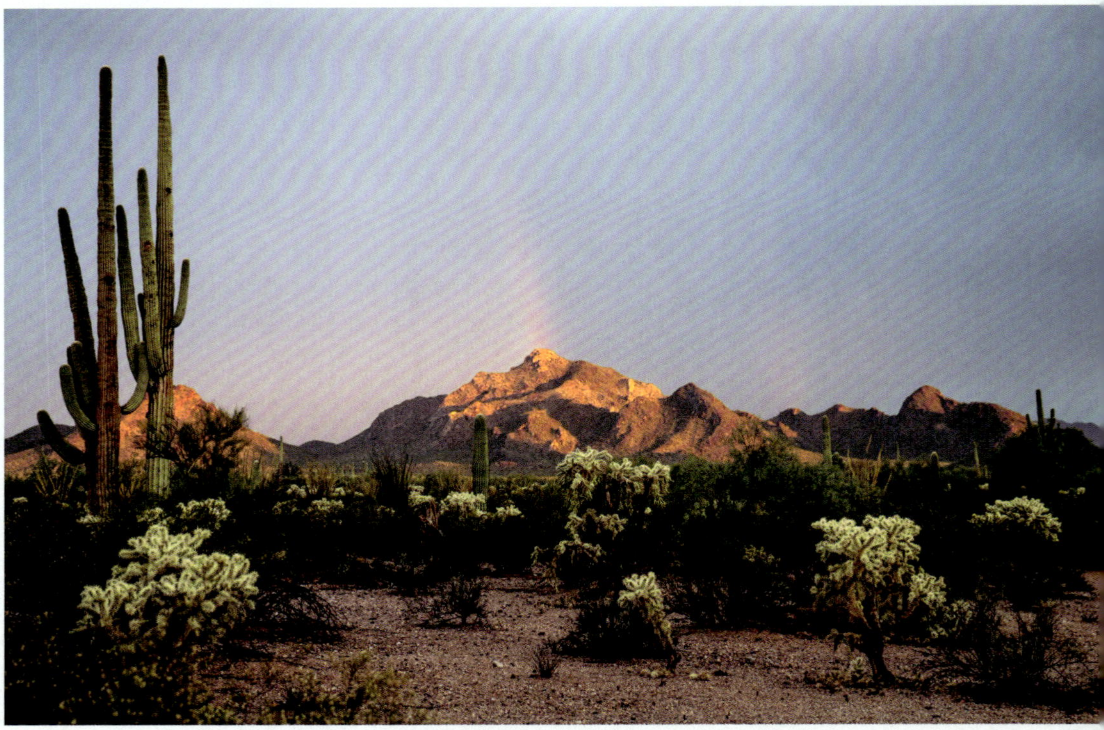

The Sonoran Desert is a biologically rich hot desert that meets the needs of several woodpecker species, including the Gila and Ladder-backed Woodpeckers and Gilded Flickers, which in turn create cavities used by many species of animals, including several small owls, such as the Cactus Ferriginous Pygmy-Owl.

to its later cacti-nesting neighbors. Gila Woodpeckers tend to carve cavities low (3 to 23 feet), inversely related to the height of the cacti, and sometimes even on the undersides of limbs on multistemmed cacti. Gilded Flickers excavate larger holes between 7.5 and 10 feet from the top of saguaros.

Sympatric Species Competition

Woodpeckers may be more aggressive toward sympatric species (related species overlapping in range) so that the territory or range of one utilizing a similar niche largely begins where the other's ends. These often fluid boundaries vary seasonally and geographically and are influenced by differing foraging and nesting preferences, which affect when each species might expend more energy toward the exclusion of the competitor.

From southeastern Canada through the eastern United States and parts of Texas, three members of genus *Melanerpes*—Red-headed, Red-bellied, and Golden-fronted Woodpeckers—are extremely territorial toward one another where they overlap in range, with the Red-headed often dominant over the other two.

This also occurs to the west, when the Lewis's Woodpecker's range encounters one of two *Melanerpes* relatives in their respective regions. The Lewis's Woodpecker overlaps with the Acorn Woodpecker in western and southwestern oaks and with the Red-headed on the western Great Plains. In these places the larger Lewis's retains the most open areas, where fly-catching is most fruitful, while in each region the Acorn and Red-headed are relegated to more wooded areas, which their more flexible feeding strategies allow them to exploit.

In the mountains of the southwestern United States and Mexico, three species nest in adjacent habitat (see the "Sierra Madre" sidebar in chapter 3, Life in the Nest). With a gain in elevation and corresponding change in habitat, the Ladder-backed Woodpecker gives way to the Arizona Woodpecker, which in turn is replaced by the Hairy Woodpecker. During fall and winter, Hairy and Arizona Woodpeckers forage downslope, and Ladder-backed Woodpeckers sometimes move upslope, with more tolerance for one another where their territories overlap, a situation that appears to be the year-round norm in some places in Mexico.

From southeastern California to northwest Mexico, similar niche partitioning occurs among the *Dryobates* Ladder-backed, Downy, and Nuttall's Woodpeckers where their ranges overlap.

Cuba has one wintering and four breeding woodpeckers, including two Cuban endemics and one Caribbean endemic. With fairly distinct feeding niches, conflicts erupt most ferociously at the scarce palm snags that serve as potential nest sites for the four species. Where territories overlap, there is a definitive pecking order, with the endangered Fernandina's Flicker facing pressure from sympatric Northern Flickers as well as the Cuban Green and West Indian Woodpeckers, with the latter also predating and pilfering their nests.

Intraspecies and Gender-Based Competition

Woodpecker competition peaks when the birds are competing for the same resources at the same time, which occurs most often between members of the same species during spring. Population density is positively correlated with the supply of food and nesting resources.

Woodpecker territoriality is expressed toward rivals and sometimes toward mates, with males of many species, including White-headed, Hairy, Downy, Red-cockaded, Strickland's, and American Three-toed, enforcing a feeding niche. In many cases, the male occupies the higher part of the foraging strata, which could be a result of his need to do more long-distance signaling behavior.

In some areas the lack of interspecies competition for food may have encouraged each sex to develop different foraging strategies, resulting in differing bill lengths. Hispaniolan, Guadeloupe, Cuban Green, Puerto Rican, and Golden-fronted Woodpeckers all live in areas with few, if any, competing species, and perhaps in response to this opportunity, males have developed longer bills, ranging from up to 15 percent on male Golden-fronted Woodpeckers in Texas to up to 21 percent on Hispaniolan Woodpeckers. The longer (up to 20 percent) bill of the male Guadeloupe Woodpeckers is the easiest way to visually distinguish between the two genders.

THE COURTSHIP CYCLE

Most woodpeckers retain bonds with their surviving mates from the previous year, especially when that nesting attempt was successful and the birds did not migrate. Since most woodpecker studies take place during the breeding season, there is still much to learn about woodpeckers' behavior throughout the rest of the year.

Some woodpeckers that breed in colder areas may fly dozens to hundreds of miles north or upslope from wintering areas to breeding locales, and males may arrive several days to a couple of weeks before females. The longer and more consistent these movements from year to year, the more they are considered migratory. Some migratory woodpeckers may return to renew their courtship with the same mate, in the same territory, and sometimes in the same cavity as the previous year. These pairs may benefit from familiarity and usually begin their nesting earlier.

Long winter movements increase the chances that woodpeckers can be injured or killed or lose their way, impacting their ability to retain mates, the timing of their breeding cycle, and how many young they might produce. Migratory woodpeckers and those that experienced nest failure in the previous year are the most likely to pair with new mates and look for them in adjacent or nearby territories.

Individuals in many species seem to forage independently of their mates within their home ranges until the male begins advertising in preparation for breeding. Most remain close enough year-round to quickly reunite with their mates in the late winter or early spring. Within some species, pairs maintain constant communication throughout the year and may even forage together. This practice is more common in tropical areas, with species such as the Lineated, Pale-billed, Cuban Green, Smoky-brown, Strickland's, Golden-olive, and Puerto Rican Woodpeckers, as well as the Antillean Piculet. In Canada and the United States, Pileated Woodpeckers not only stay in contact throughout the day but also sometimes roost in separate cavities in the same tree. Woodpeckers that live all or part-time in loose colonies communicate with their mates throughout the year, although it can be difficult for observers to distinguish paired birds from competitors outside of breeding season.

The behavioral differences between mates that stay in contact year-round versus the woodpeckers that seem to avoid one another are subtle and differ from year to year. For several years I have followed Pileated Woodpeckers in a park near my Seattle home. It took me several months to realize that a pair remain in contact most of the day year-round. They often foraged in sight of each other or on the same tree. Even when they appeared to be alone, within a few hours a call and response resulted in a sudden appearance of the mate nearby to remind me of the familiarity. It was remarkable how far calls carried and how quickly this large woodpecker would freeze when it heard his mate and how she would then arrive from several hundred yards away within seconds. With such sizable wings, the Pileated Woodpecker can cover significant distances in a short time, making a large territory effectively much smaller. This factor, combined with their reliance upon locally abundant available foods, such as wood-boring larvae, carpenter ants, and termites, allows these woodpeckers to stay together on their home territory year-round and may allow them to mate until one dies.

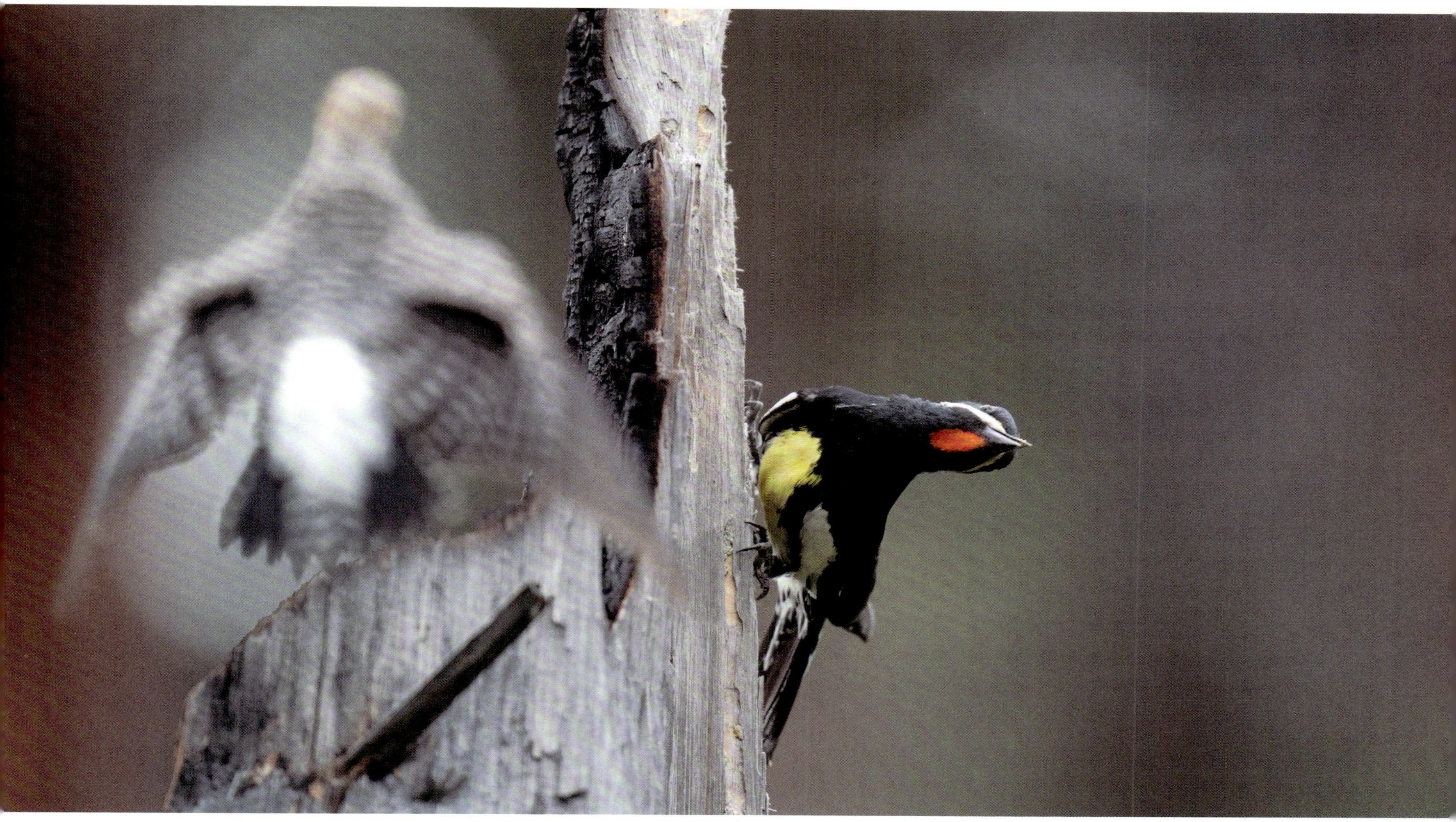

In a "bobbing-wagging" display, a male Williamson's Sapsucker, with crest raised, plumage flattened, breast puffed out, and bill pointing up, raises and lowers its head or swings it from side to side while emitting a chatter call. If a female is sufficiently impressed, she may flutter toward him.

In contrast, in that same park I only occasionally find North America's smallest woodpecker, the Downy, and rarely find a pair together between summer and late winter. The smaller wings and body of the Downy relegate it to repeatedly flying short distances between nearby trees, making small distances seem greater. Lacking the excavating power of the Pileated, it is more of a generalist that expends a lot of energy gleaning and probing opportunistically for temporary concentrations of prey. Even when foraging brings a pair closer to each other, the male will enforce separate foraging niches, keeping them apart. When food is scarce, these behaviors pull mates farther (relative to their size) from spring's nesting and roosting sites during fall and winter, although they also often retain their mates.

COURTSHIP BEHAVIORS

Trying to arrive at the zenith of courtship behaviors in a diverse woodpecker habitat reminds me of attempting to visit a wildflower meadow at its peak, when each flower reveals its most brilliant, fully opened blooms. In both cases the showiest attributes of each species are expressed over hours, days, or at most weeks, with few, if any, displaying their optimum extravagance at the exact same moment. Some overlap, others eclipse, and still others stand alone before or after the climax. And, if you arrive a few weeks early or late, you may miss it all.

The peak of courtship in northern areas with fewer woodpecker species can be so narrow that you might arrive a week too early and

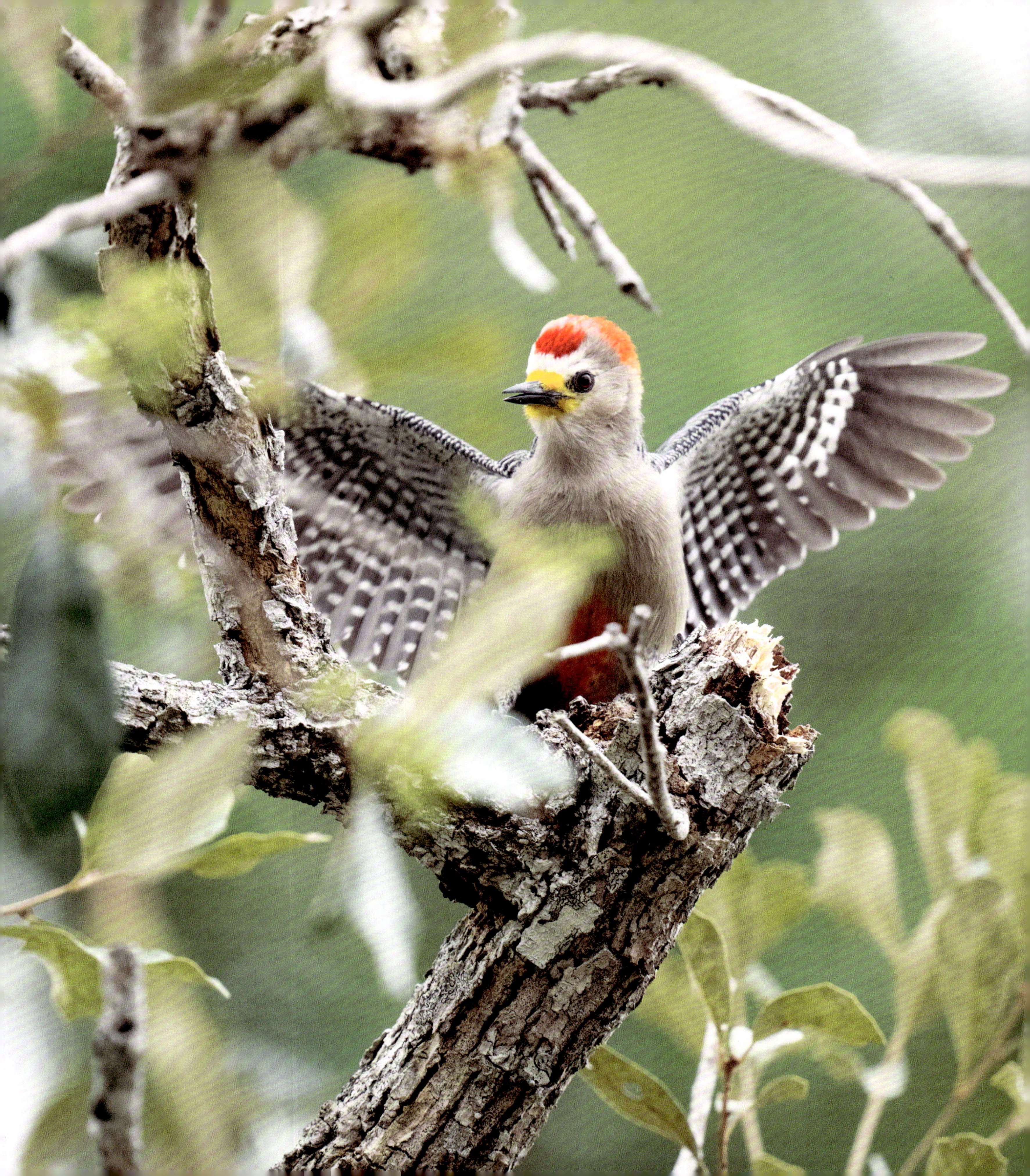

hear only wind in your ears and see icy sleeves on tree branches, or arrive two weeks later to find scattered woodpeckers foraging as they do much of the year. Within that narrow seasonal window, woodpecker activity peaks just after sunrise, after which it is often quiet until just before sunset, when individuals engage with one another again.

In all areas, the breeding cycle for each woodpecker species has evolved so that the young are fed at the very time their favored foods are at their greatest abundance. In northern and high-elevation woodpecker habitats, spring and summer are condensed into a smaller window of sufficient warmth to trigger and sustain buds on plants, sap flow in trees, the life cycles of insects, and the production of fruit, seeds, acorns, and nuts.

As you move south and downslope, spring expresses itself earlier and extends for far longer since buds, fruit, and insects are accessible over more months, allowing staggered nesting and multiple broods. Early-nesting Pale-billed Woodpeckers may begin courtship in southern Mexico in December, while Yellow-bellied Sapsuckers in northern Canada and Alaska may not begin carving cavities until June.

The dry conifer forests of the western United States feature a great diversity of woodpecker species, and in high-elevation habitats at the northern part of their range, a rich variety of calls, displays, and behaviors superficially feel as though a wave of breeding birds has erupted into courtship at the same instant. If you watch for a longer period, you can tease out that each is at a slightly different stage.

Opposite: A male Yucatan Woodpecker raises the yellow and red feathers on his head and holds out his wings as a potential mate arrives nearby.

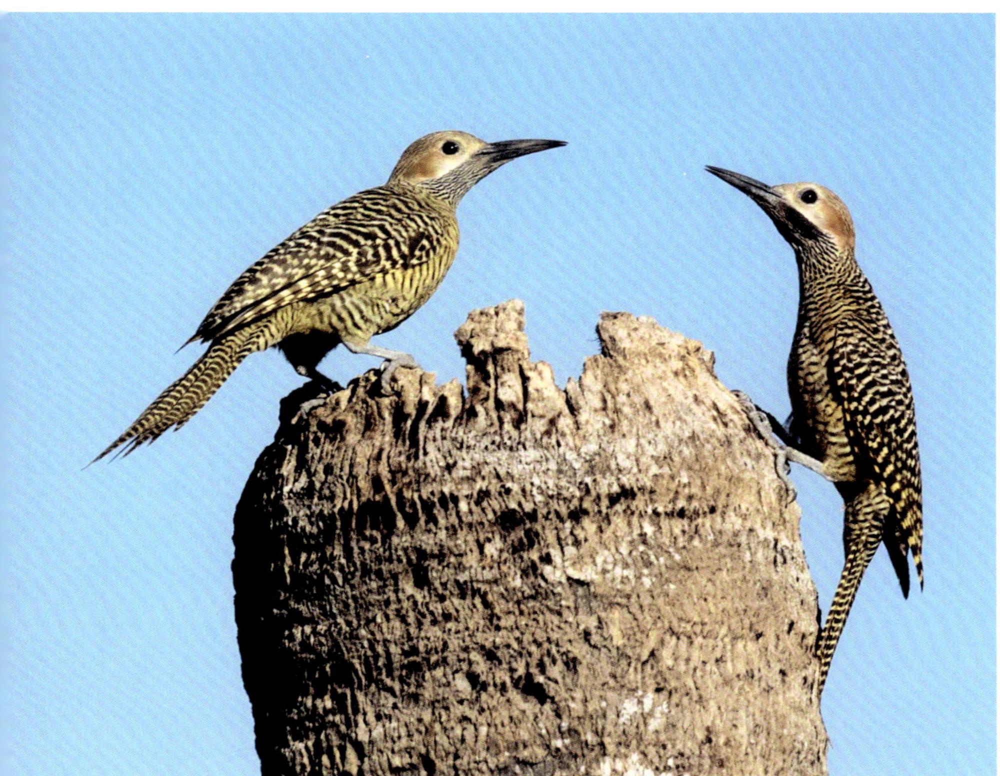

A pair of Fernandina's Flickers pause in their courtship bill-pointing atop a palm snag that will become their nest site.

Late April seemed like a good time to try observing the most species in courtship in a mixed conifer-aspen stand in central Oregon. The hurried contact calls between an established pair of Hairy Woodpeckers indicated that courtship had finished, but Williamson's Sapsuckers were drumming within a small area with females nearby, hinting that they had formed or renewed pair bonds and might nest in that stand.

Red-breasted Sapsuckers were a bit behind, still seeking mates, as I could hear three males moving about and drumming. One made a particularly outrageous appeal to a female. Over several minutes he drummed in spare, Morse code–like cadence while screaming from his favorite snag until she arrived. Instantly, he raised his crown feathers and began bobbing his head, stretching his neck high, then quickly retracting it, pumping up and down, over and over. He reminded me of a feathered wind sock expanding with the shifting breeze—up to the right, down, straight up, down, up to the left, and so on.

Male Northern Flickers were at an earlier stage as they moved broadly about in same-sex groups, reminiscent of teenagers as they called and pounded out a monotonously urgent drum for females, only to be joined by other males. Each flashed his colors and asserted his dominance by assuming a higher perch on a snag and flying at his competitors with spread wings or, rarely, lunging to displace

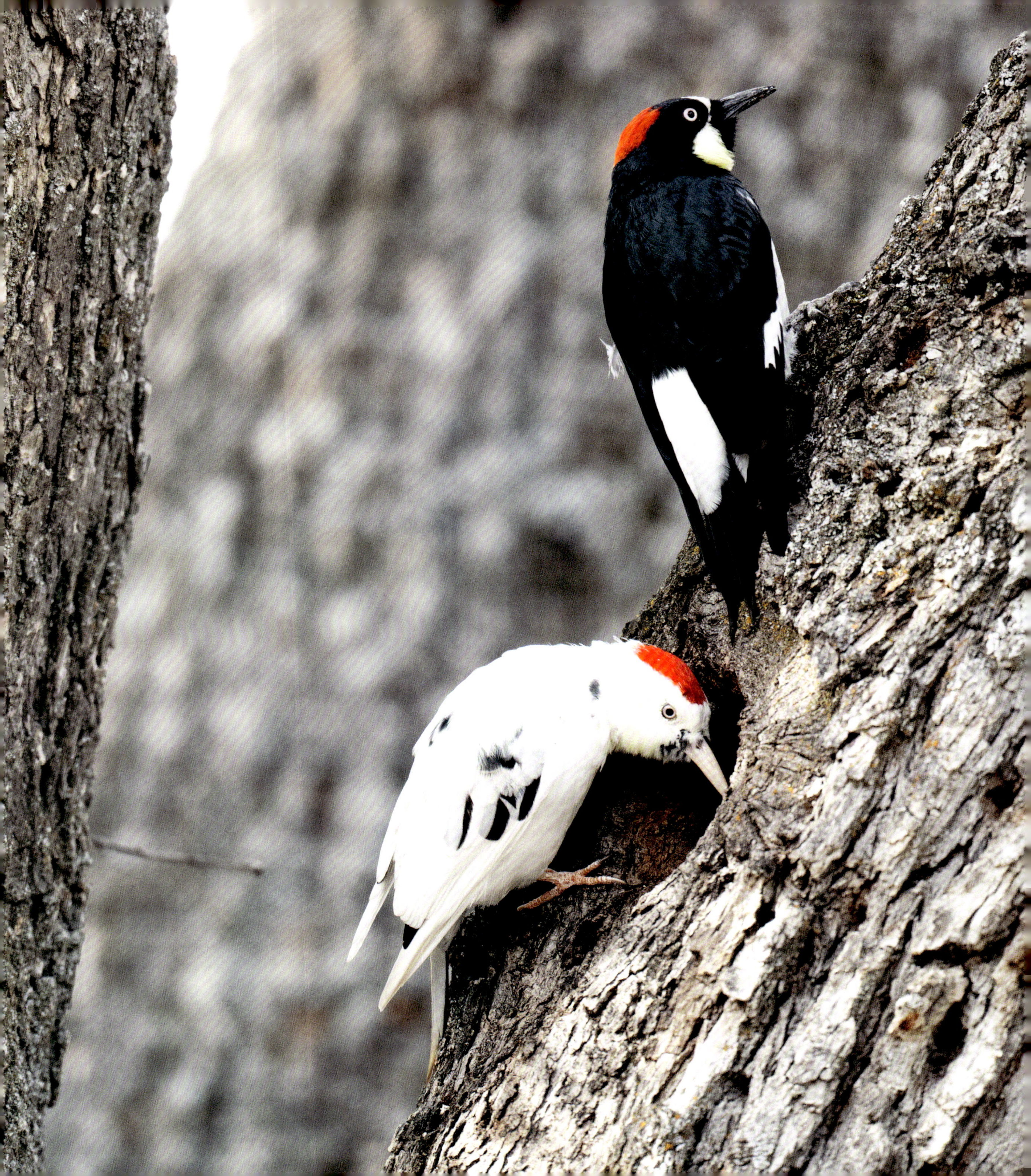

his rivals. Each subtle movement and sound were perfectly synchronized, as two birds froze into a face-off with bills pointed skyward. Responding to a signal known only to them, they simultaneously reengaged in a rhythmic and synchronized bobbing of heads in loops as if in a choreographed sword fight, while emitting a forceful *wicka-wicka*. As their heads bounced so that their bills drew circles and figure eights in the air, they danced and poked toward one another, frequently leaning from side to side and turning their tails up and forward to flash their fiery-red undersides like flags. When a female landed next to the two competitors, the intensity, pace, and amplitude doubled. Several seconds later the opponents froze, one leaning forward and the other with its bill in the air, before their bodies relaxed into apparent ambivalence, only to reengage thirty seconds later. Moments after the second bout ended, all three flickers flew in the direction of another calling male.

White-headed Woodpeckers and Red-naped Sapsuckers would begin their clamor here shortly, with Lewis's Woodpeckers beginning their own courtship a few weeks later.

COMMUNICATING WITH COLOR BADGES

Intraspecies (within a species) and interspecies (between two species) interactions are common year-round, especially near food sources and prospective nest sites during courtship. These encounters normally begin with attempts to discern which individual is dominant and thus retains a given space or resource. Since males are generally dominant over females, being able to identify gender is important, making displaying and recognizing color badges critical to reducing conflict.

Watching woodpeckers in spring can be confusing because it is difficult to discern amorous from confrontational encounters. Both are intense and involve similar calls and movements, and courtship only marks the lessening rather than the elimination of aggression. When displays occur between birds of the same gender, they are usually territorial, whereas confrontations between birds of the opposite sex more likely involve courtship or pair bonding.

Unlike most songbirds, male and female woodpeckers often look very much alike, with complex and colorful markings on their head. Such color would make open-nesting birds vulnerable to predation, but female woodpeckers can get away with this since they nest unseen within cavities. In all but four species, males differ from females in having more color in the form of brightly colored feathers on their heads, referred to as *sexual badges* or *color badges*. This obvious sex difference reduces much of the conflict between birds of the same gender. Male woodpeckers retain these gender-specific color markings year-round, in contrast to some colorful birds that only retain obvious color differences during courtship and breeding.

The color and placement of badges are relatively consistent within genera but in most species involve having more red or yellow on the throat, nape, crown, or mustache (malar). The male Red-cockaded Woodpecker possesses a color badge of red feathers that is normally hidden but can be displayed at will, which may be a useful way for this cooperative breeding species to avoid frequent conflict. In one study, color bands were placed on male Red-cockaded Woodpeckers, and when those color markings were red, the nesting group displayed more friction and raised fewer offspring.

Red-headed Woodpeckers, Lewis's Woodpeckers, Guadeloupe Woodpeckers, and Red-breasted Sapsuckers are the four woodpeckers in North America (as defined in this book) and worldwide in which males and females appear identical and lack gender-specific badges. On the other extreme, male and female Williamson's Sapsuckers sport such different base plumage color that ornithologists considered them separate species for twenty-two years.

In early spring, woodpeckers investigate many potential nest sites with their mates and may communicate preference by tapping with their bills at specific cavities. The male of this Acorn Woodpecker pair is leucistic, an unusual genetic condition causing white, pale, or patchy coloration of feathers (but not eyes as in albinism).

DISPLAYS

Displays often begin or end with noisy chases before woodpeckers assert their gender and relative dominance through visual displays, especially when close to a mate or rival. Participants attempt to establish their sex, vigor, size, or boldness as quickly as possible by calling forcefully as they raise their head and neck feathers, accentuating their color badges. They also raise, spread, fan, and flaunt other colored or white feathers on the head, neck, wings, belly, and rump and underneath the wings and tail. Once paired, mates of some species may engage in the touching of bills, which is thought to be a form of symbolic feeding.

Regardless of intent, each participant normally begins a confrontation with its bill pointed up, before pointing toward the other with its tail and bill parallel to the ground. While swinging or bobbing their heads and swaying their bodies in a jerky motion in rhythm with vocalizations, sparring partners will suddenly and unexpectedly freeze at precisely the same time in a possible stand-off before resuming.

Each of these movements can occur in isolation or in combination and are somewhat consistent across species except that, whereas most species will sway from side to side on a horizontal plane during bill pointing, members of genus *Melanerpes*—including Red-headed, Lewis's, Red-bellied, and Gila Woodpeckers—tend to bow and bob their heads on a vertical plane with accompanying calls that are much louder than those of other species. Some sapsuckers bob straight up and down and side to side with much less bowing than *Melanerpes* species. Antillean Piculets deploy a basic display that combines bouts of calling with jerky head-swinging displays usually directed at rivals. The impressively crowned Pileated, Lineated, and Pale-billed Woodpeckers raise their crests and spread their wings to emphasize their most imposing features. Red-headed Woodpecker courtship displays are simpler and involve a horizontal pose with compressed plumage, neck extended forward, and shoulders lowered, as well as a "hide-and-seek" behavior in which each individual peers at its mate from the opposite side of a tree or snag, then hides on the other side as they both spiral up the tree.

During the most aggressive encounters, many woodpeckers raise their wings in defense and flick them just before stabbing at or supplanting (replacing) their opponent. In the uncommon instances when birds do strike, the bill and feet are their primary weapons, and this sometimes results in a noisy tangle of feathers falling to the ground before they fly their own ways.

Woodpecker displays can involve more than two birds. When three or more birds, including two genders, confront one another, the birds of the same sex confront one another while the bird of the opposite sex looks on. The audience of a potential mate always rachets up the intensity of the calls and displays. An unusual exception to this behavior is the Golden-fronted Woodpecker, which defends territories against both genders. They usually chase invaders to the edge of the defended territory.

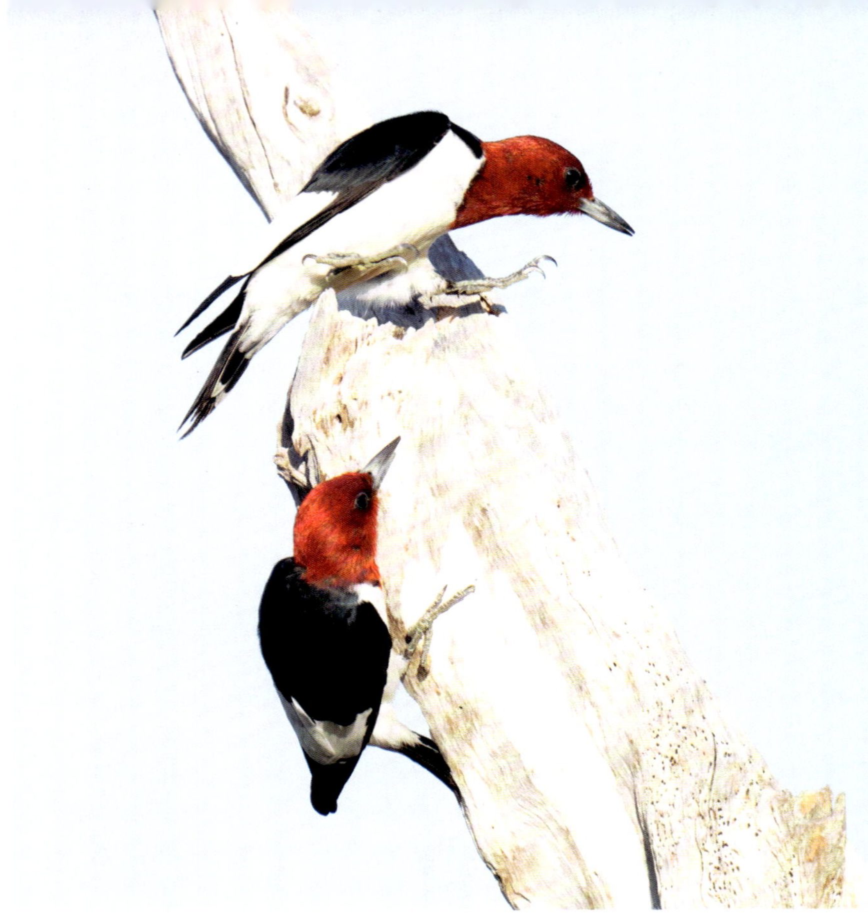

Chasing one another in flight or on tree trunks is a component of courtship for some woodpecker species, including these Red-headed Woodpeckers.

Opposite: Raised orange feathers on the crown and near the bill help a male Red-bellied Woodpecker attract a female's attention as he flies to a drumming and displaying perch.

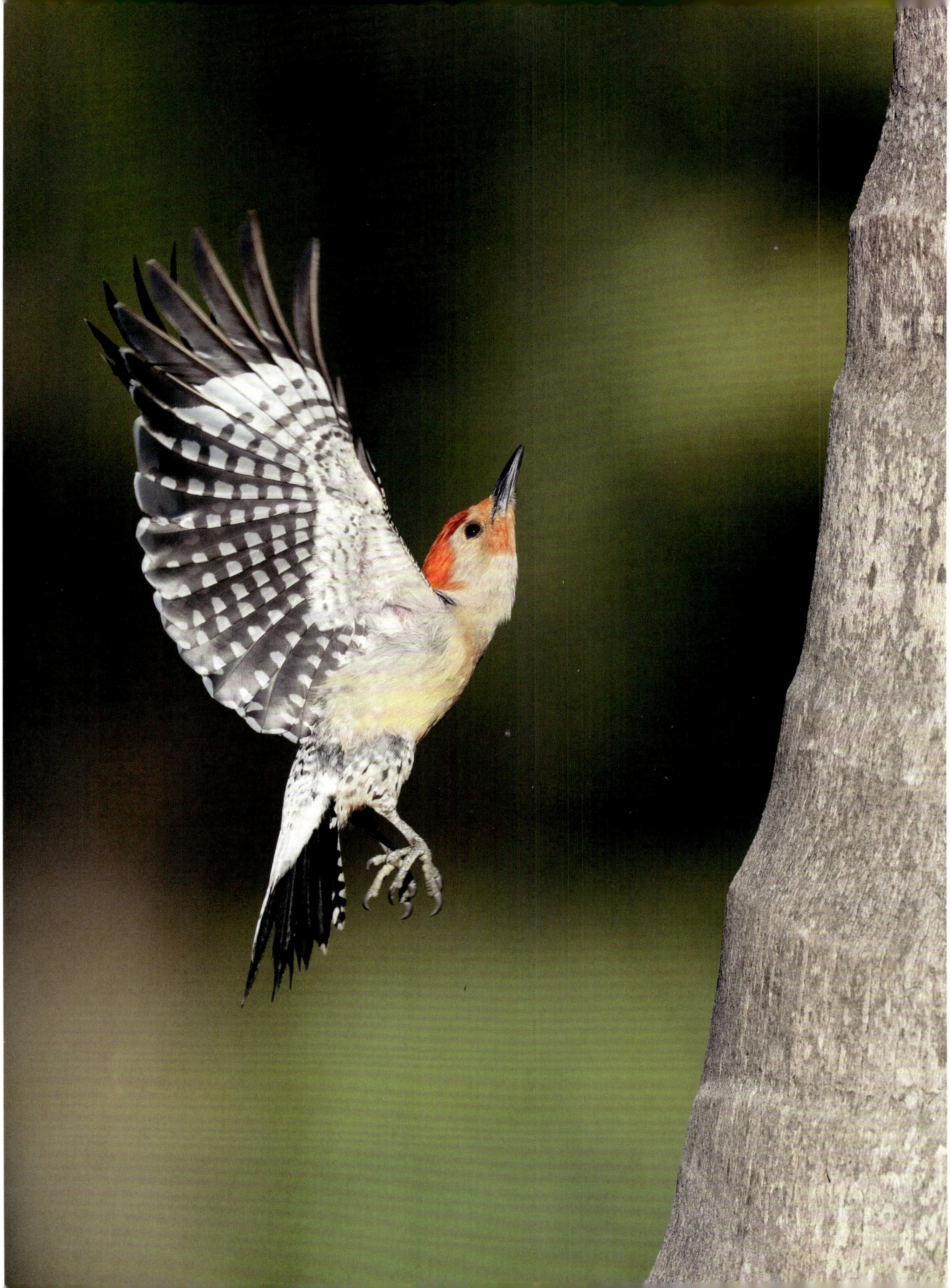

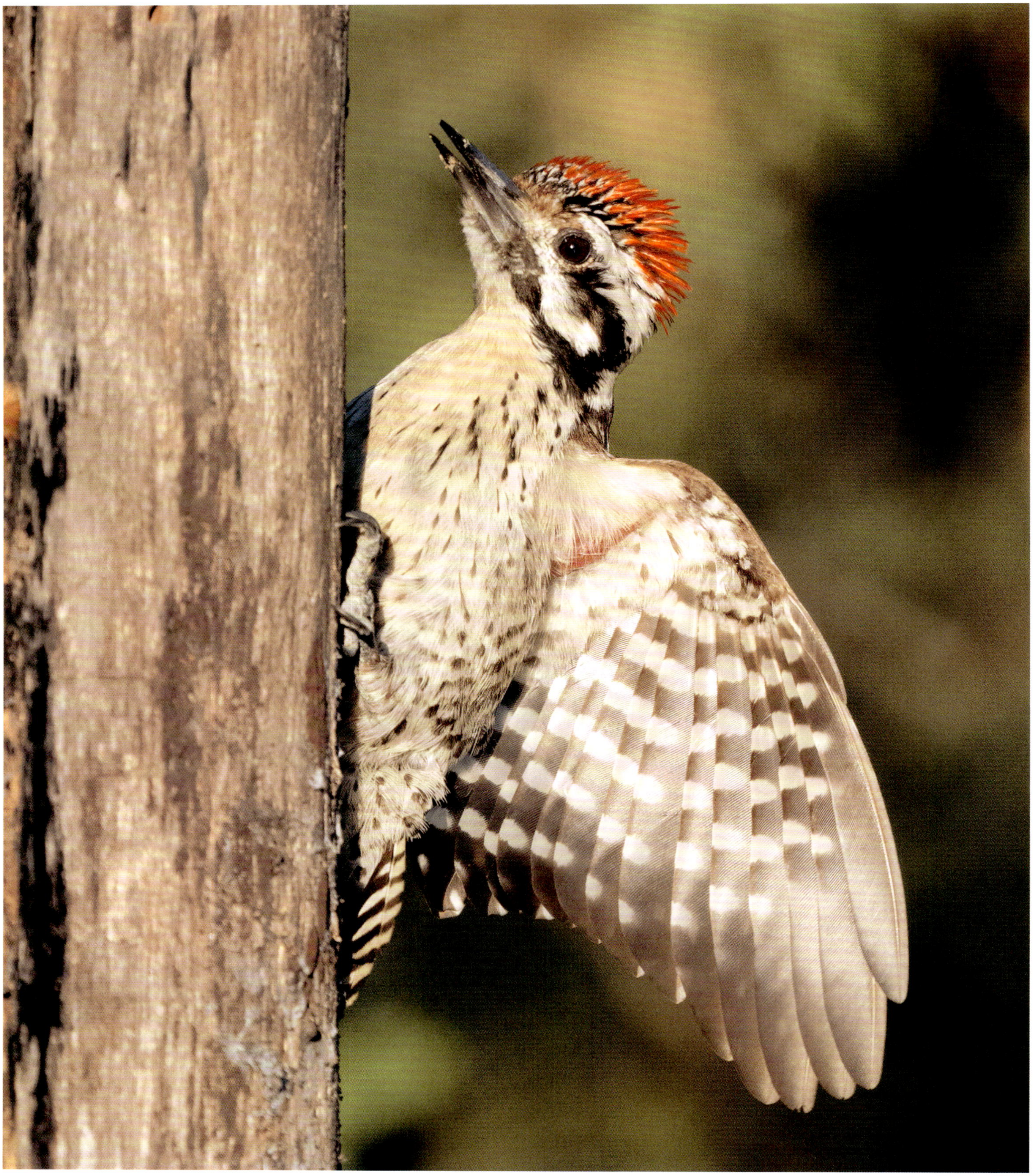

One April morning in Seattle, I watched as a male Northern Flicker sprayed wood chips from a nest box as an agitated female landed atop the container and immediately began fanning her spread tail toward another female that had just landed beside her. The two females faced each other, bills up, bobbing their heads as they jerkily punched circular patterns in the air, waving the red undersides of their spread tails, and screaming their *wicka-wicka* calls. The male in the cavity perched in the opening, looking up and watching with seeming indifference as the females bobbed, swayed, jabbed, and tilted their bodies, but occasionally he emitted his own *wicka* call and bobbed his head, which intensified the pace and fervor of the sparring females. Finally, the agitated female flicked her wing at the intruder before lunging at her, causing the interloper to fall back and into a retreating flight.

The interactions of mated pairs might appear as aggressive as those between same-sex rivals, with noisy displays and chases, and though normally it is safe to assume that aggressive interactions between birds of the opposite sex are courtship, there are exceptions. I had the privilege of watching White-headed Woodpecker courtship, including the male's flutter display as he arrived at a nest cavity, the female's invitation posture, and subsequent copulation, but was surprised by another interaction on that same day. A male flicked wood chips from a completed nest cavity when a female alit atop the snag. The male reacted noisily, screaming as he flew up toward the female, who in turn zipped to a thin branch, the male close at her tail. The two confronted each other in a spiraling chase from opposite sides of a branch, the male flicking his wings at her and lunging with bill and claws before the female retreated. It is likely that the male was protecting his mate, who was about to begin laying eggs, a behavior that has been reported with other woodpeckers.

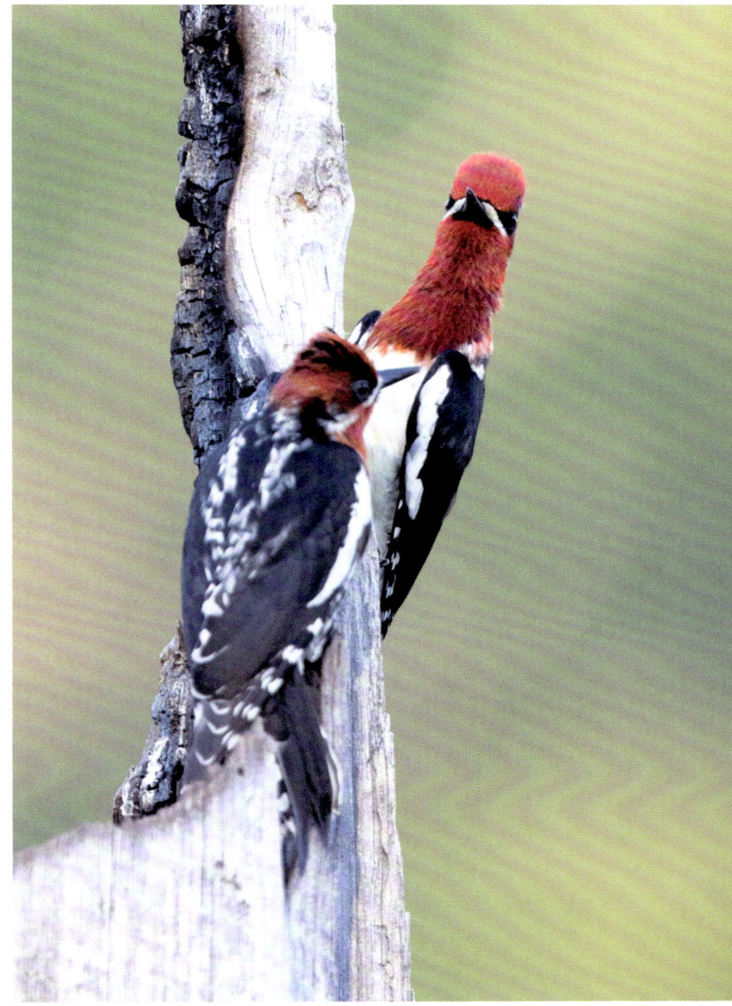

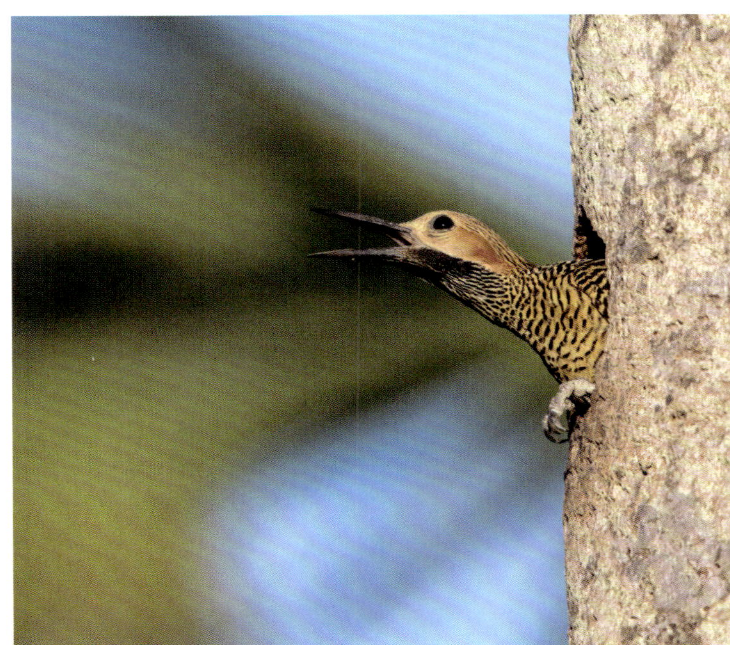

Top: The courtship behaviors of Yellow-bellied, Red-naped, and Red-breasted Sapsuckers like these two are virtually indistinguishable. The male drums and calls, and as a female approaches, he expands the red feathers on his head and pumps his head up and down as he sways from side to side.

Bottom: A male Fernandina's Flicker calls from a potential nest cavity as his mate approaches to relieve him of his excavation work.

Opposite: A Ladder-backed Woodpecker raises the red feathers on his crown and spreads his wings as he spirals up a snag to impress his mate or intimidate a rival.

FLIGHT DISPLAYS

I had begun to accept woodpecker movements as extremely ritualized, deliberate, and even predictable. So one April morning in the southern Cascades, I was anticipating the usual behaviors while watching a male Williamson's Sapsucker drumming to attract a mate. I was not expecting the aerial dance that was about to unfold. I positioned myself in sight of a snag I had seen him use. After waiting for three hours with no success while hearing sapsuckers drumming at other snags, I was trying to convince myself to wait longer when a tan, barred female arrived at the snag. She quickly began to execute a softer version of the Williamson's Sapsucker's signature staccato drum. Almost immediately I heard an unfamiliar and frenzied rattling call. Before I could identify the sound, a small quavering shape fell from above. With wings spread above his head, the male fluttered mothlike on rapid but shallow beats toward the female. I was captivated by how he managed to display spotted underwings, yellow belly, and enlarged red throat all at the same time. The strange rattle call demanded attention to these impressive features.

Upon landing, the male's display continued on the opposite side of the snag from the female. He raised his bill skyward and erected the feathers on his head and the red feathers under his throat. He pushed forward the yellow plumage of his belly, while compressing the black of his wings, and occasionally swung his head and breast backward and upside down, away from the snag, to accent his colors. Each bird then assumed a stiff, upright posture as they pointed their bills and bounced their heads in a jerky circular pattern as if dotting the outline of the letter O or the number eight, all the while swinging their bodies in sync to a pulsing rhythm. Suddenly both birds made rattling calls and took flight.

On another sunny early spring day in Seattle, I was struck by two large, white, quivering backlit shapes following each other in the canopy like tropical butterflies. The bright sun was shining through the wings of the birds and into my eyes, so it took me a few moments to realize that these were Downy Woodpeckers cavorting among the mossy limbs above me. One trailed the other with its wings spread and held high above its body, flapping softly with buoyant, shallow wingbeats.

Males and even females of other species of woodpecker utilize some version of a flight display that includes hovering, flying in a circle, and tail spreading; raised, lowered, open, or fluttering wings; and exaggerated wingbeats, all meant to draw attention to their size and color, as well as the patterns on their wings and tails. Male flickers attract the attention of their mates with corkscrewing flights culminating in perched displays, while Red-headed Woodpeckers simply chase each other through the woods.

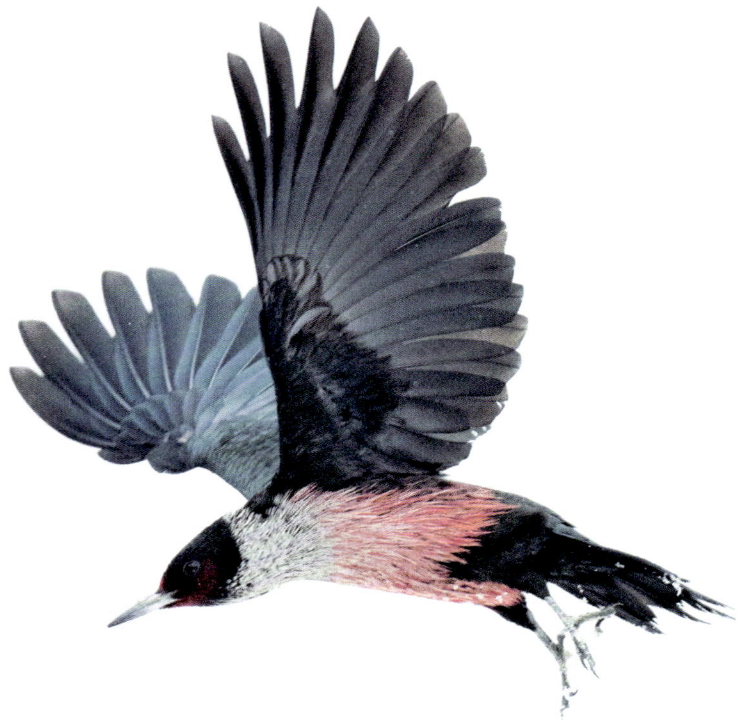

Dihedral flight displays, when wings are outstretched upward, are used by some *Melanerpes* species such as the Lewis's Woodpecker (pictured here) and the Hispaniola Woodpecker as they glide toward their mates or a rival.

All four of this book's *Sphyrapicus* (sapsuckers), both *Picoides* (American Three-toed and Black-backed), and most of the *Dryobates* (White-headed, Hairy, Arizona, and related) woodpeckers perform flight displays that make the bird conspicuous. In each case, they hold their wings fixed and open, although the presence, duration, flapping speed, and wing position relative to the body vary among species. Some hold their wings above their heads, whereas the Black-backed Woodpecker holds them more or less parallel to the body with wingtips bowed downward, and the Arizona Woodpecker holds its wings open while gliding toward a potential nest in sight of its mate.

Nest sites serve as the arena for other kinds of aerial displays from some *Melanerpes* woodpeckers. Lewis's and Hispaniolan Woodpeckers perform a *dihedral flight display* (wings outstretched upward) as they glide toward their mates at the nest. Upon landing, the Hispaniolan Woodpecker male may feed his mate, and the gesture might even be reciprocated. This courtship feeding is often preceded by a *bowing display*, in which one or both birds bobs its head with an open bill. Such mate feeding appears to be rare, but it is also known in the Guadeloupe Woodpecker. In both species the recipient of the courtship feeding is often inside the cavity working on its excavation.

DRUMMING

Percussion heralds spring for woodpeckers. Their rhythmic advertisements occur throughout the year for different purposes, but they become noticeably louder and more frequent as spring arrives and the birds need to attract or reestablish mates and delineate and defend territory. They are most intense and frequent in the early morning.

Drumming is a form of signaling that is unique to woodpeckers. Woodpeckers drum by striking their bill against a hard, resonant surface. While the head and neck move fast, the body is relatively still, which contrasts to the way woodpeckers deploy their entire body when excavating nests. Signals made by pounding upon resonant objects likely send information about the health of the drummer as well as the presence of the dead wood so important for nesting.

Each woodpecker species performs its own repertoire, varying in cadence and in number and speed of beats, as well as the space between them. Some include rolls of varying frequency or different rises and falls of amplitude. The drumming of some species sounds similar, especially within a genus, with sapsuckers using a distinctively halting series of rolls that slows noticeably toward the end before ending with a loud, demonstrative final beat or two. In contrast, Pale-billed Woodpeckers and other species within their genus can be recognized by a spare double or triple slow rap on the most resonant hollow tree that sounds like an ominous warning. Northern Flickers and White-headed and Red-headed Woodpeckers drum in an evenly paced cadence. Drumming is used by most North American species, but its deployment varies greatly by gender, with females generally drumming less frequently and more softly.

Individuals can vary slightly from the standard pattern and may have their signature recognized by others within their species. Given the sometimes slight interspecies differences and intraspecies nuances, sometimes woodpeckers respond rather indiscriminately to the percussion of other species, particularly at the hypercharged beginning of the breeding season.

Some species, such as sapsuckers, seem to have a strong preference for one or a few drumming posts, whereas others move to many within a territory, although woodpeckers restrict their drumming to a smaller area as they secure mates and select nest sites. Certain species signal from high, exposed locations while others select lower, more obscured places. Woodpeckers typically find the most resonant objects, often a dead or hollow part of a tree. Signaling to other woodpeckers is of such importance that they regularly experiment with surfaces to gain an advantage over competitors and reach more potential mates.

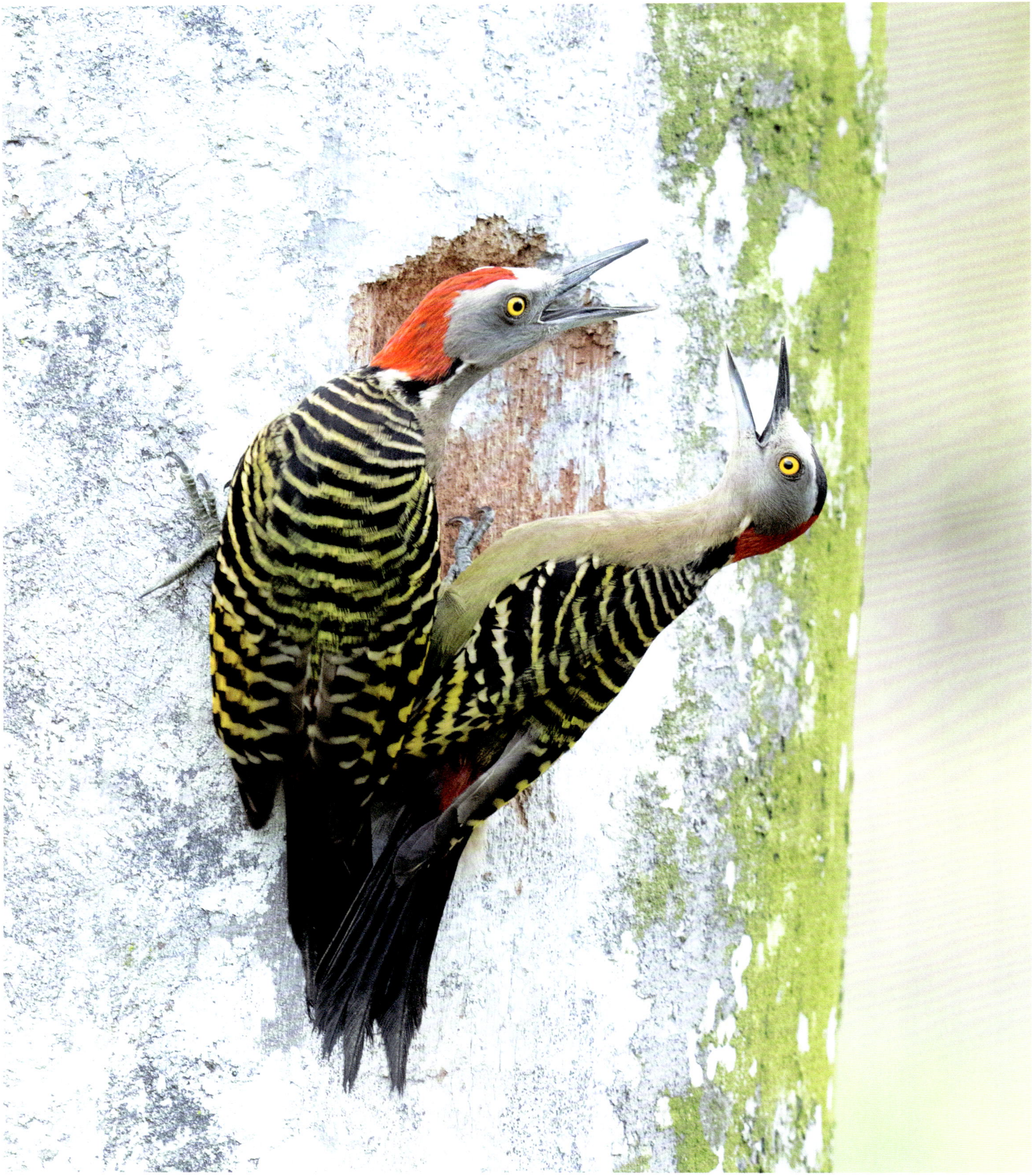

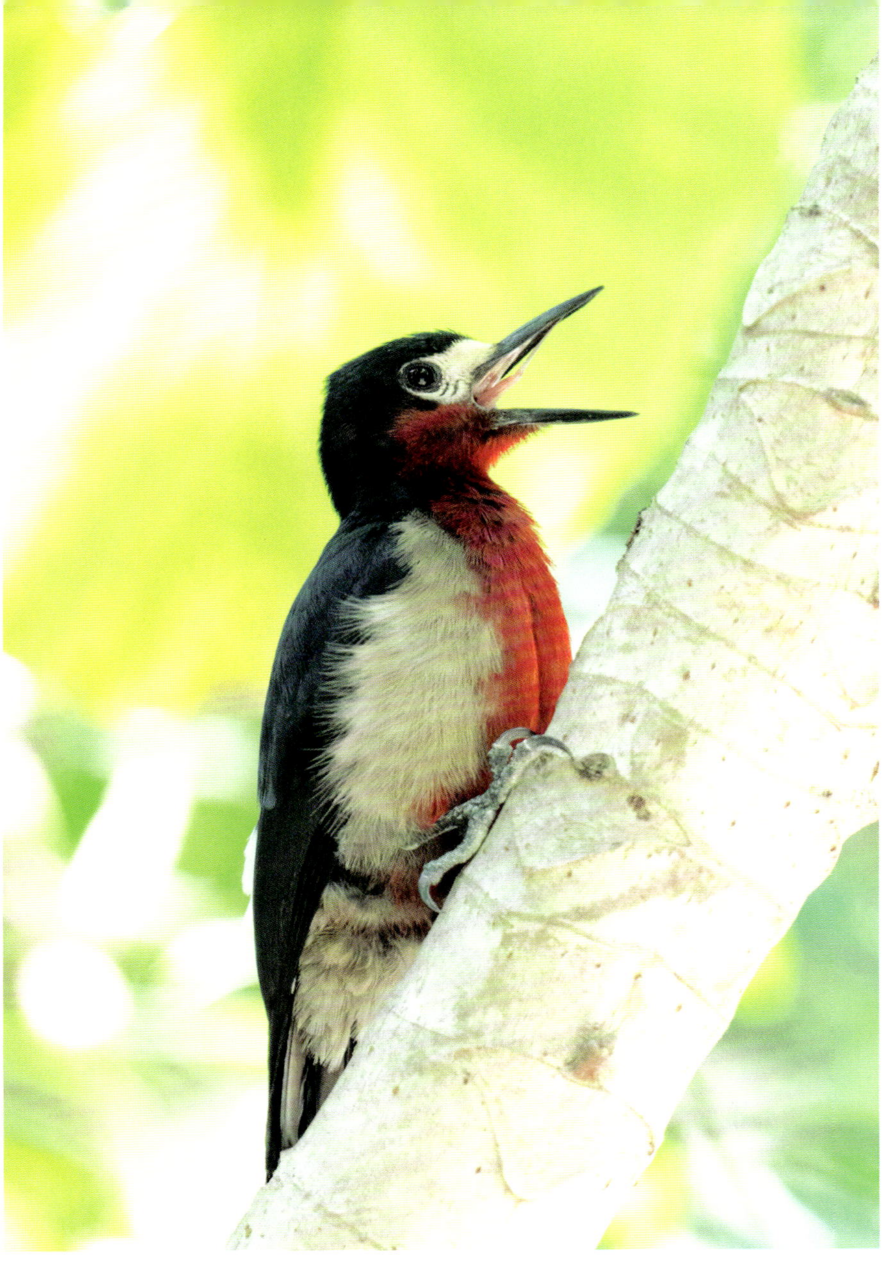

Possessing several different calls with various functions, Puerto Rican Woodpeckers are especially vocal at dawn and when reassembling a flock, yet they also have been occasionally reported to respond only to the call of their mate.

Opposite: Hispaniolan Woodpeckers rarely drum, but they engage in a variety of noisy courtship movements with their mates.

On a trip to the Jamaican highlands, I heard a man complain that a woodpecker was destroying the poles holding his family's clotheslines. To discourage the bird and protect the wood, he capped them with coffee cans. After the man left, I turned in response to a loud metallic drumming by a jaunty Jamaican Woodpecker that seemed to be enjoying the amplification provided by the louder surface. I have admired the ingenuity of Golden-cheeked Woodpeckers utilizing the metal frames of window screens, Acorn Woodpeckers going to town on signs, Northern Flickers and Red-headed Woodpeckers working the metal siding of chimneys, and Red-breasted Sapsuckers tattooing metal interpretative displays.

VOCALIZATIONS

Vocalizations are important for woodpeckers throughout the year, but especially during spring, when they are used to attract mates, warn of danger, challenge a rival, and announce their location. These calls are used in conjunction with drumming, and some are emitted during flight.

Woodpecker calls are simpler than those created by songbirds but cover a wide range, including single high-pitched notes, trills, twittering, rattles, and even screams. The more social the species, the more diverse its calls. Regardless of the species, it is relatively easy to hear similarities between calls with similar functions, particularly those in the same genus and between closely related genera. Whenever I visit a new habitat, I can always recognize the genus of the caller well before I identify the species.

The best way to learn woodpecker songs is to spend time in the field or listen to widely available recordings, as each of us and each reference book writes out sounds using often contradictory mnemonics. I generalize here about some basic types of calls utilized by many woodpeckers. These vary by species, season, and context, with social species possessing the widest range of vocalizations.

Call notes are basic and often click-like and can function as a location call between mates, parents and offspring, or rivals. The speed, length, and intensity of these calls vary with the level of excitement. The complexity ranges from the single note of the Hairy Woodpecker and double or triple note of the White-headed Woodpecker to the double *tchur* of the Red-headed Woodpecker.

Long calls such as the wild laughing *cuk-cuk* calls of the Pileated Woodpecker, the similar *kik-kik kik-kik* of Northern Flickers, or the rattle calls of *Dryobates* species are loud, rhythmic series of similar notes, sometimes preceded by introductory notes. They help maintain contact between members of a pair and serve as territorial advertisements. Many *Dryobates* woodpeckers utilize versions of their rattle call for these purposes, including the Downy Woodpecker's shrill, descending "whinny," but some deploy different versions of rattle calls as interaction calls.

Interaction calls are used by many species during bowing, swinging, or wing-spreading displays. The Northern Flicker is one of several species that emits a *wicka-wicka* call, while others utilize a rattle or even a sputter call. Black-backed Woodpeckers deploy a "scream-rattle-snarl" call while in the "hunched-head swinging display," which is targeted at woodpeckers of its own or other species when establishing a territory.

Quee or **kweek** calls are functionally similar to the songs of other birds and are typically used only during the breeding season, especially by males from elevated perches. They are long and squeaky, expressed in a series, and used to both attract mates and defend territories, sometimes in combination with drumming or flight displays, although those that drum more utilize this call less frequently.

Screams are sometimes uttered in the most aggressive struggles with predators or competitors or when being handled by humans.

Intimate calls are soft sounds between mates that can be heard only at close range and are expressed during copulation or when engaging each other near the nest.

Nestlings emit a buzzing, rasping, or twittering, which peaks in intensity as an adult arrives and more so when it enters the nest.

NEST CREATION

All the woodpeckers in North America nest in cavities, which could be in trees, cactus, agave, yucca, termite nests, fence posts, house siding, utility poles, nest boxes, or, rarely, in other natural or human-made cavities. Their custom-made cavities protect their eggs from the talons, teeth, and tenacity of most predators, and keep rain, snow, and wind out and in some cases reduce temperature extremes. Birds that nest in cavities are 25 percent more likely to fledge young than those with open nests.

Woodpeckers are usually primary cavity excavators in that they excavate not only cavities in which they nest but also roost cavities in which they escape the elements. Birds nesting in new cavities typically have more breeding success than those that reuse nests, particularly nests that were used the prior year. Primary cavity excavators have the advantage of customizing their nests to serve the size of the parents, making it less accessible to predators, and when a variety of options are available, they can choose a substrate that is less likely to fall apart and that provides more insulation. Nests that were used the prior year can become infested with lice and mites that could weaken or kill nestlings.

Cooperative breeders are more likely to nest in the same cavity each year. The Red-cockaded Woodpecker may do so for several decades unless its cavity is enlarged or usurped, and the Acorn Woodpecker does so as much as half of that time, although not in consecutive years. The difference between the two is explained by the inordinate time the Red-cockaded puts into creating each nest cavity in live trees so it can maintain a

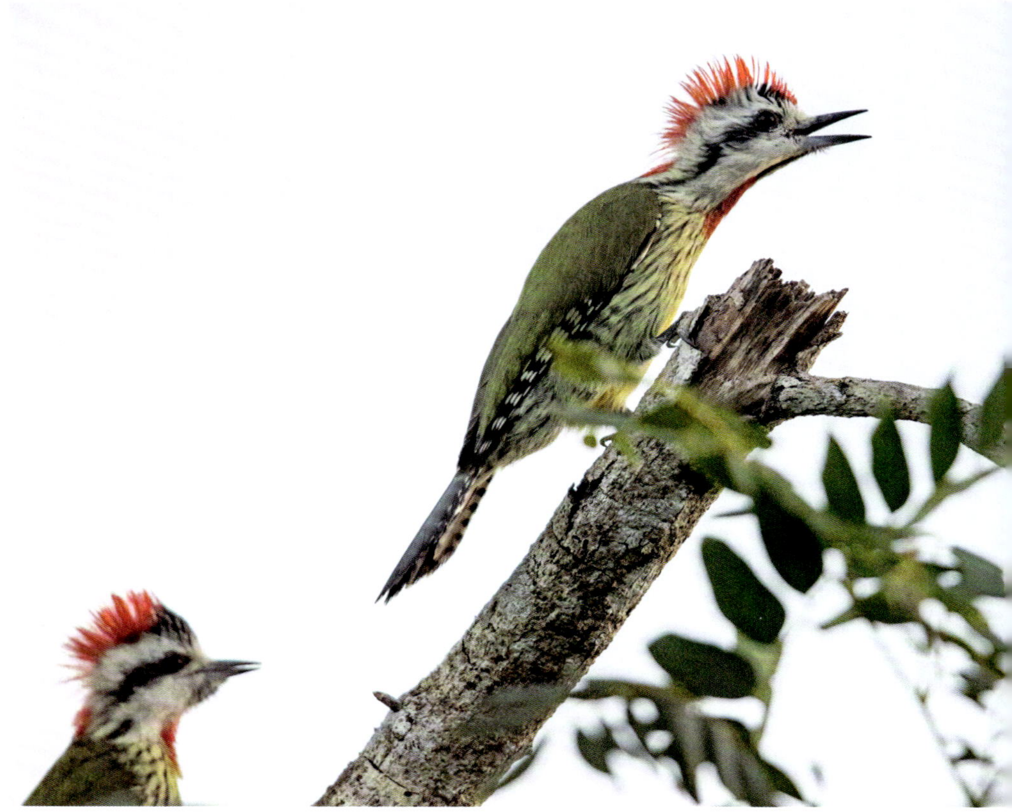

Highly territorial, Cuban Green Woodpeckers are surprisingly aggressive toward other woodpeckers and will even raise their crests, call at, and pursue larger interlopers.

continuous flow of protective sap around its nest, in contrast to the Acorn Woodpecker, which excavates in dead wood.

Woodpeckers are more likely to use existing cavities when their first nest failed and the remaining breeding window is small. Sapsuckers may reuse their own cavities regardless of the timing, as will White-headed Woodpeckers, although much less frequently. Some weak cavity excavators, such as the Antillean Piculet, *Colaptes* species such as flickers, and *Melanerpes* such as Lewis's, Red-headed, Red-bellied, Gila, Guadeloupe, Golden-fronted, and Acorn Woodpeckers, more frequently reuse cavities from an earlier year, or they might tidy up one previously used by another species. Red-headed Woodpeckers can be particularly aggressive, enlarging and taking over cavities created by Red-bellied, Downy, and even Red-cockaded Woodpeckers. Lewis's Woodpeckers may enlarge cavities created by flickers or Hairy Woodpeckers. Though not weak excavators, Pileated Woodpeckers also enlarge nests created by other species. When woodpeckers do reuse nests or utilize natural cavities, they often peck at the cavity walls to create a fresh layer of wood chips at the bottom of the nest.

Red-headed, Lewis's, and even Puerto Rican Woodpeckers are known to customize natural cavities such as those formed by tree rot. Northern Flickers and Lewis's, Red-bellied, and Golden-fronted Woodpeckers are the most likely to utilize nest boxes. Where snags are hard to come by, such as in the plains or deserts, some woodpeckers will carve nest cavities in utility poles or make use of nest boxes. In Puerto Rico, I observed a woodpecker family that had nested in a hole within a concrete utility pole, as snags were scarce in the years after a major hurricane.

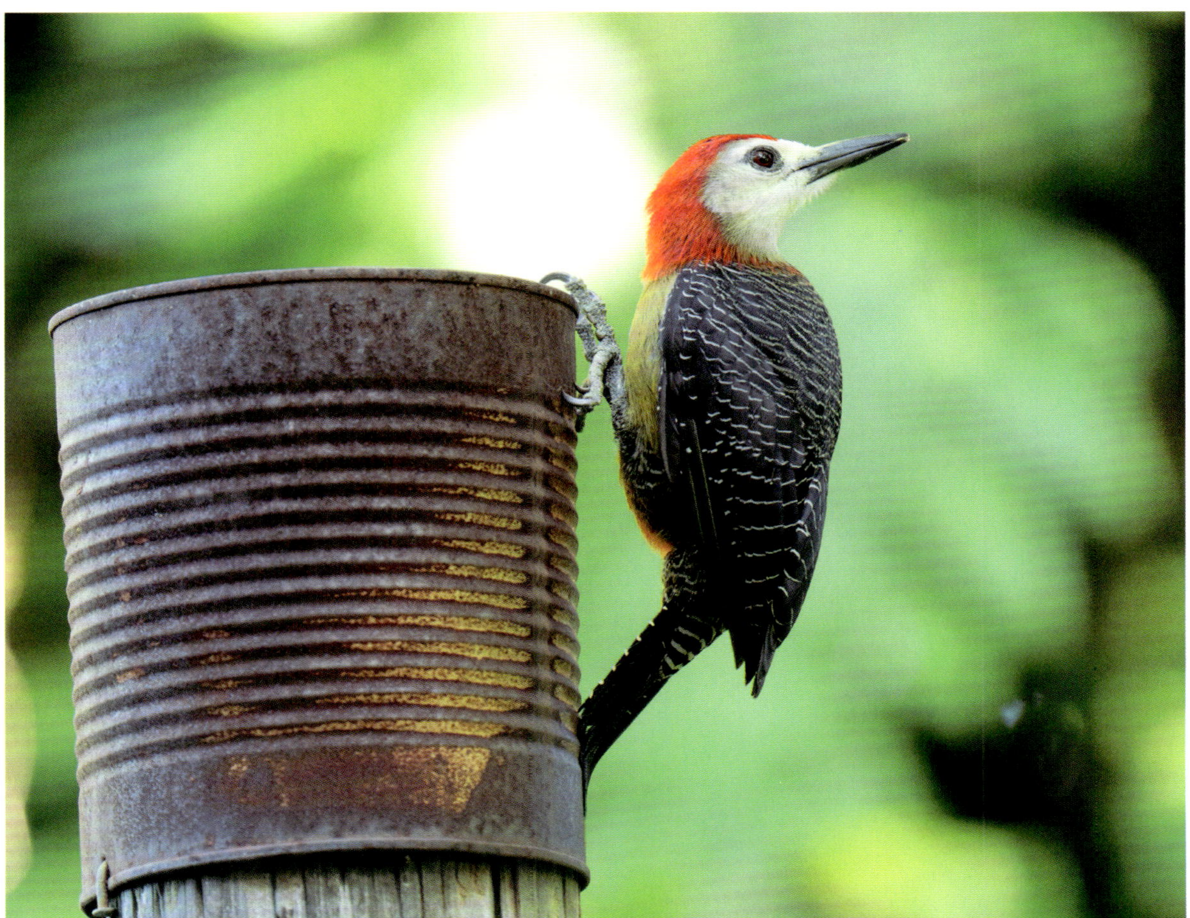

The pursuit for the most resonant drumming post for courtship led this male Jamaican Woodpecker to a clothesline post "protected" from woodpeckers by an old coffee can.

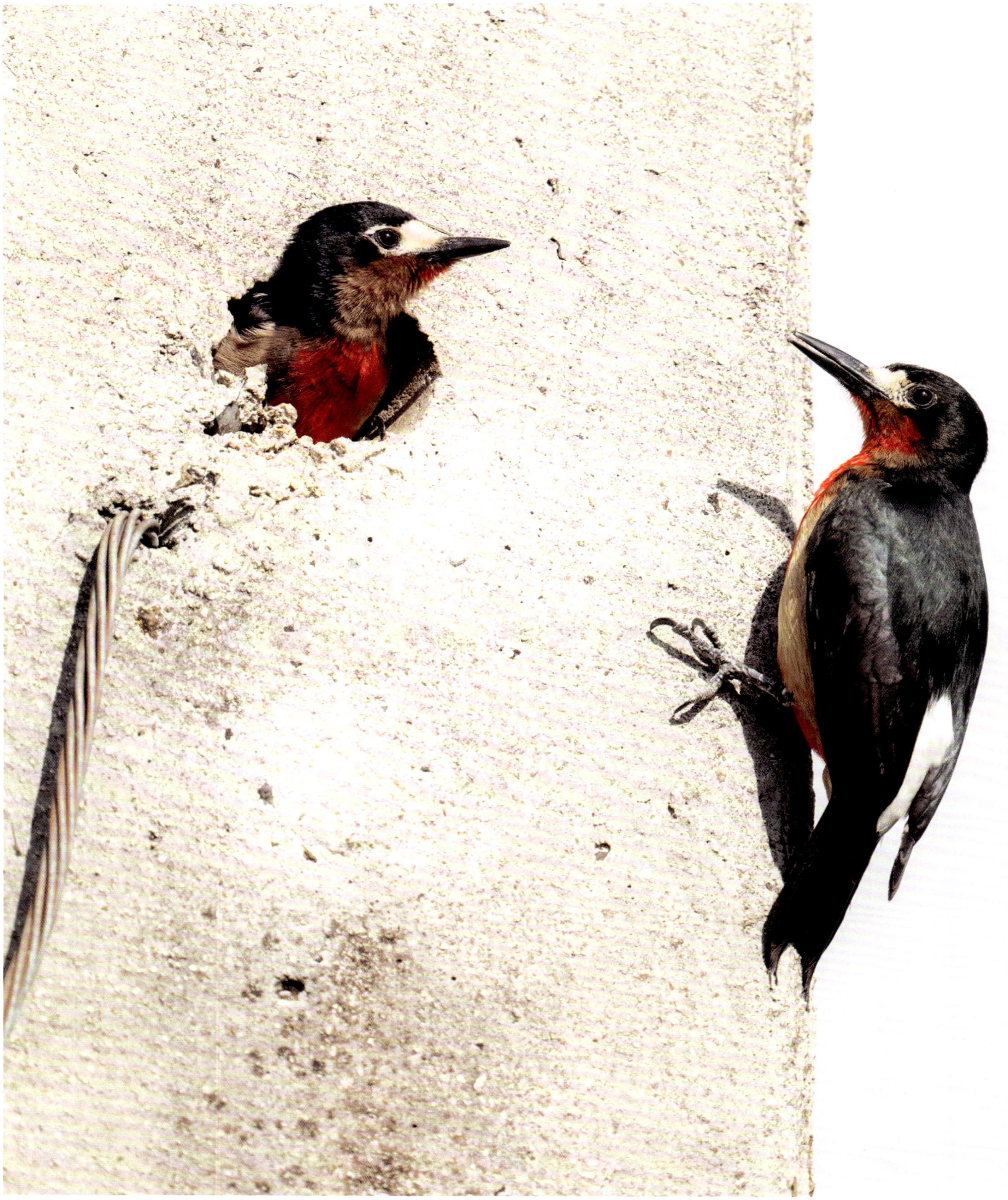

SITE SELECTION

Most woodpecker species create cavities in dead trees, dead parts of living trees, or living trees infected with a sapwood fungus or heartrot. I have witnessed Cuban Green Woodpeckers carving a nest cavity in a termite nest on a tree. Strickland's and Ladder-backed Woodpeckers frequently nest in stalks such as agave or yucca, and cacti are used by Gila and Gray-breasted Woodpeckers, Gilded Flickers, and to a lesser extent Ladder-backed, Hispaniolan, and Golden-cheeked Woodpeckers. So prized are agave stalks that I know people in both Oaxaca and Arizona who put old agave stalks in the ground as decorations only to have Ladder-backed Woodpeckers carve holes in them within days!

Excavating is costly in the time it takes a woodpecker to search for the right spot and carve the hole, as well as the energy spent guarding this valuable resource, especially when cavities are in short supply. First the bird must assess the hardness of the exterior and the softness of the interior to make sure it can withstand the onslaught of predators and weather, but not so hard as to be beyond the woodpecker's excavating capabilities.

Some species return to the same stand of trees, and sapsuckers often select the very tree they used the previous year, although they will usually make a new cavity above or below the prior nest, consistent with the direction of the rot within the tree or snag. It is not uncommon for the roost cavity of the male or an abandoned, partially finished nest from a previous year to become the new nest. This is most common when snags or trees with rot or fungus infections are scarce. In such situations, sapsuckers are more likely to share their nest tree or snag with other species, including other woodpeckers, and may in rare cases share the snag with others of their own species.

When suitable snags or trees are lacking, woodpeckers are also more willing to nest in close proximity to their own predators. I have seen this when American Three-toed and Black-backed

Opposite: The factory-made hole in a utility pole was chosen by this pair of Puerto Rican Woodpeckers since nest snags were in short supply the spring after a major hurricane. The male waits for the female to investigate the cavity before leaving.

During early spring, Pileated Woodpeckers demonstrate many typical woodpecker courtship behaviors including crest raising, wing-spreading, and head swinging displays. Surprisingly, they even engage one another on the ground!

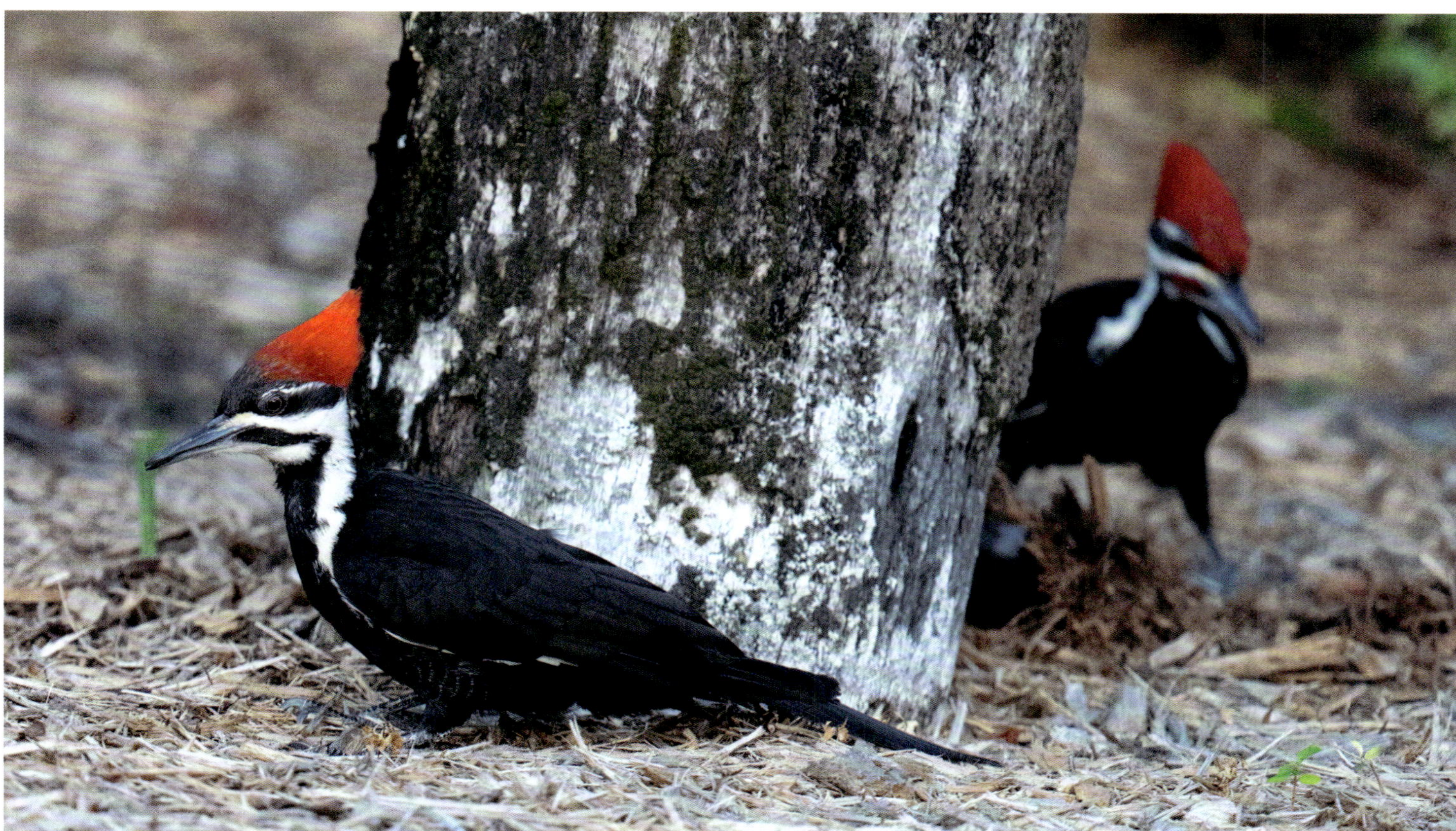

Woodpeckers nested close to one another and to their predator the American Goshawk, when Lewis's Woodpeckers and Hispaniolan Woodpeckers nested close to American Kestrels, and, amazingly, when seven woodpecker species nested in the same stand as one of their most dangerous foes, the Northern Pygmy-Owl!

While watching Williamson's Sapsuckers at this stage, I was struck by their diligence. I sat waiting at a stand of gray, barkless ponderosa pine snags pocked with perfectly round ping-pong-ball-size (1.6 inch) nest holes from previous years lined up vertically, each appearing finer edged than the older one below. Periodically I heard a subtle scratch as the male landed on the smooth, hard bark. He would hitch up the face of the tree in quick, jerky spurts, irregularly pecking and pausing. Unlike drumming, excavation involves a woodpecker slamming his entire body, led by his bill, against a tree. This male did just that in various places on three trees for several hours each day until the outline of three cavities began to form. Each hole faced mature ponderosa pine habitat. Over the first two days, the male appeared to spend equal time at the three cavities, one only eight feet from the ground that he seemed to have an easier time excavating, and the other two above the prior year's cavity. He spent enough time at each to fully carve an entrance, but on the third day, he focused his attention on the highest hole, some fifty feet above ground level. This would eventually become his nest.

The Williamson's was likely evaluating subtle changes in resonance that indicated wood density in the interior of the snag. Pecking at these areas and starting holes likely helped determine the thickness of the exterior wall. It is probable that woodpeckers prefer the maximum hardness they can excavate from the most leveraged position on the outside of the cavity with their bill striking at a right angle to the tree. At the same time, trees with more rot are easier to carve from the less ideal digging postures inside the cavity when they need to remove more wood. Although the lower cavity was easier to excavate, the higher cavity likely offered more security and perhaps a less predator-penetrable surface.

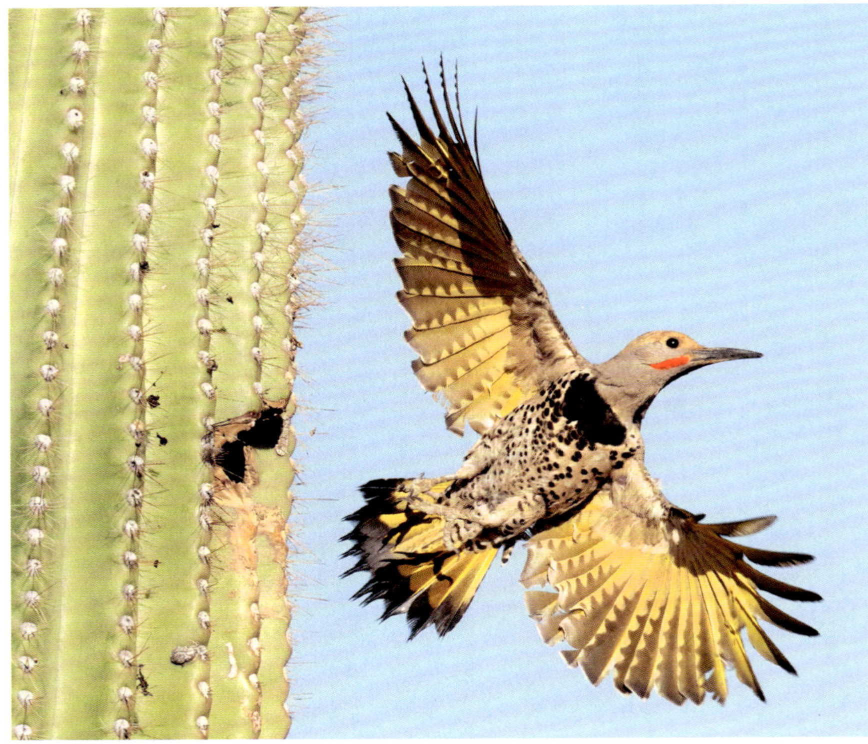

A male Gilded Flicker flies from a nest cavity in a saguaro. During early stages of courtship, a male woodpecker will often introduce his mate to several nest options. The golden feathers under the tail are the easiest way to distinguish Gilded from Northern Red-shafted Flickers.

Opposite: The normally hidden red crown feathers (cockade) of the male Red-cockaded Woodpecker are raised and visible during territorial or courtship encounters. This streak of color in the crown is the only way to distinguish between sexes in the field. Group members regularly peck around the cavity entrance to keep sap flowing and deter predators.

Suitable dead wood usually limits a woodpecker's nesting options, making it difficult to discern a woodpecker's ideal nest location. Individual species may have preferences for height of cavity, type of substrate, surface and internal density, orientation to foraging habitat, orientation to the horizon to collect less precipitation, and direction in relation to the sun.

Each of these preferences varies by species, habitat, region, and availability. In cold areas, southern exposure can be advantageous for warming the nest in the mornings. When the first nest changeovers begin, and in hot deserts, facing north and west can help prevent overheating of nests. Red-cockaded Woodpeckers prefer to nest just below the lowest branch of live trees, and Pileated Woodpeckers often choose the underside of leaning trees and snags. Red-bellied Woodpeckers often utilize the underside of a limb, while Northern Flickers and Lewis's Woodpeckers nest anywhere on trunks. Red-headed

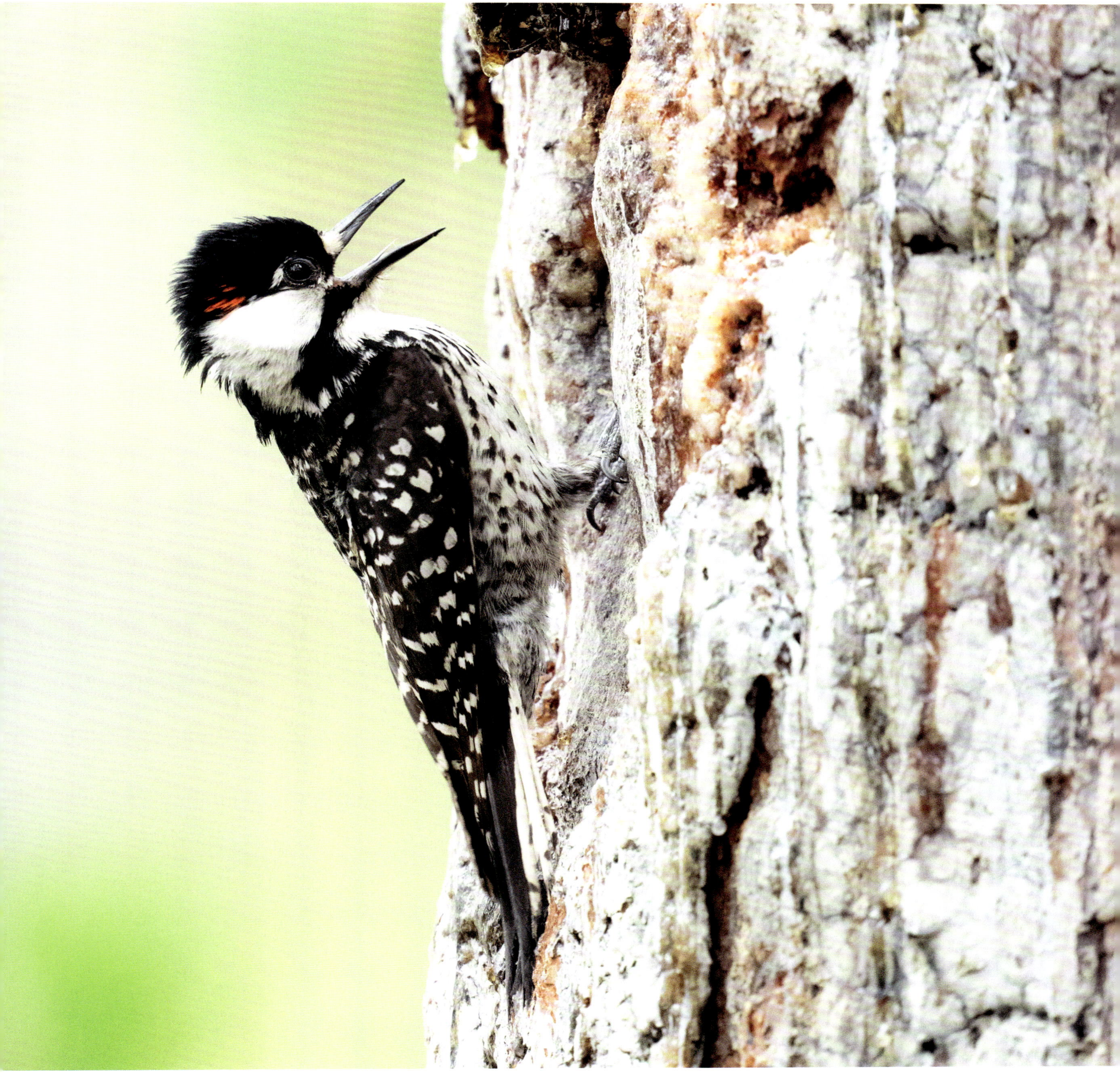

and White-headed Woodpeckers often pick barkless snags.

Black-backed, American Three-toed, Hairy, and Pileated Woodpeckers generally select trees with hard exteriors, whereas Downy, Lewis's, and White-headed Woodpeckers, as well as flickers, prefer softer surfaces that are easier to excavate, although there can be much overlap between the species, especially when options are limited. Such preferences can reduce conflict during the nesting season.

EXCAVATION TIME

Red-headed and White-headed Woodpeckers are fairly typical among woodpeckers in that it takes them one to two weeks to create a new nest cavity in a snag or the dead part of a live tree. They might accomplish this in as little as a few days in a well-decayed snag, or it could take them as long as several weeks in a harder tree.

Woodpeckers that nest in cactus likely spend a similar amount of time actually excavating, but it takes more time overall because they need to remove the spines and then periodically stop to let the pulpy insides dry and harden before excavating further.

The relatively small Williamson's Sapsucker often uses live aspen with heartrot or aspen snags but is capable of carving through a few inches of sapwood. It normally invests three to four weeks excavating, the same timeline as the larger Black-backed Woodpecker. The Pileated Woodpecker typically requires three to six weeks to complete its nest.

Extraordinarily, the small Red-cockaded Woodpecker may require several years to more than a decade to carve a nest cavity in a live pine since it must first get through 2.5 to 3.5 inches of sapwood to reach the softer decayed heartwood, the prime reason they live in family groups. The entrance tunnel begins upward before making an abrupt downward turn, deterring would-be predators that might get past the constant flow of sticky gum (sap) flowing around and below the entrance.

CAVITY SIZE, SHAPE, AND HEIGHT

Most newly carved woodpecker holes are round or oval and range from about 1 inch by 1 inch for the Downy Woodpecker to roughly 4.75 inches by 3.5 inches for the Pileated Woodpecker. Height ranges from ground level to more than 165 feet, although most are between 6 and 82 feet. The lowest reported nests belong to the Northern Flicker; the White-headed, Ladder-backed, Golden-fronted, and Black-backed Woodpeckers; and the Antillean Piculet. The highest belong to Pileated, Lewis's, Puerto Rican, and Red-cockaded Woodpeckers, as well as Red-breasted and Williamson's Sapsuckers. These heights may reflect available rotted wood more than species preferences.

The ideal finished cavity is just wide enough at the opening to accommodate the larger of the woodpecker pair, whether that be the male or a gravid (with eggs) female, with no extra room for a predator to invade with its body, bill, teeth, or talons. Once past the constricted opening, the hole opens up and quickly curves downward into an increasingly wider chamber that is vase-, egg-, or gourd-shaped. Wood chips left over from excavation sit at the bottom, but no material is added to the nest from the outside.

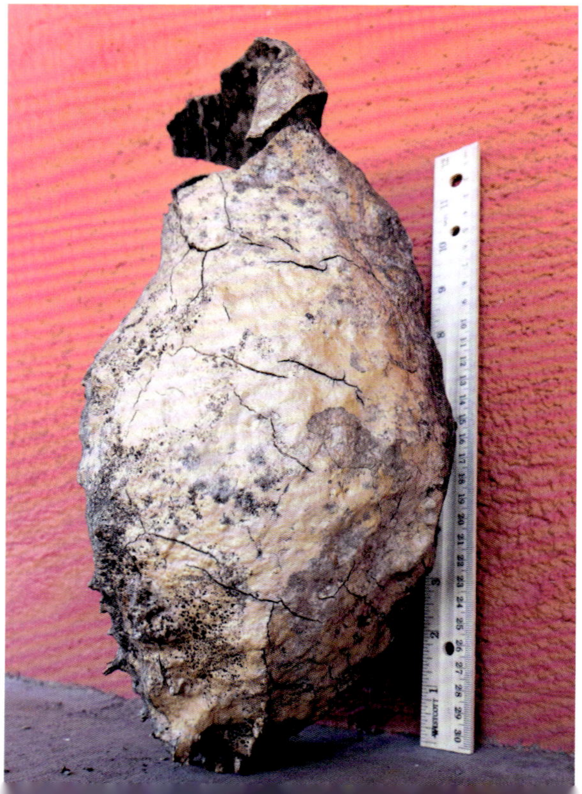

Gila Woodpeckers and Gilded Flickers carve cavities in live cacti. The pulpy walls of the cavity eventually harden into a protective and insulating shell, which remains intact even after the cactus rots. These forms, known as "saguaro boots," were historically used by Native peoples to carry water.

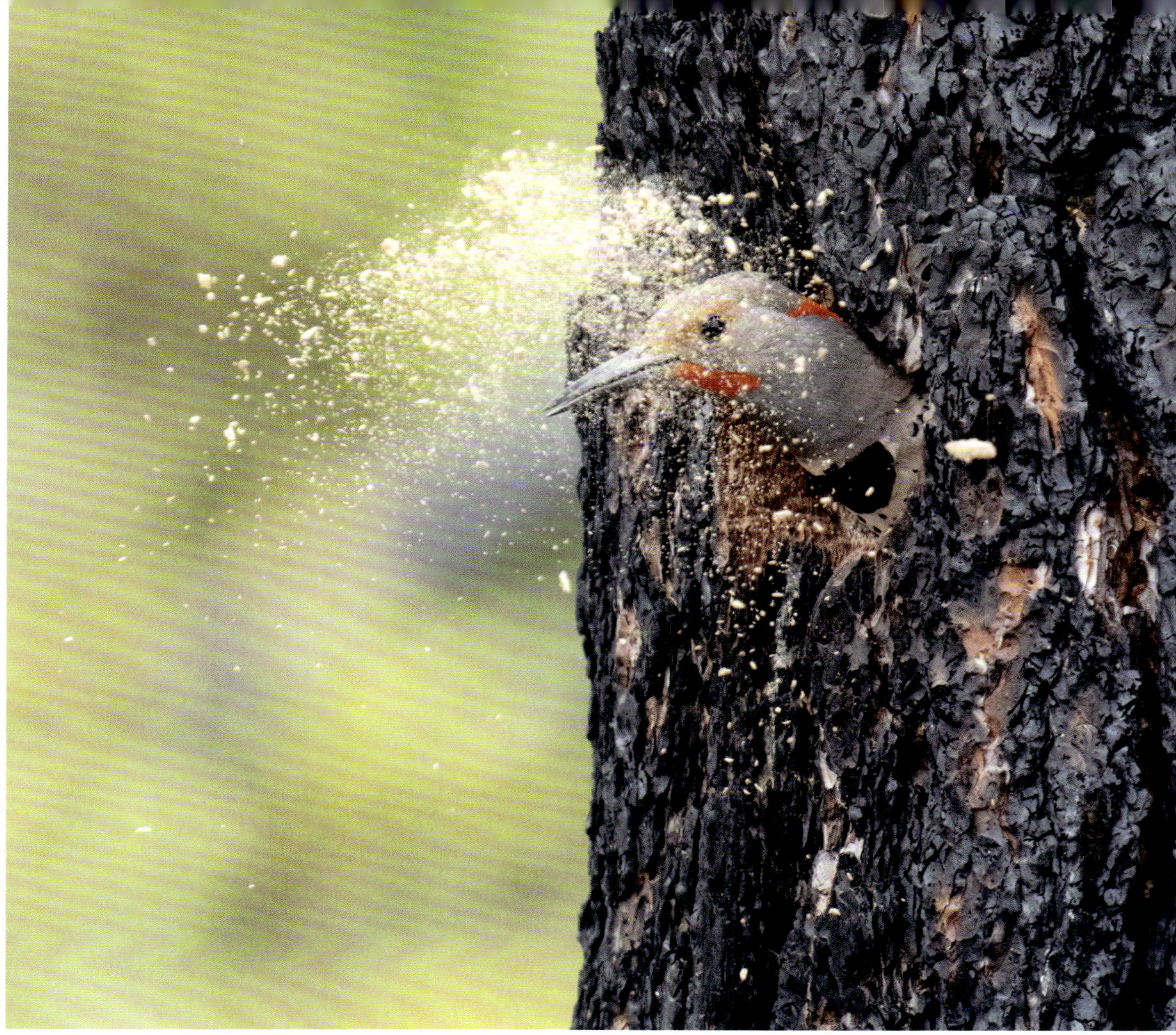

The softest trees, such as burned trees with rot, can be excavated by Northern Flickers, who are frequently secondary cavity nesters. Notice the red malar typical of the Red-shafted and the red nape crescent typical of the Yellow-shafted, evidence of interbreeding of the two subspecies.

THE EXCAVATION PROCESS

Initial excavation takes place from outside the cavity, with the male first pecking at the substrate so that wood chips spray in all directions. As the cavity deepens, he begins to lean in to break up the wood and then take mouthfuls of chips that he tosses to the side with vigorous flicks of his head. As he digs deeper, the male fully enters with his entire body. He usually continues to flick the chips out of the cavity but may sometimes fly away with the biggest mouthfuls, perhaps to hide his work from potential predators.

A lot of mystery still surrounds the exact process for selecting a cavity. Firsthand accounts suggest that the male often introduces the female to the partially excavated nest cavity and nest cavity options, and in many species the female makes the final choice. The male may suggest a cavity by tapping near the entrance with his bill, and the female appears to accept or reject his suggestion with taps of her own. It has been reported that sometimes the pair may disagree, with each woodpecker pecking near its chosen cavity, trying to entice its mate to agree. If they disagree, they sometimes move on to another option.

Once a pair decides upon a cavity, the male uses that cavity as his overnight roost so he can guard the nest, and later the eggs, day and night. Red-headed Woodpeckers have been captured on video waking and guarding their cavities at night, so it is likely that other species do as well. The female spends her nights in her own roost cavity, perhaps one of the rejected nest sites.

After the female has accepted a potential nest cavity, she will periodically relieve the male of his excavation duties, but she usually does not excavate with the same enthusiasm and throws out far fewer chips. One possible explanation is that the male will be the one roosting in the hole overnight until the female lays her eggs and thus needs to perfect it to his needs.

Pileated Woodpeckers make loud drumming double taps before they let their mate take their own turn excavating, perhaps confirming their continued interest in the site. Mutual tapping, in which pair members tap in roughly synchronized fashion at a potential nest site or nearby perch, probably functions in strengthening pair bonds.

I have noticed that Pileated Woodpecker females do as much as one-third of the excavation, which is significantly more than I noted with other species. It has been suggested that for smaller species, eggs take up a large proportion of the female's body weight, so she needs to conserve her energy. Since there is more sharing of excavation duties with Pileated Woodpecker pairs, they communicate more around the nest during this time. I observed that if one bird was inside excavating and heard the other land at the nest tree, it would tap to confirm its presence, and the bird outside would tap to confirm its interest in trading places, a request that was not always granted.

A pair continues to make minor improvements in the nest until the eggs are laid, perhaps occupying the cavity to prevent others from co-opting it. Regardless of the species, if an initial chosen nest fails, by necessity both mates seem to share excavation duties more evenly at a new site since the breeding window is narrowing. During these situations the pair becomes more likely to use an existing cavity or create a cavity in softer substrate.

One spring at over 7,000 feet in a pine-oak forest in the mountains of southern Mexico where the trans-Mexican volcanic belt joins the Sierra Madre Oriental, I spent a week watching Strickland's Woodpeckers finish and guard a nest cavity. Each day the woodpeckers threw fewer wood chips from the cavity as the nest neared completion, but they rarely left the hole unattended. At all times, one of the woodpeckers, usually the male, foraged on or near the nest tree and in the early morning would spend an hour or so removing just a few chips. The female would relieve him briefly in the morning but otherwise foraged elsewhere. It was clear that they were guarding this rare resource, a finished cavity in an area with few snags of the proper size. The birds were extremely quiet throughout the day unless another Strickland's called or drummed, in which case the male and sometimes both birds quickly returned to the cavity, drummed, called, and perched close to the entrance. Each night the male roosted in the cavity, leaving just before sunrise every morning to forage.

By the end of spring, the most colorful, demonstrative, and interactive period of a woodpecker's year is complete, and energies turn toward producing and protecting the next generation.

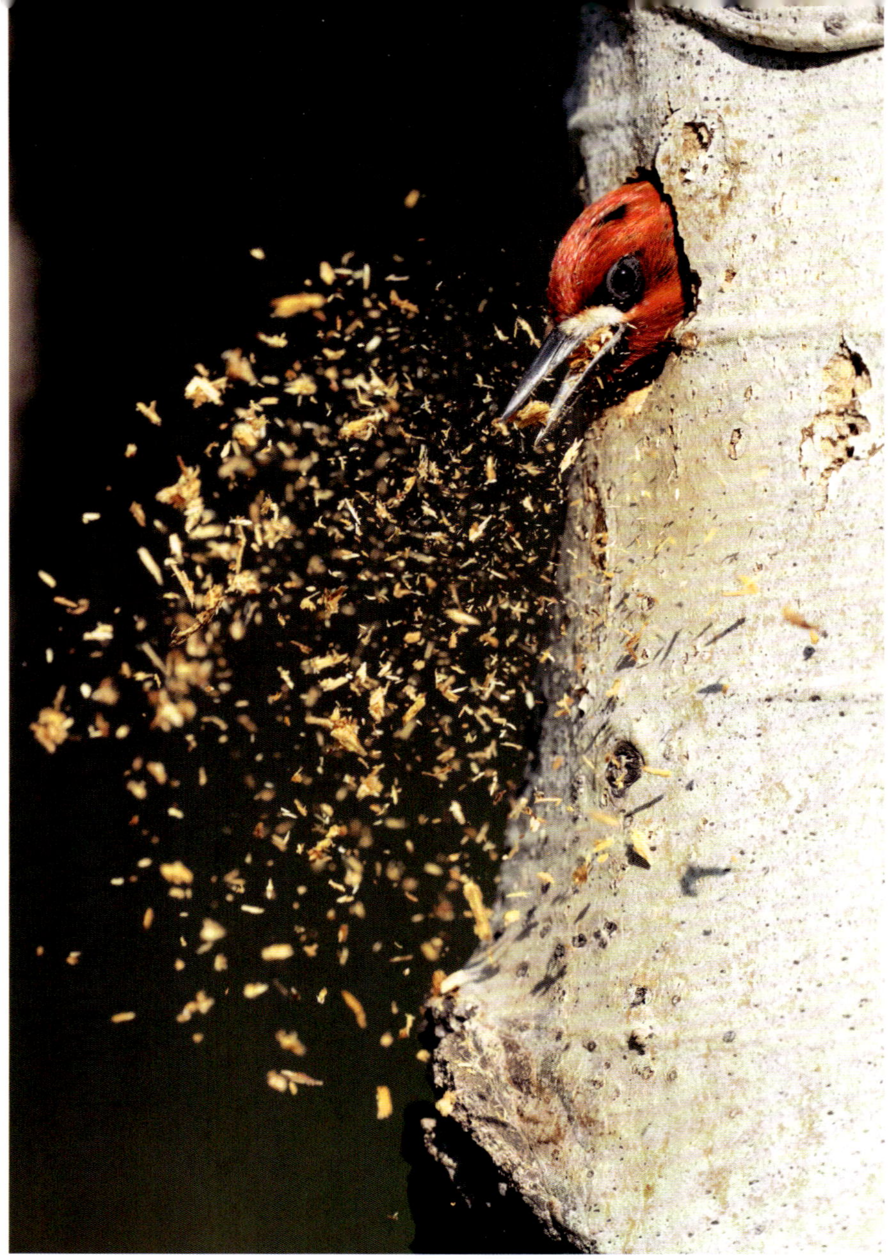

Sapsuckers, like this Red-breasted, utilize rot-infected quaking aspen for their nest cavities more often than other tree species. Sometimes they will nest in the same tree in consecutive years.

Opposite: Once a nest cavity has been selected, male woodpeckers, such as this Strickland's, guard it overnight until the young are near fledging. After eggs are laid, the male and female alternate incubating during the day.

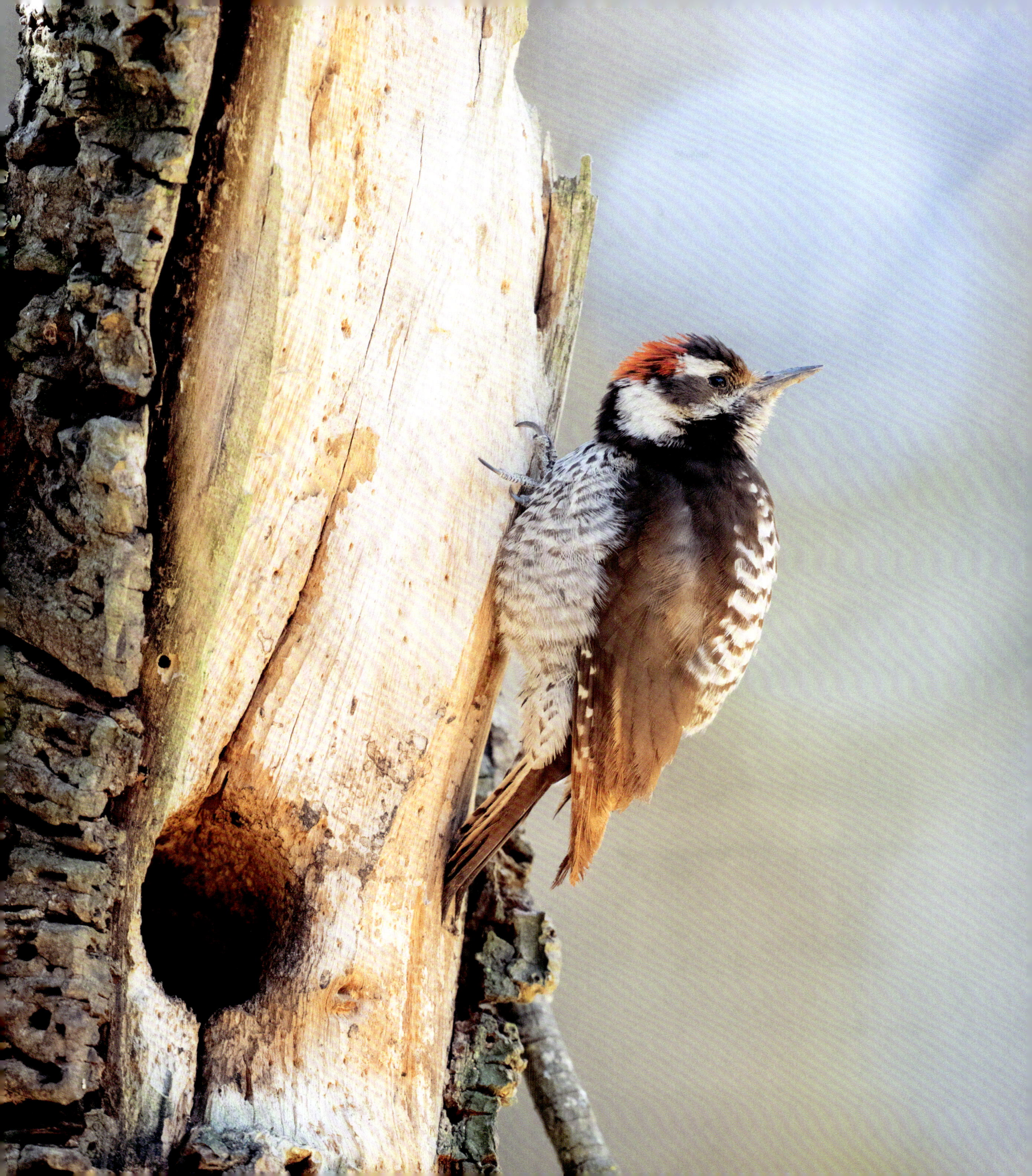

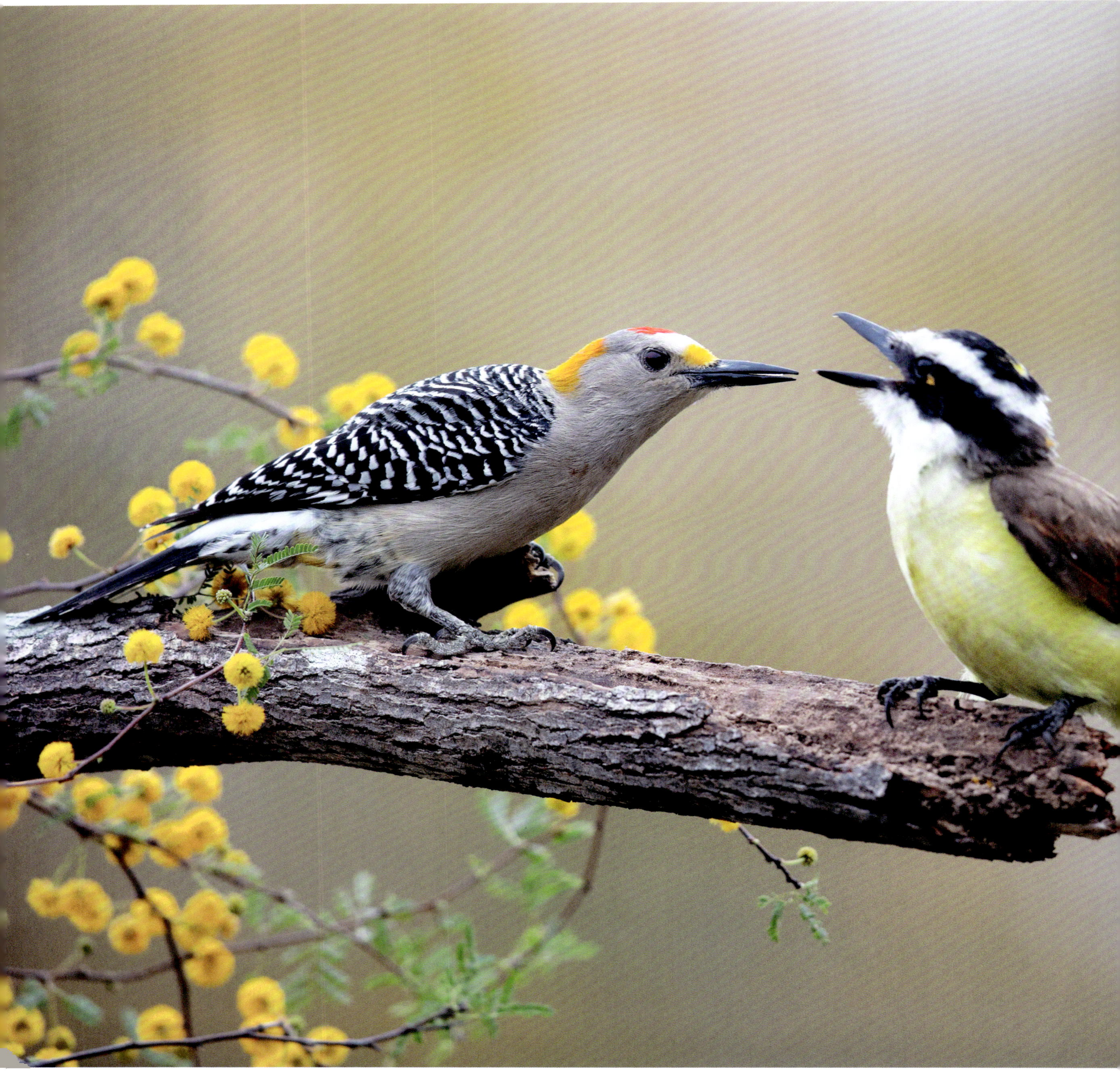

Left: Golden-fronted Woodpeckers will sometimes confront other species, such as this Great Kiskadee; yet at other times they may forage in close proximity.

Below: Colonial nesters such as the Hispaniolan Woodpecker are noisy and interactive, and mated pairs frequently work together to drive off competitors. Dominant birds often occupy the highest nests.

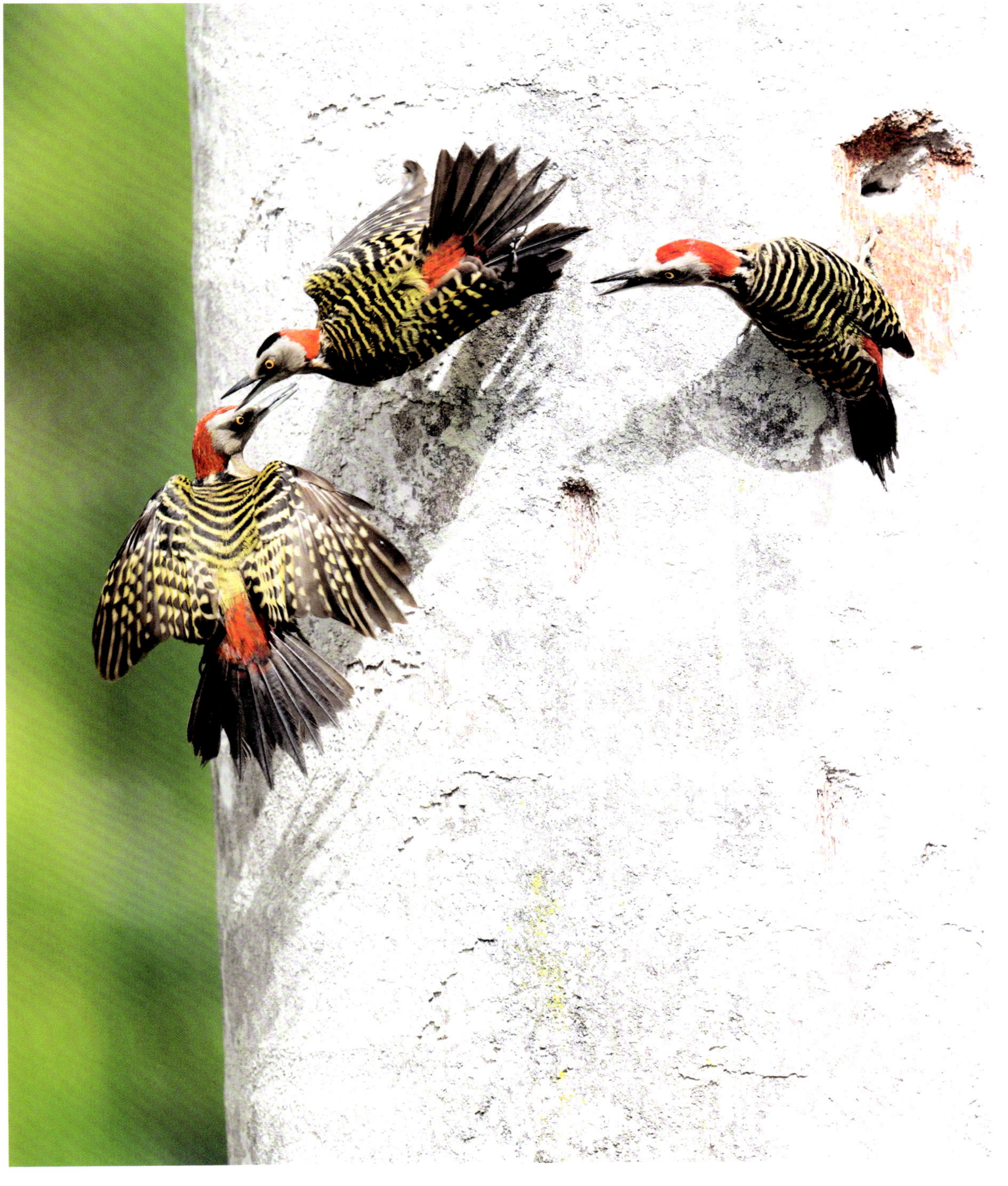

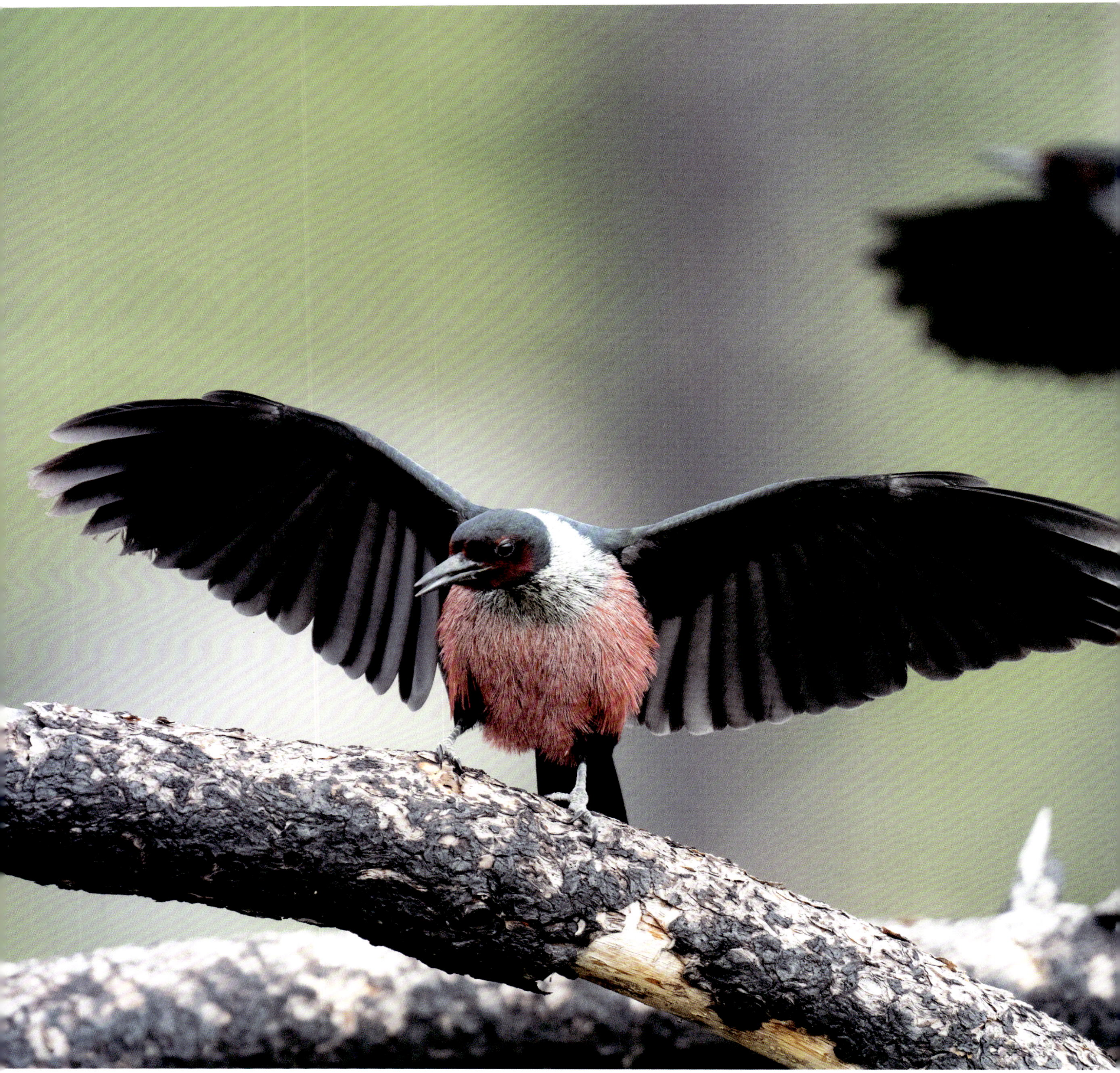

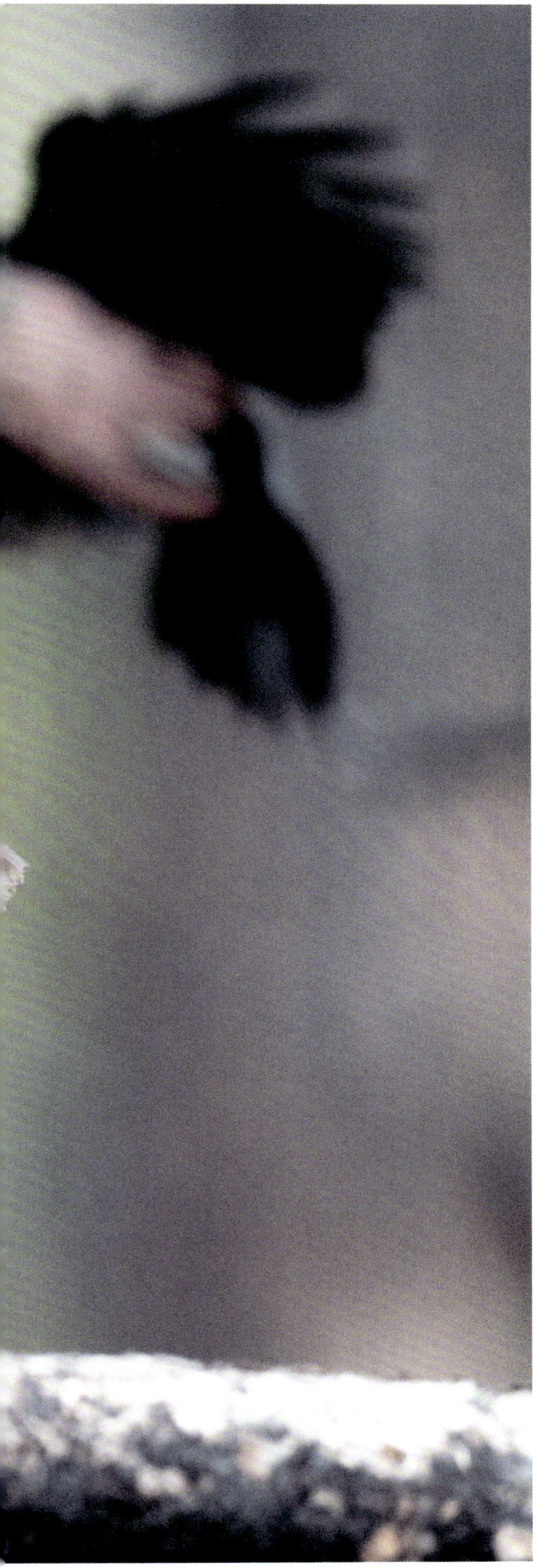

Left: In the "wing-out" display, done from a perch, a Lewis's Woodpecker holds its wings above the body, exposing the pinkish-red feathers of the abdomen; puffs out the gray feathers of the throat and neck; and depresses the head. This display can be directed at mates or rivals.

Below: The West Indian Woodpecker is a close relative of the Red-bellied Woodpecker and possesses a similar vocal repertoire, including a version of a churr call, often described as a scream.

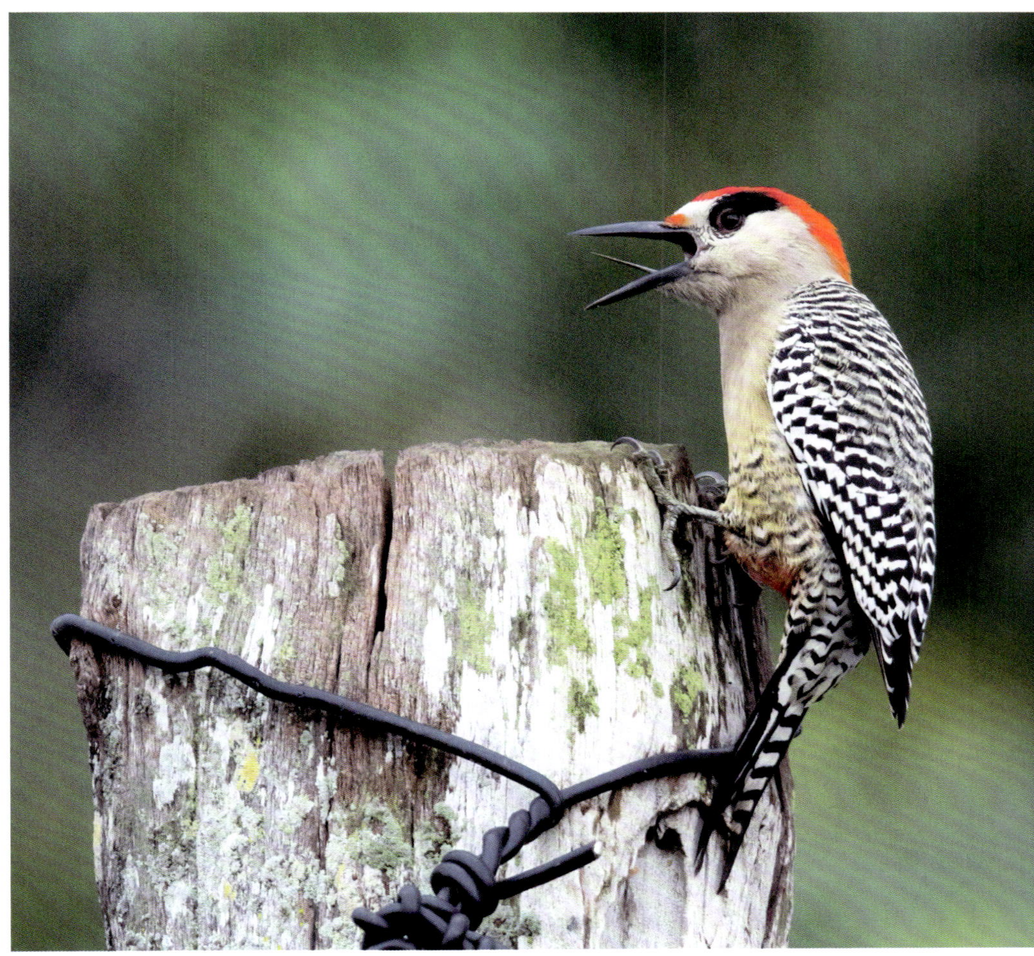

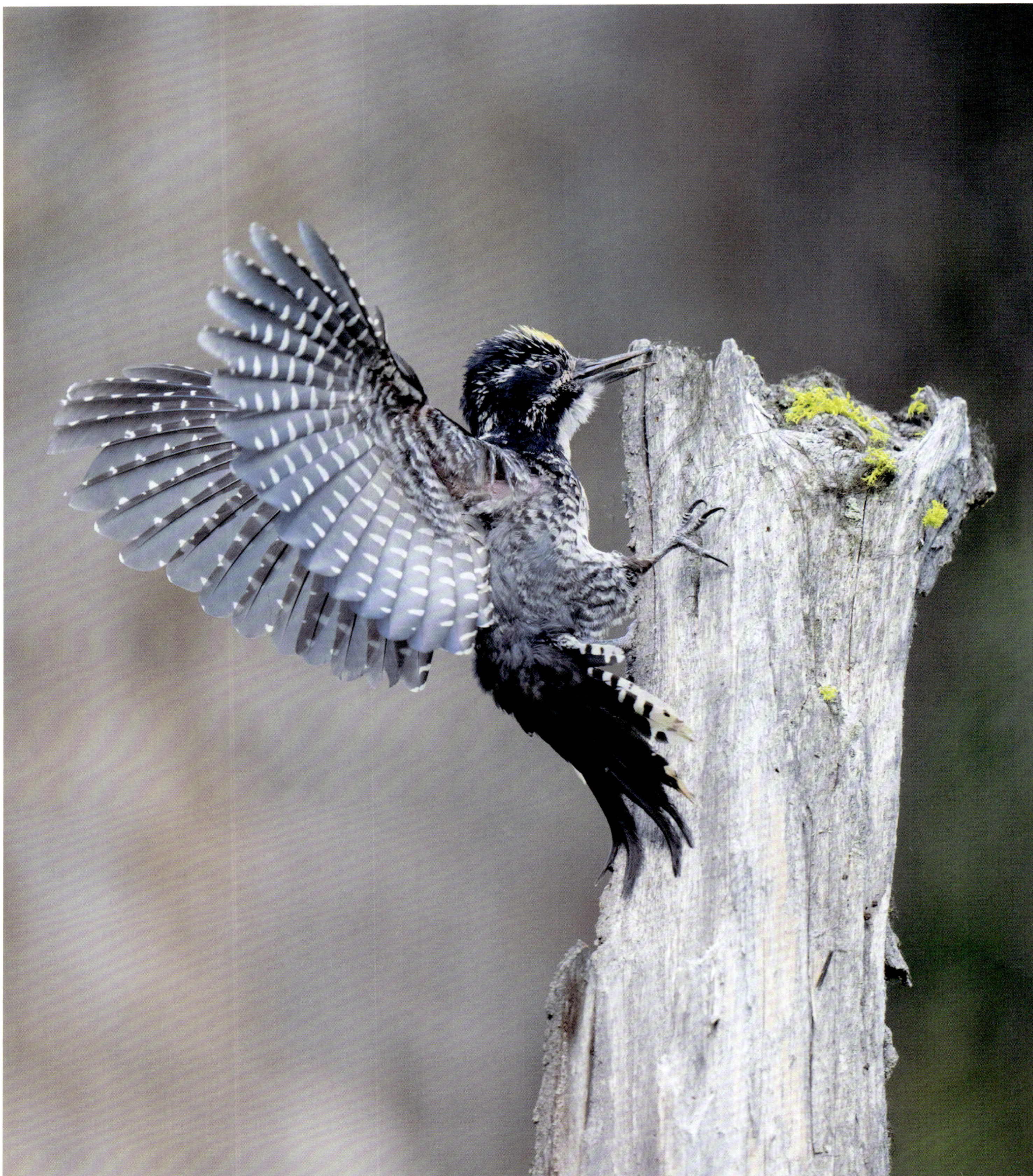

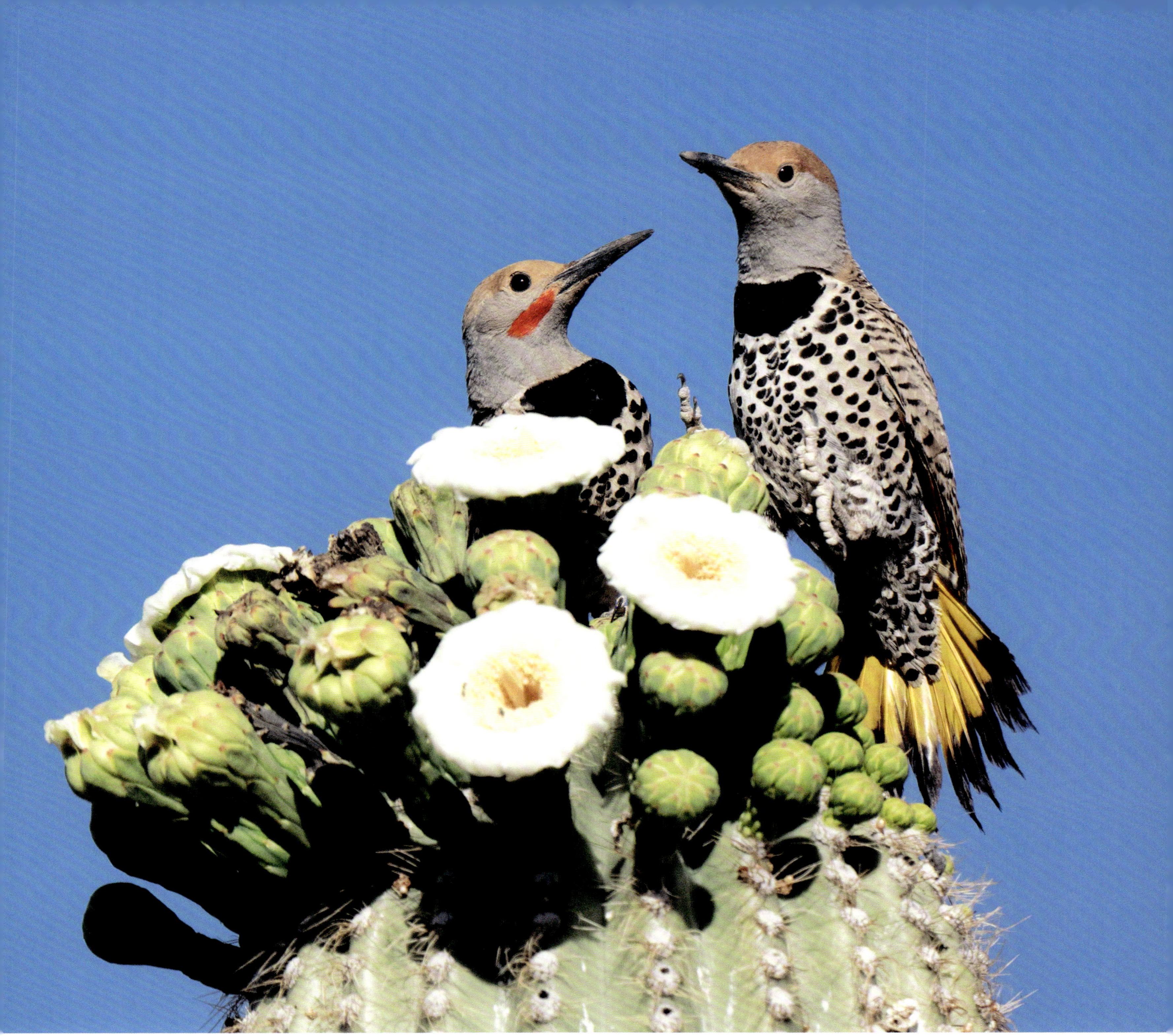

Opposite: An American Three-toed Woodpecker raises its wings as its mate approaches—a common woodpecker greeting during courtship or confrontations.

Once a pair bond has been established during the spring, individuals such as these Gilded Flickers acknowledge one another with the raising of bills and spreading of tails but much less vigorously than during courtship.

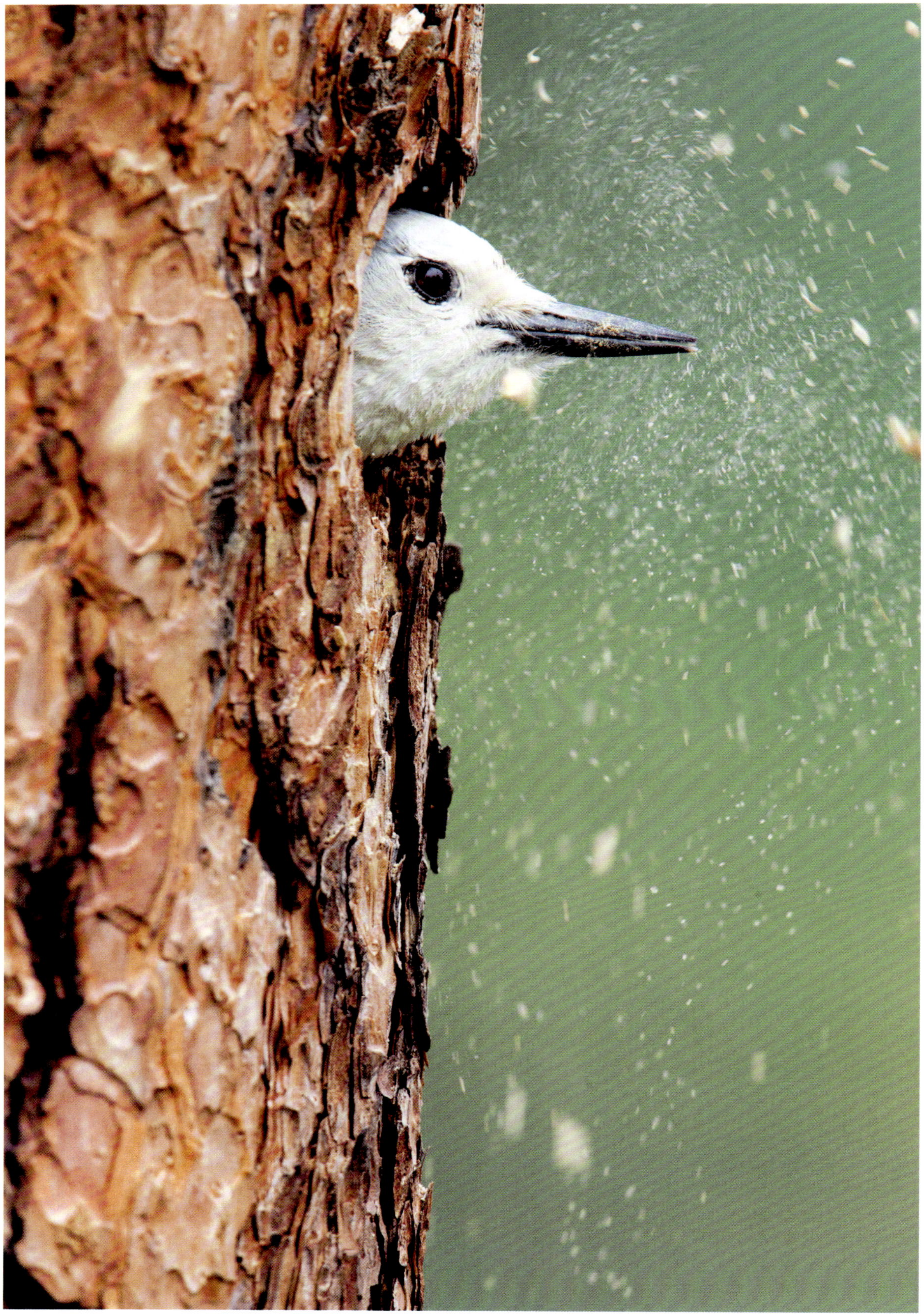

When rot infects trees, it makes them the preferred choice of woodpeckers looking to dig nest cavities. White-headed Woodpeckers often choose trees such as this ponderosa pine that are in a well-advanced stage of decay and leave behind cavities that benefit a range of other species from Flammulated Owls to Pygmy Nuthatches.

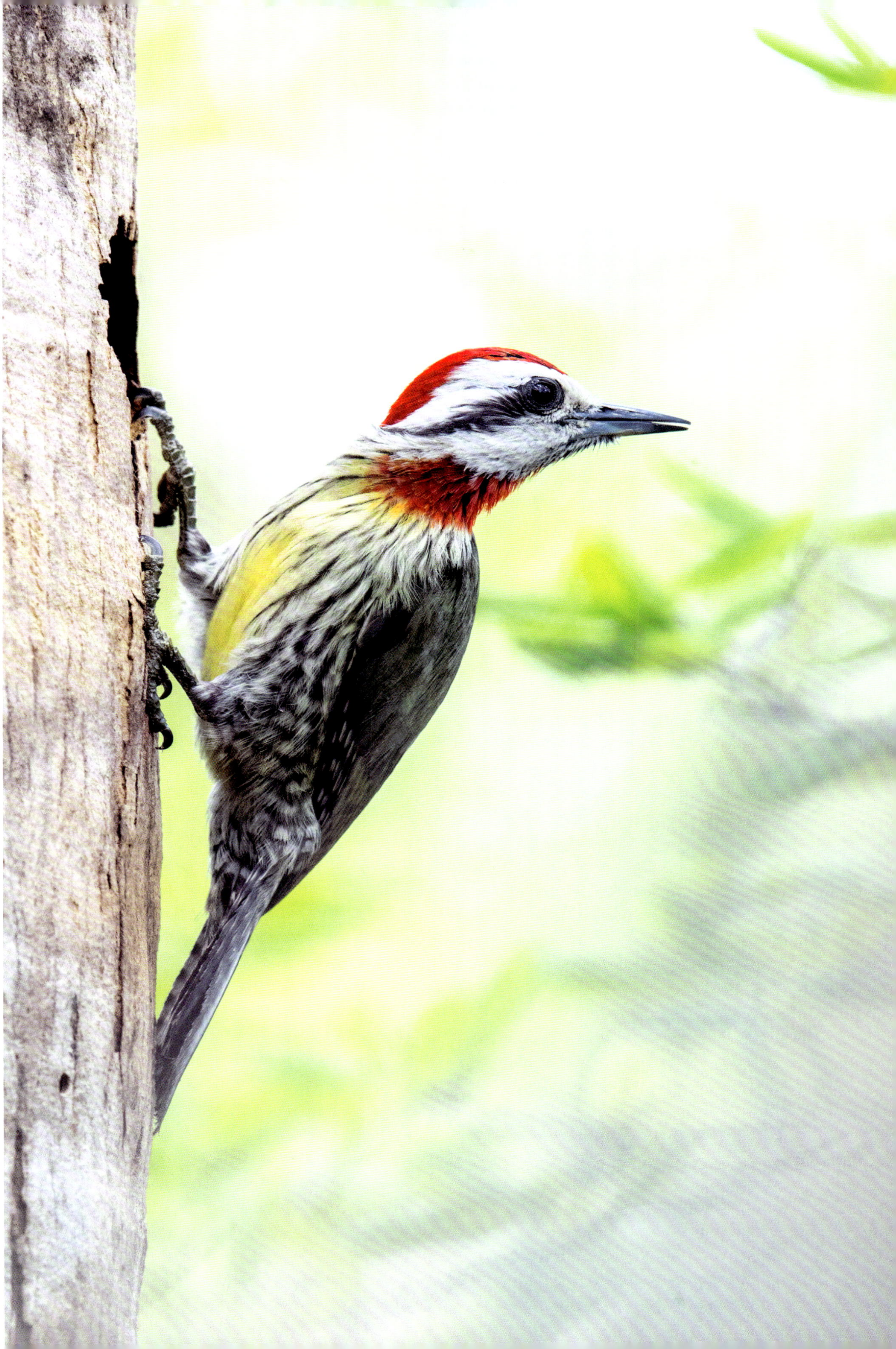

A Cuban Green Woodpecker waits for his mate to investigate a potential nest cavity. This Cuban endemic species is the world's only member of the genus *Xiphidiopicus*.

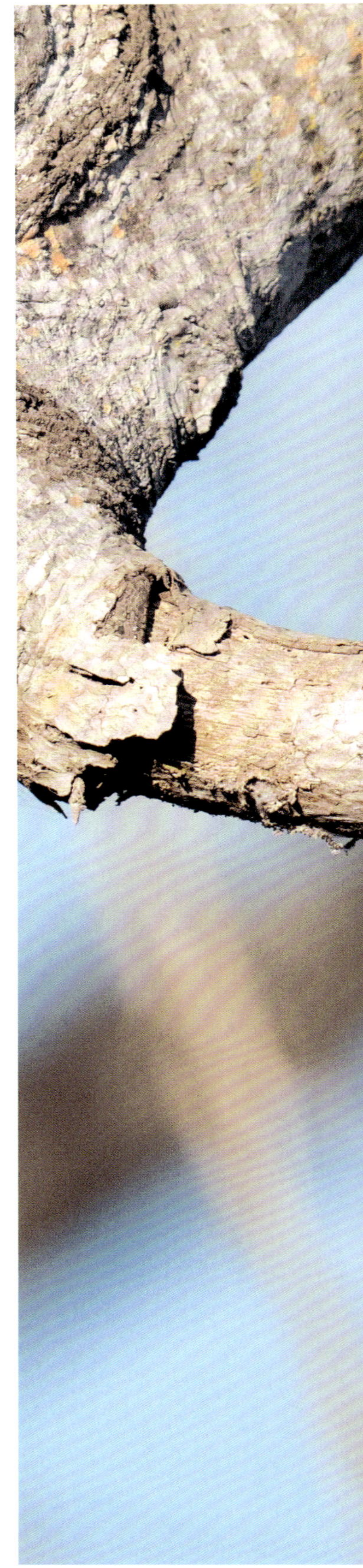

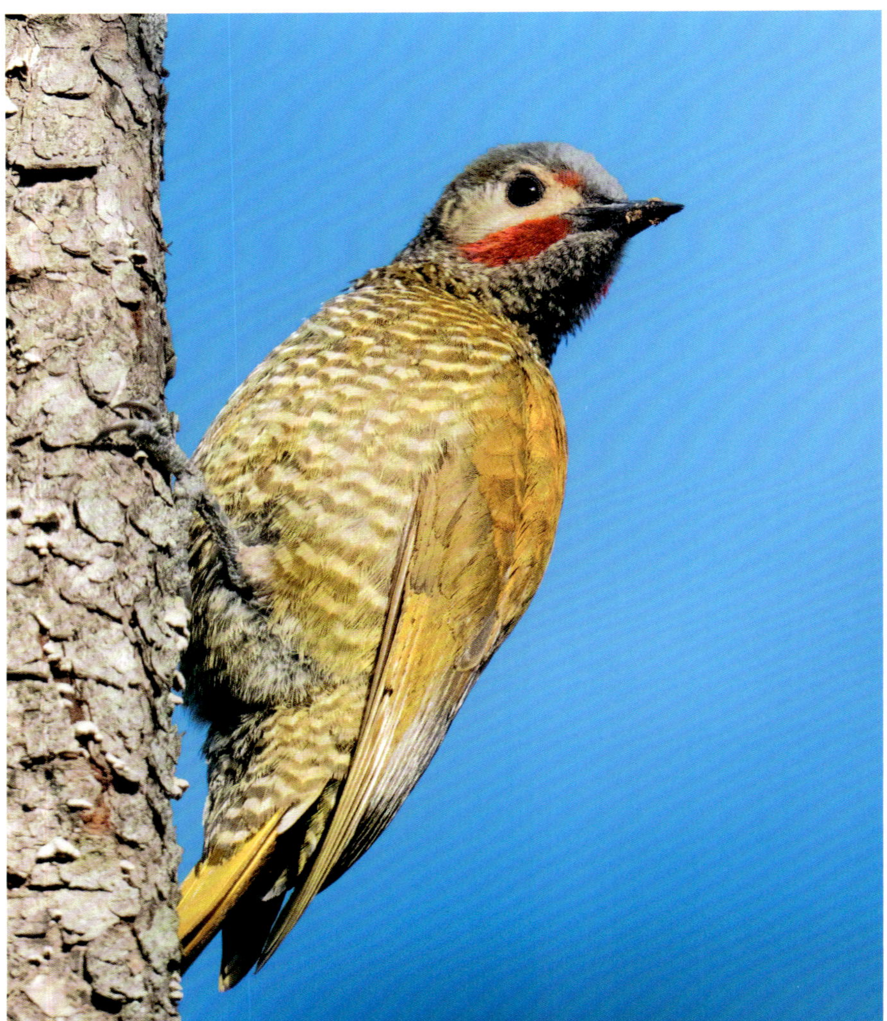

Endemic to the middle- to high-elevation forests of the Sierra Madre in western Mexico, the Gray-crowned Woodpecker is vulnerable to habitat loss. This species was once considered a subspecies of the Golden-olive Woodpecker.

Opposite: A male Lineated Woodpecker flies from one thick branch to another while foraging. Observations suggest that females often forage upon smalller-diameter branches than those used by males.

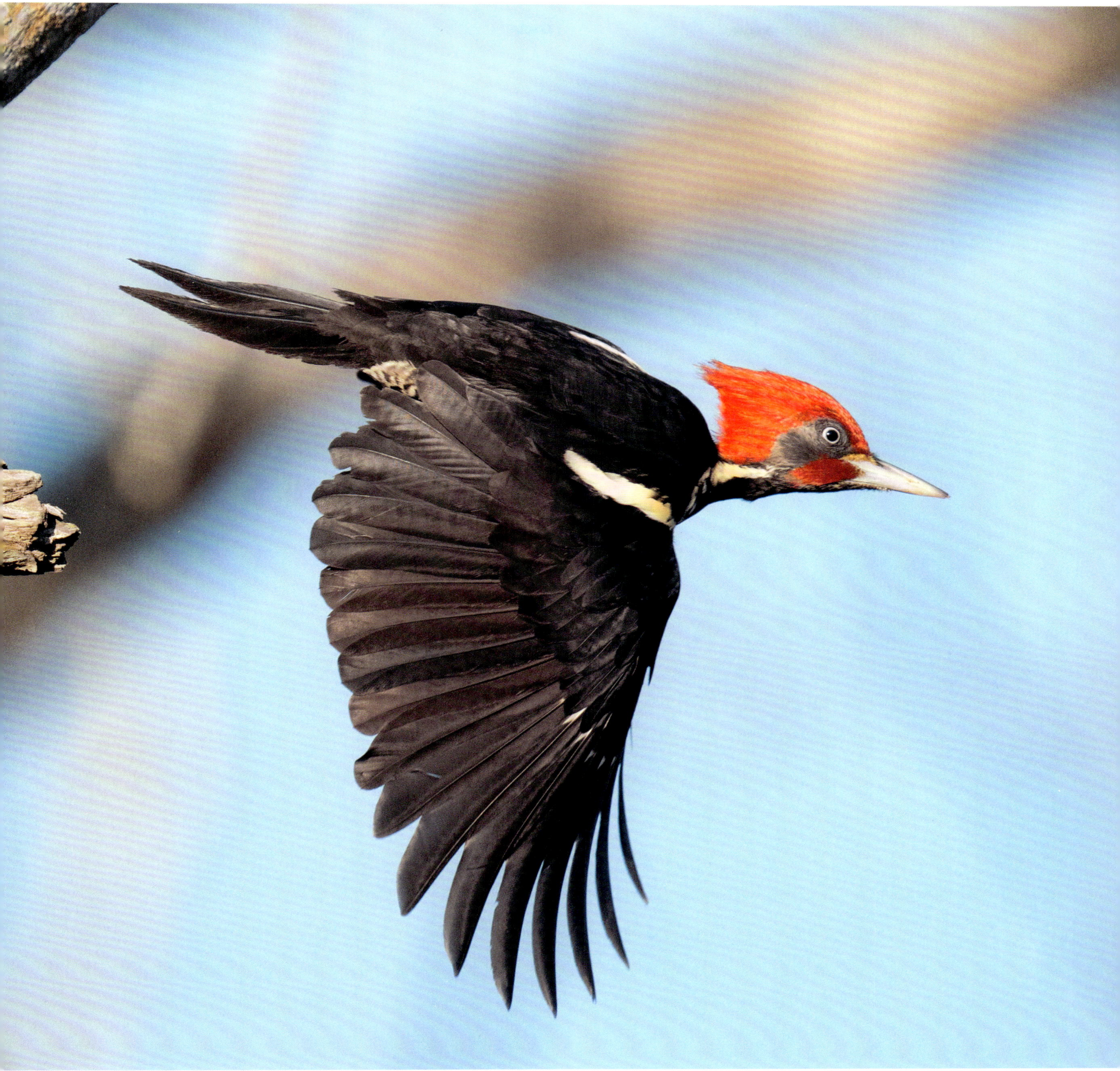

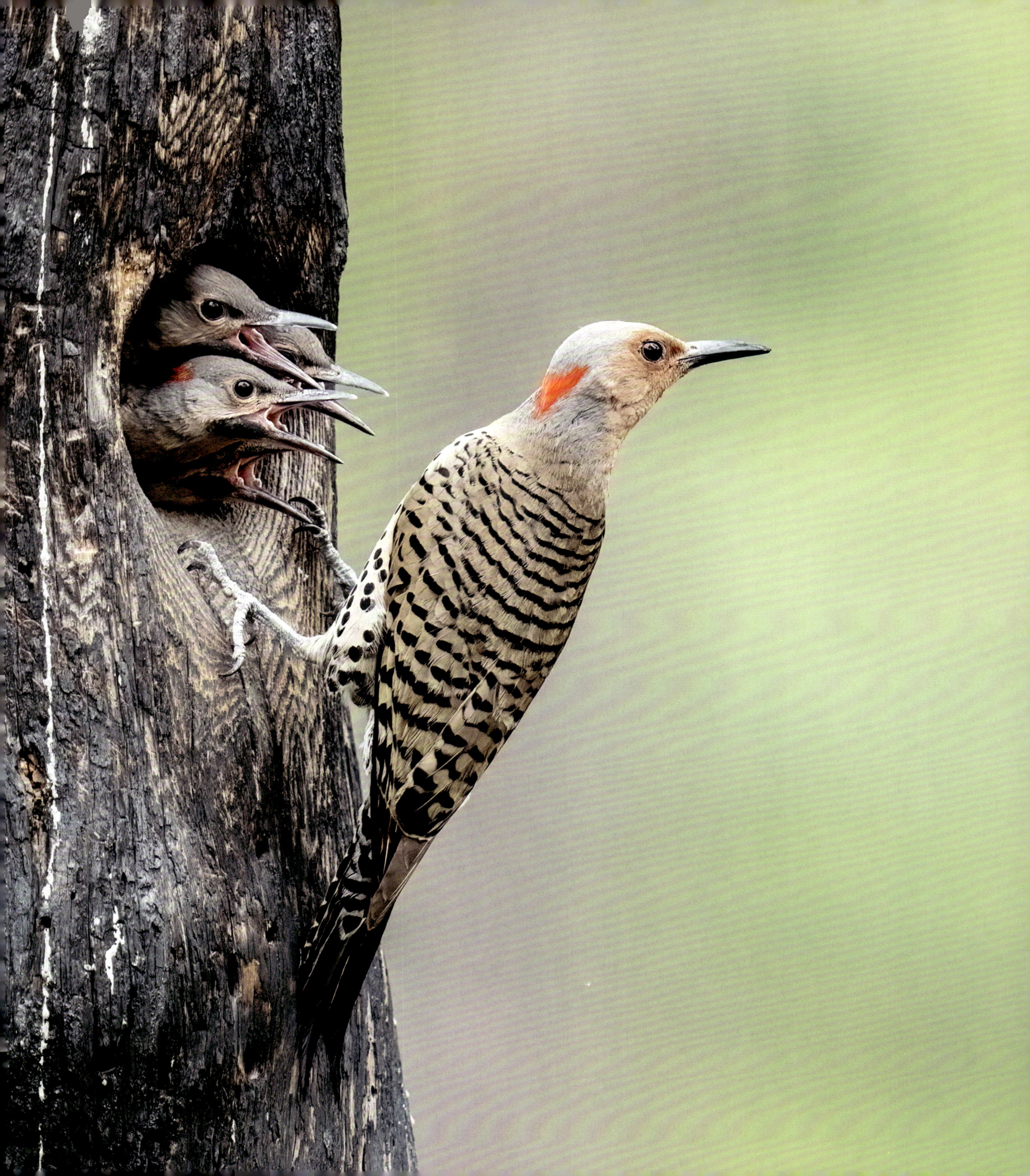

3

Life in the Nest

The silvery frost of a May morning dissolved as the sun rose over the central Oregon Cascades. Years earlier, fire had consumed shrubs and small trees, leaving behind an open woodland of red ponderosa pine trunks accented by the occasional blackened snag. With the extravagance of courtship concluded, all was hushed, save the churring of late-courting Lewis's Woodpeckers and the soft taps of a White-headed Woodpecker chipping the final slivers from its nest wall inside a barkless pine snag.

As fledging nears, nestlings look very much like adults and are constantly hungry. This female Northern Yellow-shafted Flicker prepares to take flight after feeding her four young.

I shivered as I stepped out of the hot rays and brittle soil of the burn and headed downslope, across a stream, and under the voluminous canopy of the giant quaking aspen, Douglas-fir, and ponderosa pine that still held the morning's chill against the icy ground. These mere steps seemed to move me miles into a distinct habitat where the rooflike mesh of upper branches had trapped cool air and protected the last of the winter snow.

Rays of sun through openings in the canopy crossed one another, dropping narrow shafts of light to create changing patterns on leaves, branches, and downed logs. Within the shifting shadows and light, several species of well-camouflaged black-and-white woodpeckers were moving into another stage of their lives.

Breaking twigs and crackling needles under my feet masked all faint sounds until I stopped and stood still. Only then could I make out various taps and scrapes, soft pecking, and the occasional sound of wings or squeaky calls of the new life stage underway. Now that the fittest birds had secured mates and nest sites, the priority shifted from performance at drumming sites to secrecy at nest cavities, where each pair would lay their eggs and raise their young. Hairy Woodpeckers were already guarding their nests in an aspen snag, while Williamson's Sapsuckers were quietly finishing their excavation and tossing the last pale wood slivers to the breeze.

I chose my next steps carefully as any noise might change the woodpeckers' behavior, wasting their precious time and energy at the threshold of the most vulnerable time of their year, when they attempt to mate, retain a nest cavity, produce and incubate eggs, and switch places at the nest, all while striving to remain hidden from predators or competitors that could render these efforts futile. I decided to leave them alone for at least two weeks.

Woodpeckers begin their most reticent period once territories, mates, and nests are secured. During this time, one woodpecker is almost always hidden in the nest while the other must at once feed and be alert for competitors and potential predators before taking its own turn at incubation.

NESTING AND BREEDING STRATEGIES

Most woodpeckers practice *monogamy*, mating with one mate during a single breeding season and occupying one nest that they defend against potential predators and other cavity nesters, including their own kind. Pileated and Lineated Woodpeckers are among the species that may maintain their pair bond for several years, sometimes for as long as both survive. Members of these two species are known to stay in contact year-round, even during winter, a time when most bird pair bonds are weak. Such genetic monogamy is rare among small land birds.

Some birds deploy *multiple-nest polygyny*, in which one male mates with two females and maintains two nests. This approach is likely extremely rare among woodpeckers since males almost always roost with the eggs or young in the cavity at night, and nests would likely fail without this nocturnal protection.

A third strategy is *multiple-nest polyandry*, in which the female mates with and lays eggs in the nests of more than one male. As much as 5 percent of Northern Flickers in one British Columbia population were documented to utilize this strategy, enabled by the uncommon occurrence of inordinately more females than males and slightly staggered breeding. In these instances, the female alternated incubation with the male at one nest as she was laying eggs at another and later incubated at the second nest while provisioning young at the first. This has also been recorded in at least one West Indian Woodpecker and a Williamson's Sapsucker but is likely a rare practice overall. Given the constant attention to eggs and young required during the breeding season, it is difficult for young to survive in these situations.

COLONIAL NESTING

Colonial nesting occurs when woodpeckers tolerate nesting close to others of their species and live in loose colonies. This arrangement is not the norm for the practicing species and is seen most often when there is a concentrated abundance of potential nest sites proximate to prey-rich landscapes. Pairs defend a small area around the nest but disperse widely when foraging. The participants benefit from the alarms of their neighbors and, in some species, may receive feeding assistance from an inadvertent and sometimes unwelcome helper at the nest, but these contributions appear to be rare. In North America this tolerance is found in several *Melanerpes* woodpeckers such as Lewis's, Red-headed, Hispaniolan, and Gray-breasted Woodpeckers, as well as in Fernandina's Flickers.

The most extreme example can be found in Hispaniolan Woodpeckers, which nest far from others but may sometimes nest in colonies of up to twenty pairs within a few trees and, remarkably, twenty-six individuals have been recorded in a single tree! In these situations, with this species, the highest holes are believed to be occupied by the most dominant individuals. There is some evidence to suggest that helpers assist at these sites.

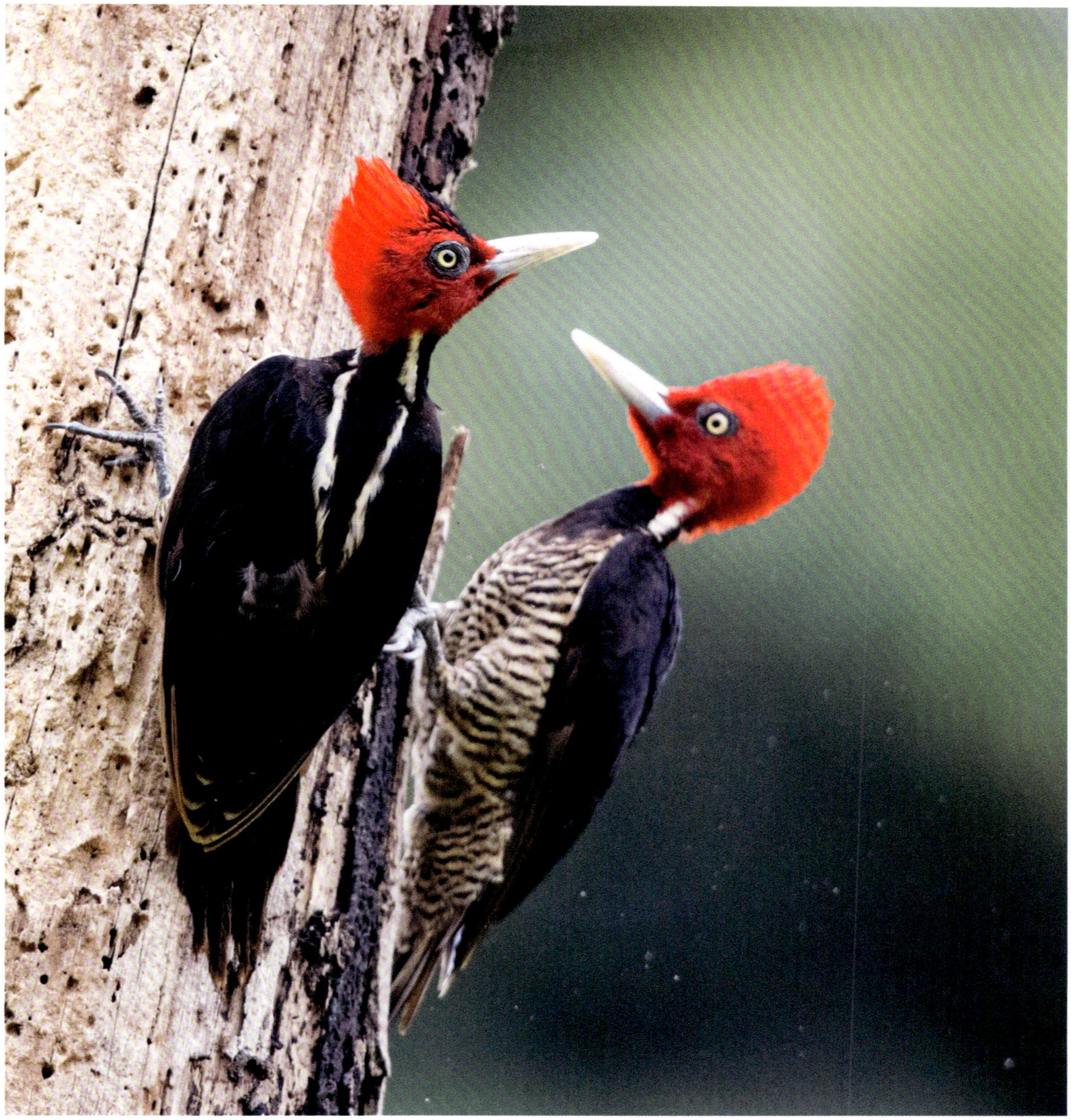

A female Pale-billed Woodpecker (foreground) prepares to take flight after the male arrives to take his turn brooding the young.

I have watched up to a dozen pairs of Lewis's Woodpeckers nesting in burned woodlands in Oregon and Washington with many of the nests less than fifty yards from one another. In both situations I witnessed birds other than parents attempt to feed nestlings, and there have been reports of several birds feeding nestlings at a single nest and of multiple nests in a single tree, although this is likely a rare occurrence.

Over subsequent years, the brilliance of this strategy became apparent when the easily excavatable snags either fell or were engulfed by the trunks and canopy of the regrowing young trees. Every year, fewer Lewis's stayed to nest. Being able to tolerate high nest density allows this weak excavator to take advantage of temporary boons of soft snags after fires, floods, or other disturbances.

At a fire-maintained oak woodland in Minnesota with a plethora of soft snags and a high concentration of Red-headed Woodpecker nests, I witnessed an adult deliver food to nestlings that were not its own.

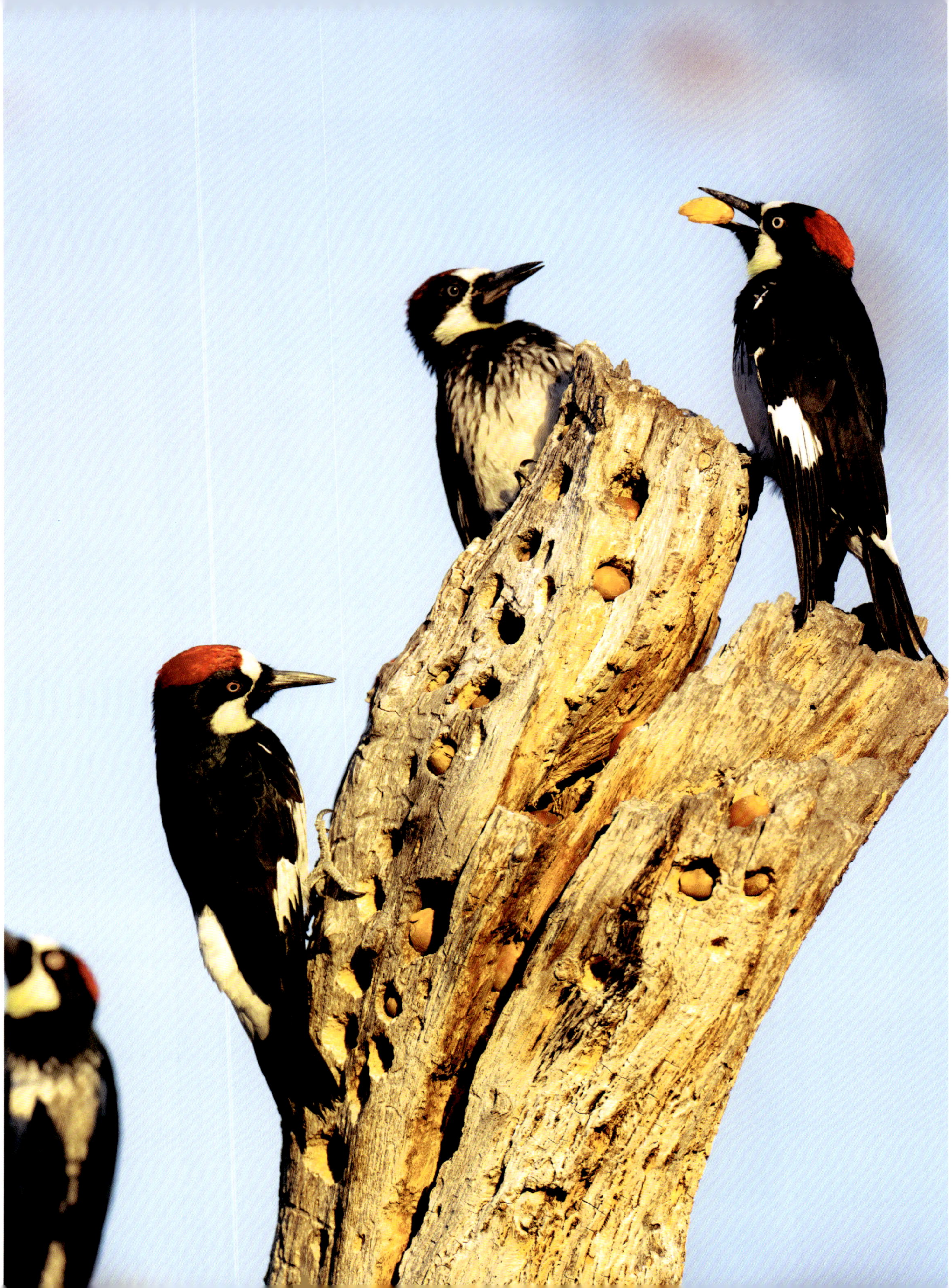

COOPERATIVE BREEDING

Cooperative breeding, in which one or more birds of each gender contribute to the nesting effort, is usually the rule for two species, making it the most common exception to the typical breeding strategy. This approach requires the efforts of multiple birds to raise the young and involves parents as well as helpers, nonbreeding birds that may assist breeding adults. Nests with helpers produce more fledglings than those without.

The Red-cockaded Woodpecker and the Acorn Woodpecker have developed their own unique approach to cooperative breeding to better take advantage of a scarce resource: secure nest sites in the case of the former, and a critical food source within a short window of time in the latter. In both cases this arrangement provides a better chance of securing resources through cooperation. The Gray-breasted Woodpecker is said to be both a colonial and an occasional cooperative breeder.

Red-cockaded Woodpeckers are a relatively small species living in areas with high levels of predation. Excavating in live trees allows birds to carve tiny resin holes that will fill with sap and keep a sticky barrier flowing around the cavity entrance, which prevents predators, especially rat snakes, from entering. Maintaining this flow requires tremendous effort, most efficiently undertaken by a group. Even with heartrot, a mature live pine is a difficult substrate to excavate and requires several years to complete, even with the work of multiple woodpeckers.

Red-cockaded Woodpeckers are monogamous cooperative breeders. Each breeding group includes one breeding male and one breeding female. Breeding groups may have one to four nonbreeding helpers, which are usually related young males from previous years. The roost sites of the individuals of the group are close to the nest, and combined with the nests are called *cavity tree clusters*. Males are dominant over females and inherit territories and nest sites. Since creating new nests and roosts takes so long, it may be smarter for males to wait for a breeding vacancy in their natal territory or a neighboring territory, even though this too will usually take several years. Striking out to find a vacancy also means risking predation while trying to locate or create a cavity and recruit a mate. For females, it is clearly advantageous to disperse upon fledging.

Acorn Woodpeckers maximize their nesting success by amassing as many acorns as possible during a narrowing window in fall, a challenge best met by a group. These woodpeckers, which have the most complex social system of all birds, are cooperative breeders that at times practice *polygynandry*, in which multiple females mate with multiple males in the same nest. Incest almost never occurs, as breeding males are related to one another but not the breeding females, who are also related to each other, and sons do not replace fathers, and daughters do not replace mothers in these groups.

Typical Acorn Woodpecker nesting groups include four to five birds, and many groups include only a single breeding pair. Add in the nonbreeding helpers that are young from previous years, and a nesting group may include up to eighteen birds.

The participation of multiple members of each sex results in some unique breeding behaviors. Females are secretive about copulating, and some mate with more than one male, creating uncertain parentage, which motivates all males to participate in the raising of young. Females know that any eggs in the nest prior to their own egg laying are not theirs and will remove them, often placing them in a hole as if an acorn. These eggs are usually destroyed and eaten by members of the group, including the mother. Once all females are laying eggs, the destruction ends, as none are certain which eggs are from competitors.

All breeding Acorn males and females contribute to the feeding of young and the storing of acorns. The breeding parents share duties regardless of genetic realities and are assisted by the nonbreeding helpers, enabling them to success-

Family groups of Acorn Woodpeckers often help raise the young. Here, a female (note the black forecrown) prepares to feed a piece of an acorn to a juvenile (note the dark eye) as a male (red forecrown) looks for another acorn to offer the youngster. A second female can be seen on the far left.

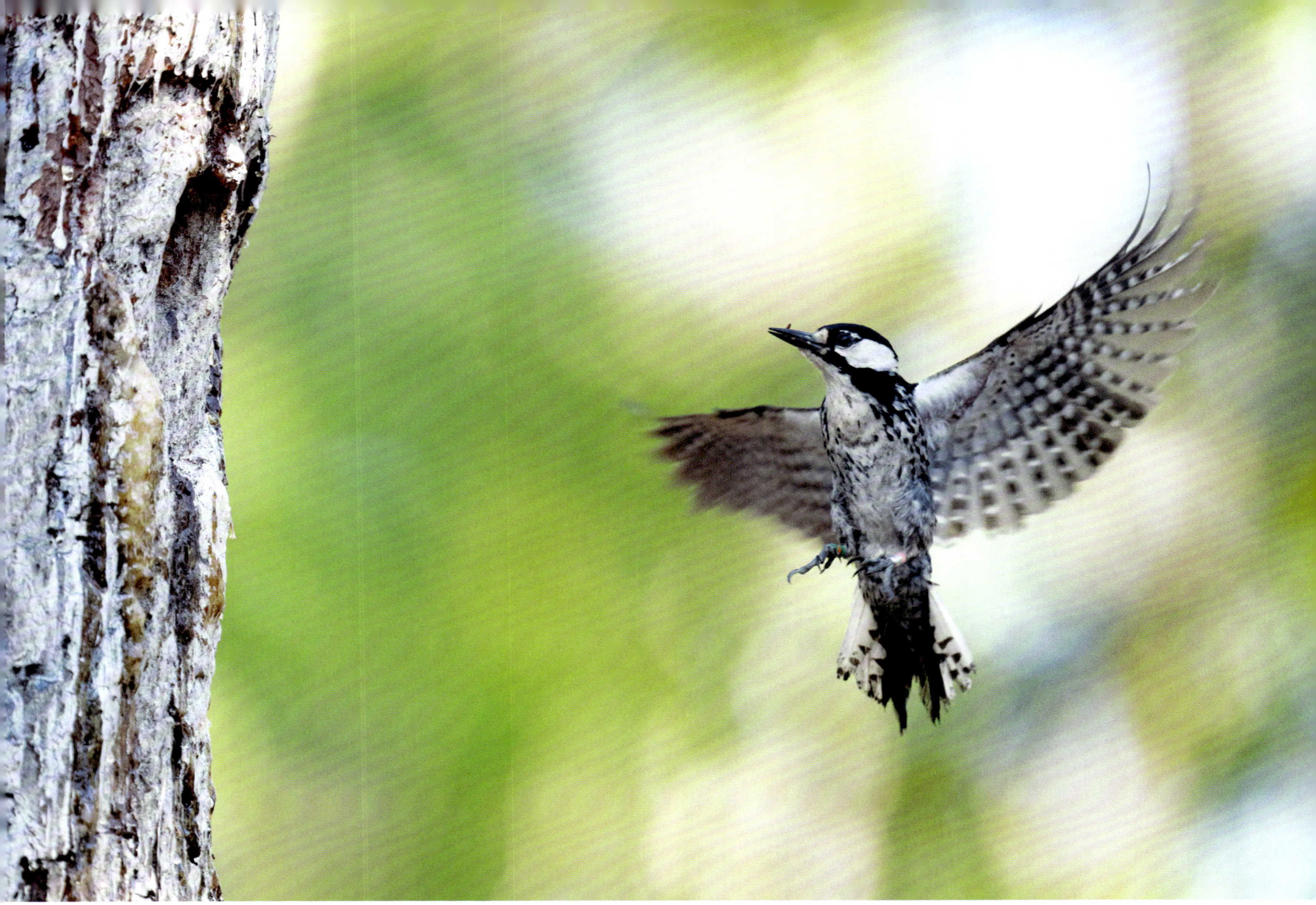

fully fledge more woodpeckers and, at times, produce multiple broods in a single year.

Nonbreeding Acorn Woodpecker helpers benefit from the security, roost sites, and abundant food stores within a group but will not have a chance to breed unless all the breeders of the opposite gender in another nesting group die or disappear. These openings are always filled with unrelated birds from outside a group to the exclusion of related helpers within the group. As many as twenty nonbreeding helpers from external groups may fight, usually in single-sex sibling groups, in battles that may last for days or weeks and involve vocalizing, chasing, drumming, fighting, and perhaps even death. These struggles may involve either gender but are usually most severe in attempts to fill female vacancies. Territories with more acorn storage capacity have more longer struggles and more invaders than lower quality territories. Normally, the larger the group of collaborating siblings, the more successful the effort. Those that win earn the right to share breeding roles in the territory while losers return to their role as nonbreeding helpers within their natal groups.

There is a record of three Red-headed Woodpeckers cooperating to raise young in two consecutive years at a site in Illinois. In this case, the three adults truly cooperated, incubating eggs, brooding young, and defending the nest. This occurred in an area with an unusual abundance of snags and mast (nuts, seeds, and fruit, including acorns and corn).

A Red-cockaded Woodpecker flutters to its sap-surrounded nest cavity in a live pine tree. Family groups help keep the sap flowing by pecking at the area surrounding the nest entrance and thus help deter predators such as snakes and rodents.

SIERRA MADRE

Formidable Montezuma pines tower among sacred fir and over lichen-encrusted Mexican alders at 10,400 feet in the trans-Mexican volcanic belt, which marks the southern extent of the Sierra Madre mountains and is the transition zone between Nearctic and Neotropical pine-oak forests. The habitats on the slopes of the Snowy Mountain Range, as this area is also known, include isolated mountain forests strewn between the Sierra Madre Oriental to the east and Sierra Madre Occidental to the west, which terminate to the north in the scattered sky-islands of southeastern Arizona, southwestern Mexico, Texas, and northern Baja California. These pine-oak habitats host several species of woodpeckers, including the Madrean endemic Arizona and Strickland's Woodpeckers and Williamson's Sapsuckers from the north, which migrate here during winter.

The *life zone* concept first described by C. Hart Merriam suggests that each thousand feet gained in elevation is the biological equivalent of traveling north three hundred miles. In some parts of this area, as you ascend from deserts to grasslands and dry shrublands, you are also ascending from tropical to temperate habitats, finding oaks around three thousand feet and reaching oak woodlands at four thousand feet. Just below five thousand feet, Sierra Madrean pine-oak woodlands emerge, which then give way to pine forest, then mixed-conifer forests such as pine-cedar above about nine thousand feet, and, finally, fir above about ten thousand feet. Three *Dryobates* woodpeckers help illustrate this concept—Ladder-backed Woodpeckers occupy foothill desert scrub before being replaced by Arizona Woodpeckers as oaks become dominant, which are in turn replaced by Hairy Woodpeckers as pines take over.

The Madrean pine-oak woodland is one of Earth's richest and most endangered terrestrial ecoregions, hosting a quarter of all Mexico's plant species, many of them found nowhere else, including roughly 40 percent of all pine species and 30 percent of all oak species, many endemic mammal and bird species, and up to 500 million monarch butterflies. This ecosystem also hosts the greatest woodpecker diversity in the Sierra Madre. Depending upon the region and season, Arizona or Strickland's Woodpeckers might be joined by Acorn, Ladder-backed, Hairy, Gray-crowned, Nuttall's, and Golden-olive Woodpeckers, as well as Northern Flickers and Williamson's and Red-naped Sapsuckers. Many birds of the Sierra Madre, including the endangered Thick-billed Parrot and Eared Quetzal, as well as the Whiskered Screech-Owl, Flammulated Owl, and Elf Owl, utilize woodpecker cavities.

The extinction of the Imperial Woodpecker, once the largest woodpecker in the world and resident of Madrean mountain pine forests, provides a warning to conserve these rich ecosystems. The primary threat to Madrean habitats is loss of forests from unsustainable logging, wood harvesting for cooking and fuel, clearing for agriculture, livestock grazing, and, the most catastrophic: wildfire.

From southeastern Arizona to southern Mexico, the Sierra Madre forms rich habitats that transition over short distances. Cacti-filled deserts give way to grasslands, which in turn become oak and then pine-oak woodlands before they change to conifer forests. You can find a wide range of woodpecker species in small areas, with one species often giving way to another as the habitat changes.

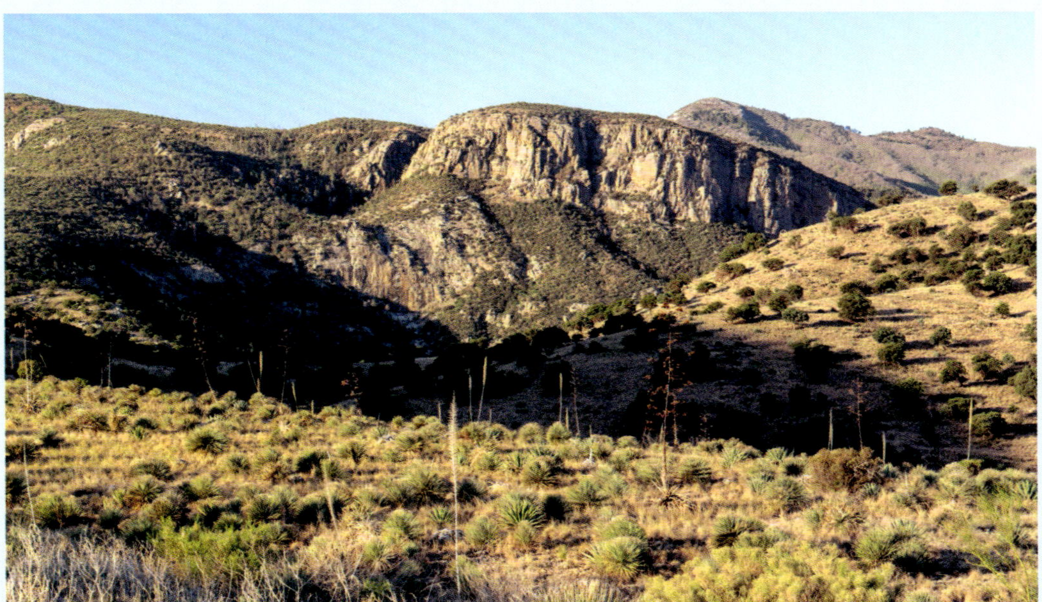

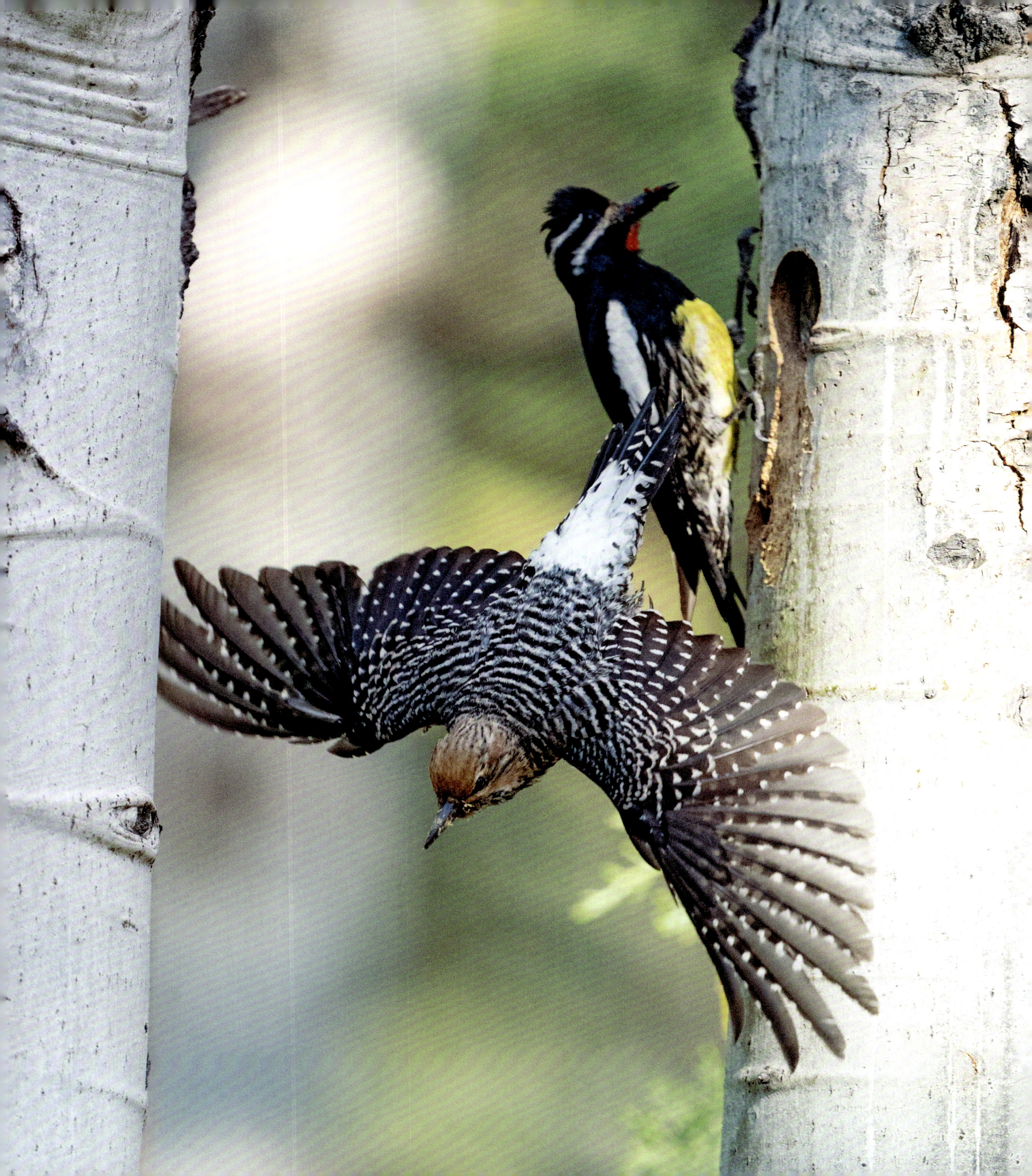

OUTLIERS

Sometimes woodpeckers feed unrelated young of their own species and of other species in a manner that could not be called "cooperative" and is not always welcomed by parents.

I have seen the more social *Melanerpes* Lewis's and Red-headed Woodpeckers from loose colonies attempt to feed unrelated nestlings but was more surprised to see a male of the less social Williamson's Sapsuckers attempt the same. Williamson's Sapsuckers and Downy Woodpeckers elsewhere have on occasion demonstrated the tendency to feed unrelated young. Unmated male Williamson's Sapsuckers may feed the nestlings of an adjacent pair and will also sometimes replace the paternal male in feeding the young when the genetic father leaves the family unit before the young fledge. One Williamson's Sapsucker was even reported to have helped two Red-breasted Sapsuckers feed young at their nest.

One early morning in a central Oregon ponderosa pine woodland, I was watching young White-headed Woodpeckers beg incessantly from within an old snag. At different times in between the parents' feedings, male Hairy Woodpecker and Northern Flicker adults moved down toward the cavity opening much as they would with young of their own species, although I was not certain if the calls were motivating paternal or territorial instincts.

COPULATION

The occasional fluttering and soft calls surrounding copulation are usually the only fleeting commotion near nest sites during this secretive period. Woodpeckers benefit from discretion at this vulnerable time, which keeps them from attracting the attention of rivals or predators and allows them to successfully breed despite the occasional high concentrations of other cavity nesters.

Woodpeckers usually start copulating at completed or nearly completed nests a week or two before the female begins laying eggs. These engagements become more frequent during laying but could continue throughout incubation and sometimes even after hatching.

Either woodpecker may initiate copulation by gaining the other's attention. The female may fly in front of the male or call to him before alighting on a horizontal branch near the nest. After initiating or after accepting a male's approach, she compresses her body crosswise or lengthwise, with her head and tail slightly lifted in what is called a *soliciting posture* or *invitation pose*. Sometimes the tail is held straight out, and at other times the head leans back.

The mate's response varies by species and situation. Sometimes the male approaches his mate quietly or may respond with displays and calls. A Pileated male begins the engagement by drumming several times, compelling the female to fly toward him and assume the soliciting posture, whereas the related Lineated Woodpecker female is reported to move to the end of a branch to assume the same invitation pose, causing the male to slowly climb the trunk toward her, copulate, and punctuate their action with its signature long call.

A Black-backed Woodpecker may fly excitedly toward his inviting mate in a flutter-flight and might even engage her in bill waving. A Williamson's Sapsucker male descends fluttering to perch beside his mate as both chatter and engage in bobbing and wagging displays before he initiates copulation. Flicker copulation is much more perfunctory and may or may not involve calls. The flicker's simple, unceremonious mounting is executed after the female arrives to relieve the male during incubation. During these few seconds, his wings are normally spread over her torso.

The sky-loving Lewis's Woodpecker often heralds copulation with an aerial flourish. One mid-May morning in a burned ponderosa pine woodland in central Oregon, I recognized the screeching calls of Lewis's Woodpeckers and was able to make out one soaring high above the trees, which surprised me as I could not think of insects it could be targeting so early in

Young nestlings must be kept warm by one parent as the other forages. A female Williamson's Sapsucker flies from the nest as soon as the male, bill full of ants, arrives to relieve her.

the season. As it glided in a circle above an isolated snag with wings held high, back arched, neck extended, and bill, feet, and tail hanging, I realized what was happening. I was not surprised to see that upon landing, it lowered and spread its wings as it approached another bird displaying the invitation pose on a branch near the top of a snag. He climbed onto her back, and they copulated just above a cavity they would both later enter.

In most species the male flaps or climbs onto the female's lowered back and quickly slides down her body until their tails align. He then usually falls slightly to her left side, bracing himself by pressing his wing against her or a branch as he wraps his posterior under her, allowing their cloacae (reproductive openings) to be in contact for three to twenty seconds before he dismounts. Wing-flapping by the male and side movements by the female are indicative of unsuccessful attempts. After successful copulations, woodpeckers might perch alongside each other and preen themselves, call and drum to each other, and fly off or fly to the nest together.

Frequent copulation during the egg-laying period better guarantees fertilization, while other copulatory encounters are likely an important part of pair bonding. Reverse mounting and pseudocopulation are common in some *Melanerpes* species, such as Lewis's, Red-bellied, and Golden-fronted Woodpeckers. Reverse mounting, which often happens after an intruder is repelled, is usually initiated by the female, who hops or flutters onto the male's back before he mounts her in the exact same way, or he may attempt copulation with his mate before she hops onto his back. Acorn Woodpeckers engage in ritualistic pseudocopulatory behavior before they enter their evening roosts, with many, if not most, birds mounting one another regardless of gender.

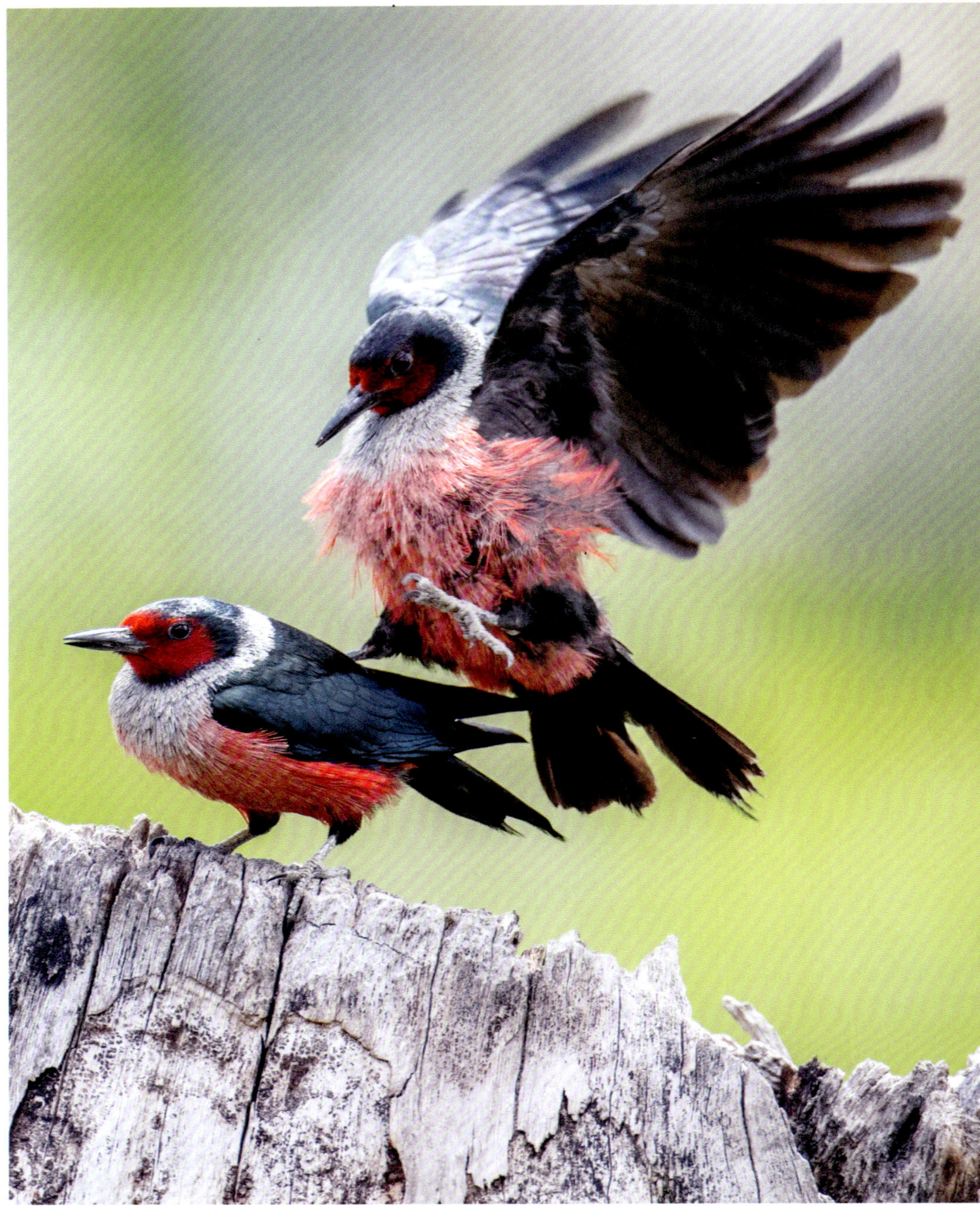

Woodpeckers copulate frequently during the early nesting season, both to better guarantee fertilization and to affirm the pair bond. Lewis's Woodpeckers are among the species that engage in "reverse" mounting, where each member of the pair takes a turn in the mounting the other.

NEST COMPETITION

From the moment a nest is chosen, woodpeckers defend it, the degree of aggression varying with the potential threat. Rival cavity nesters, especially those most likely to try to steal or predate a nest, are driven off with the most ferocity. Potential predators, such as squirrels, are driven off more aggressively, accompanied by spread wings, raised crowns or crests, rattling calls, and threatening bills. Following confrontations with potential aggressors, the birds seem to linger longer near the nest after changeovers.

Woodpeckers serve the vast cavity-nesting community through the creation of cavities as well as the restoration and maintenance of older nests, which they render clean and ready for safe occupation. Woodpecker pairs are vigilant and formidable defenders during this time. Many secondary cavity nesters wait to challenge cavity holders during incubation, when one is attending the eggs and the other is away.

Woodpeckers are infrequently the usurpers of other woodpecker cavities, especially when they share nest preferences. It is most common with weak cavity excavators but also happens with strong excavators in areas with few snags or a shrinking breeding window.

The most aggressive and sustained attacks over the possession of a cavity occur when few nesting holes for secondary cavity nesters and nesting snags for primary excavators are available. This is because of competition for the scarce resource and because it is more likely that predators are finding nests and forcing birds to try to steal a nest rather than create a new one with the breeding window closing. Secondary cavity nesters that can fit into the existing cavity and that breed a few weeks later than the architect are often the most threatening since stealing nests is reflexive.

Unfortunately, the invasive European Starling is a threat to woodpecker populations throughout the US, Canada, Mexico, and the Caribbean and steals nests from many brooding woodpeckers. Red-bellied Woodpeckers suffer as their nest phenology (schedule) is similar to the starling's, and they almost always succumb to them, forcing Red-bellieds to nest later and compete with the dominant Red-headed Woodpeckers.

Even feisty Red-headeds lose their cavities to European Starlings. In some northern areas, starlings are responsible for as much as half of Red-headed Woodpecker nest failures, and their presence can make nests four times as likely to fail!

One afternoon in Seattle I watched two European Starlings attack a female Northern Flicker brooding her eggs. She panicked in the cavity entrance as she fended off one with her bill, while the second pushed its way past her into the cavity, attacking the female from the inside, forcing her out and into the claws of the first starling. With the flicker dislodged, the starling filled the cavity entrance and would not let the female reenter before proceeding to destroy her eggs.

Starlings are unable to fit into smaller cavities such as those created by sapsuckers. Smaller secondary cavity nesters have many more cavity options and are surprisingly aggressive in their attempts to pilfer the cavities of woodpeckers. Eastern, Mountain, and Western Bluebirds; Violet-green and Tree Swallows; Pygmy Nuthatches; House Sparrows; and House Wrens all attempt to take over cavities from woodpeckers as large as flickers and as small as sapsuckers, with the House Wren being the only known avian nest predator of sapsuckers.

American Kestrels are both a cavity competitor and a potential predator that causes some woodpeckers, especially Lewis's Woodpeckers, to sometimes abandon nest sites if kestrels are nesting nearby.

Other avian predators at or near nests include Sharp-shinned and Cooper's Hawks, Peregrine Falcons, Red-shouldered Hawks, Northern Goshawks, Aplomado Falcons, and Common Ravens, as well as screech-owls and pygmy-owls, which are also potential nest competitors.

Keeping nest sites secreted at this time also helps reduce nonavian predators such as weasels, black bears, raccoons, squirrels, rat snakes, and gopher snakes. One year I watched a pair of Pileated Woodpeckers work laboriously over two months carving four potential nest cavities in different parts of a forest. As their intense activity at one cavity suddenly hushed, I realized they were incubating eggs. Two weeks later, while I was watching attentively for signs of the first feeding, the male arrived with his crest raised and wings spread as he swooped aggressively toward an eastern gray squirrel near the nest. The next morning, while I was still waiting to see the first signs of feeding, the face of a squirrel filled the cavity entrance, apparently after predating the nest.

When cavities and snags are numerous, competitors, predators, and prey can sometimes nest in virtual apartment buildings, with nests of various species each above the next. One year in central Washington in a stand of aspen within a ponderosa pine–dominated mixed-conifer forest, I witnessed Pileated, White-headed, Hairy, and Downy Woodpeckers; Northern Flickers; Williamson's, Red-breasted, and Red-naped Sapsuckers; Northern Pygmy-Owls; Western Bluebirds; House Wrens; Mountain and Black-capped Chickadees; and White-breasted Nuthatches nest within fifty yards of one another. It was not unusual to see the heads of three to four birds peering from cavities at the same time! Another year at a dry wash in the Sonoran Desert, I observed Gila Woodpeckers, Gilded Flickers, and Cactus Ferruginous Pygmy-Owls nest in cavities in a clump of four saguaros, and American Kestrels, Purple Martins, and Elf Owls nest in nearby saguaros.

THE NEST CYCLE

In most of the temperate parts of Canada, the United States, and Mexico, woodpeckers begin their most secretive period at their nests during April and May. Courtship is delayed by a month or more at higher elevations and northern latitudes. Woodpeckers of tropical areas from the southern coasts of Florida, through the Caribbean, and in the coastal and southern parts of Mexico enter this stage a month or two earlier and sometimes multiple times during a year. Some species, including Acorn and Red-headed Woodpeckers, can breed in response to extraordinary food supplies and can produce second broods in response to favorable conditions.

Sometimes woodpeckers are forced to renest, as when a mate dies or their cavity is stolen or predated. This forces them to begin using an alternate cavity or making another. If a mate dies, the remaining bird retains the territory but must recruit an unmated bird, perhaps a floater without a territory, and begin the courtship and nest selection process anew.

Each cavity is designed to be a tight fit for the particular woodpecker to limit the access for potential predators. Holes angle abruptly down from the normally vertical face of the entrance, making it a challenge to enter and depart nests. I have seen woodpeckers spend a few seconds wiggling and rotating their bodies to squeeze out of snug nest entrances. Nest cameras reveal adults entering headfirst before making the sharp turn downward, upside down with legs spread until reaching the bottom, at which point they turn around while raising their tail and place their brood patch against the eggs, fluffing up their feathers to create more body warmth, and lowering their wings. Then the incubating bird either rests its head against the wall of the cavity or tucks its bill inside the feathers on its back before closing its

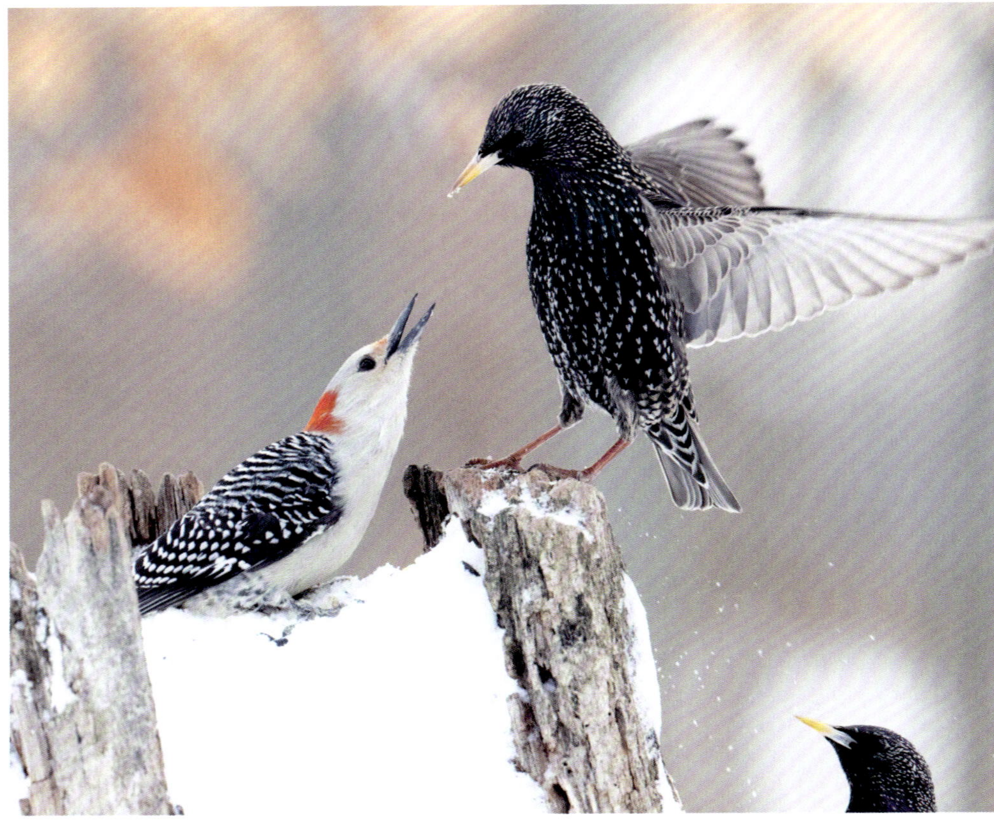

European Starlings, like these two harassing a female Red-bellied Woodpecker, are one of the biggest threats to woodpecker nests. Red-bellied and Red-headed Woodpeckers, Northern Flicker, and other species often have eggs destroyed and nests overtaken by this aggressive invasive species.

eyes and appearing to rest. If the day is particularly warm, the adult inside the cavity may roost on the nest wall.

Female woodpeckers can lay eggs beginning after their first winter, but not all do. Nearly all species lay one clutch a year, although there are exceptions among *Melanerpes* species, including Red-headed, Acorn, Red-bellied, West Indian, Golden-fronted, and Black-cheeked Woodpeckers, which may produce two clutches in a year, and Gila and Jamaican Woodpeckers, which might produce three. Rare second broods have been recorded for Downy and Hairy Woodpeckers as well as for Northern Flickers.

The size of clutches within a species is likely influenced by many factors, including the experience and health of the parents, the quality of the habitat, and the warmth within the cavity. Northern birds and species tend to produce larger clutches, while those closer to the equator usually produce smaller clutches.

In the temperate regions from Alaska and Canada through the mountains of northern Mexico, most woodpeckers lay four to five eggs. In tropical areas, such as southern Florida, the Caribbean islands, and coastal and southern Mexico, most produce two to three. Pale-billed Woodpecker clutches max out at two eggs, Lewis's Woodpeckers may produce up to nine, Red-headed up to ten, and Northern Flickers may have thirteen, although it is rare for all to fledge for most species.

Most if not all woodpeckers are believed to be indeterminate layers, meaning that if an egg is removed or destroyed, the female will lay a replacement. Many studies have been done on

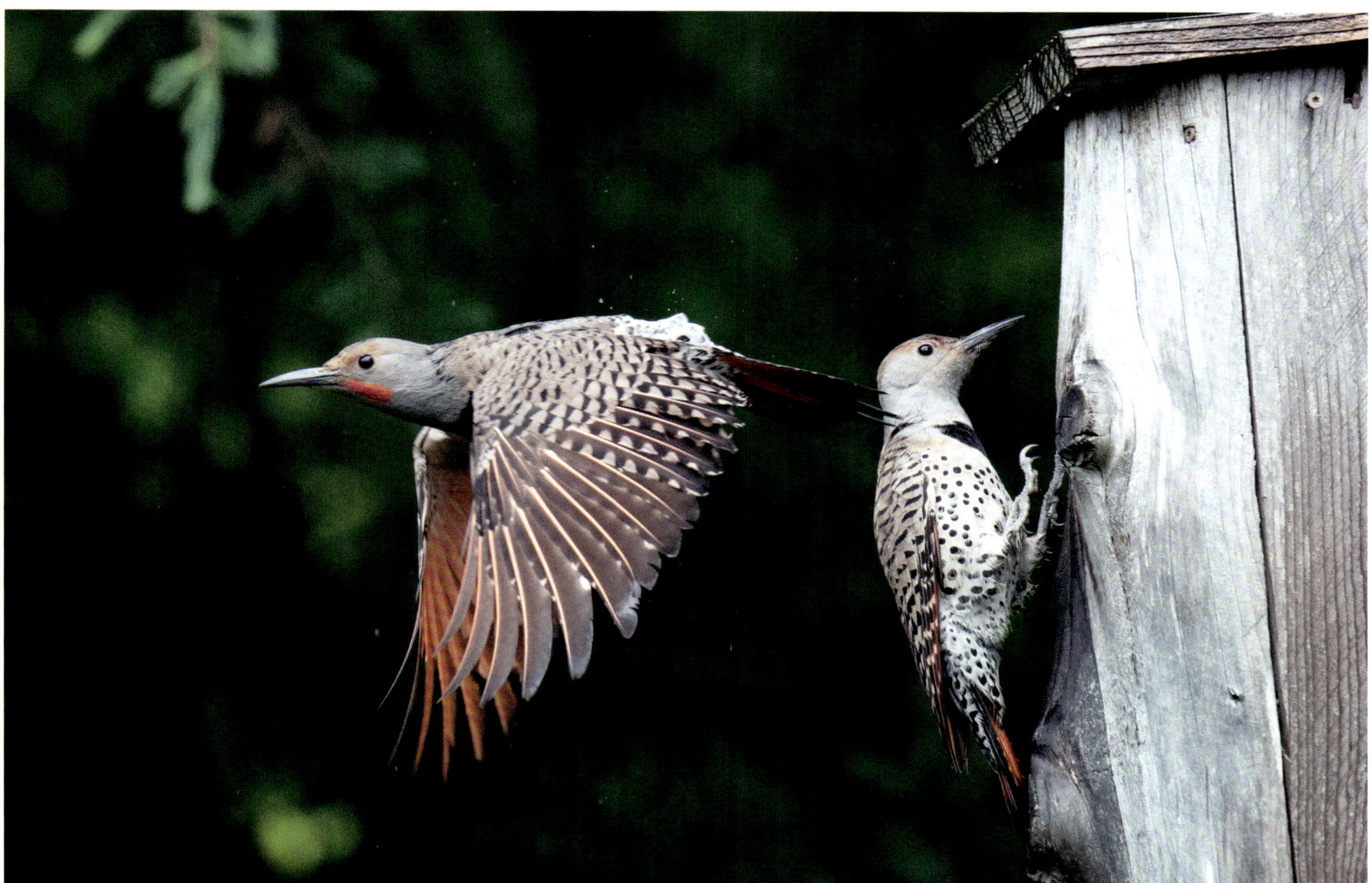

A male Northern Red-shafted Flicker flies from his nest as his mate arrives to relieve him of incubation duties in a nest box in my yard. During this vulnerable stage, one member of the pair is almost always in the nest.

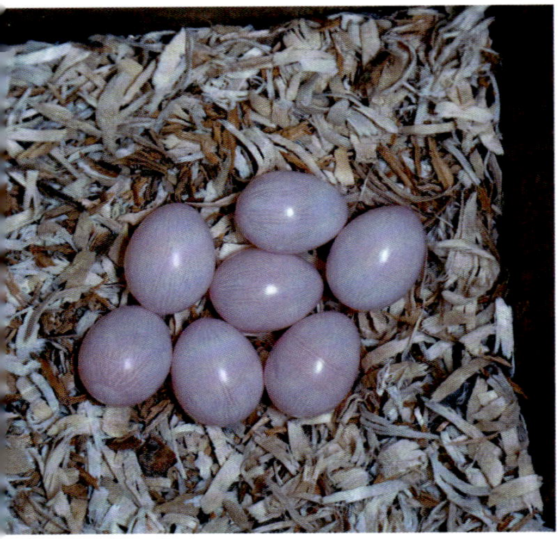

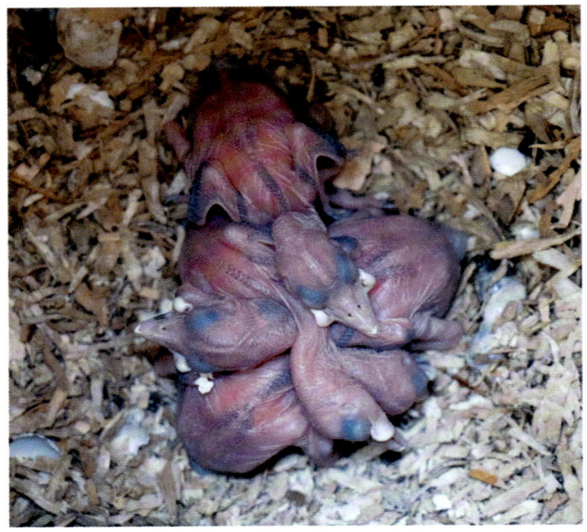

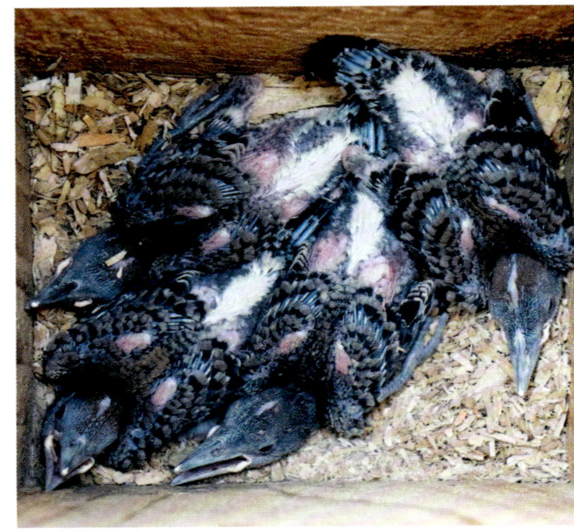

Northern Flickers, and in one case an egg was removed each day and the female continued to replace it until seventy-one eggs were laid! A similar study was conducted in which a Red-headed Woodpecker had her clutch of eggs removed six times over thirty-one days (twenty-eight eggs total) before she was allowed to raise her seventh clutch (four eggs). The cavity size does not seem to impact the size of the brood.

Woodpecker eggs vary in size, with the Downy Woodpecker likely having among the smallest eggs and the Ivory-billed Woodpecker likely to have had among the largest. The size of the smallest Downy Woodpecker eggs are about 0.75 inches long, and the largest recorded Ivory-billed Woodpecker eggs were about 1.5 inches long. All woodpecker eggs are all white, unmarked, and glossy to some degree.

During the egg-laying period, both parents develop brood patches, a featherless area of the abdomen that is dense with blood vessels. The adults place this area in contact with the eggs to keep them warm. Acorn Woodpecker and Red-cockaded Woodpecker helpers that assist parents with various duties, including incubation, have less defined, if any, brood patches commensurate with the time they invest in brooding.

Most woodpecker nesting behavior takes place deep within the cavity, hidden from view, but the more we learn about what happens sequentially inside the cavity, the better we can guess the stage in the life cycle through the behavior of the parents at and around the nest.

What follows represents the typical nest cycle for woodpeckers but varies by species, geography, and individual nests.

Nest Attendance

Woodpeckers are egalitarian throughout the breeding cycle, especially in comparison to other birds, although the relative amount of work of each member of a pair or family unit at each stage varies by species. Male woodpeckers are unusual among birds in how large of a role they play. They share the brooding of eggs and young as well as feeding the young and protecting the nest, and they usually take on most of the nest sanitation and are the exclusive brooders and guardians of both eggs and young overnight.

Typically, once the female commences laying eggs, one parent will almost always be guarding the cavity entrance from the inside or very close to the nest. The pair will continue to copulate, and the female will generally lay one egg early each morning until her clutch is complete. After the penultimate or final egg is laid, the pair will begin sharing incubation duties during the day, while the male continues to roost in the nest and brood the eggs at night as he has since the nest was completed.

Left: Woodpecker eggs are white and almost glossy, but they can appear pinkish during the first four days of incubation as the orange yolk reflects through the shell until embryonic membranes form.

Middle: During the first four days after hatching, the nestlings are featherless with closed eyes. At this stage one adult is always in the nest, and the male broods throughout the night. These Northern Flicker nestlings are roughly three days old.

Right: These Northern Flicker nestlings are roughly twelve days old. Their eyes are open, and they have begun to hold their heads up against the wall of the cavity.

LONGLEAF PINE HABITAT

The longleaf pine ecosystem is one of the most biodiverse in North America, stretching from northeast Texas across the coastal plains of Louisiana, Alabama, Florida, Georgia, and the Carolinas, and encompassing savannas, rolling hills, sandhills, and montane forests. This system once covered 90 million acres; roughly 5.2 million remain.

Lightning fires shaped the longleaf pine ecosystem. Flames, fueled by wiregrass and dry pine needles, clear out flammable woody vegetation, exposing mineral-rich soil conducive to seed germination. The resulting open forest dominated by large older pines, with a well-developed ground layer of grasses and forbs and virtually no midstory, hosts many eastern woodpeckers but is the ideal habitat for the near-threatened Red-cockaded Woodpecker, the indicator species of this system. Mature live trees, especially those utilized as nest and roost trees, are critical as they allow Red-cockaded Woodpeckers to keep the sap around the entrance holes flowing to protect the cavity's inhabitants. These long-lived trees continue growing beyond 150 years, pulling climate-changing carbon dioxide from the air and making them important for carbon storing.

Pine trees dominate a healthy fire-maintained longleaf pine habitat. Fire removes smaller pines as well as oaks and other deciduous trees and leaves behind an open, grassy understory—perfect habitat for the Red-cockaded Woodpecker.

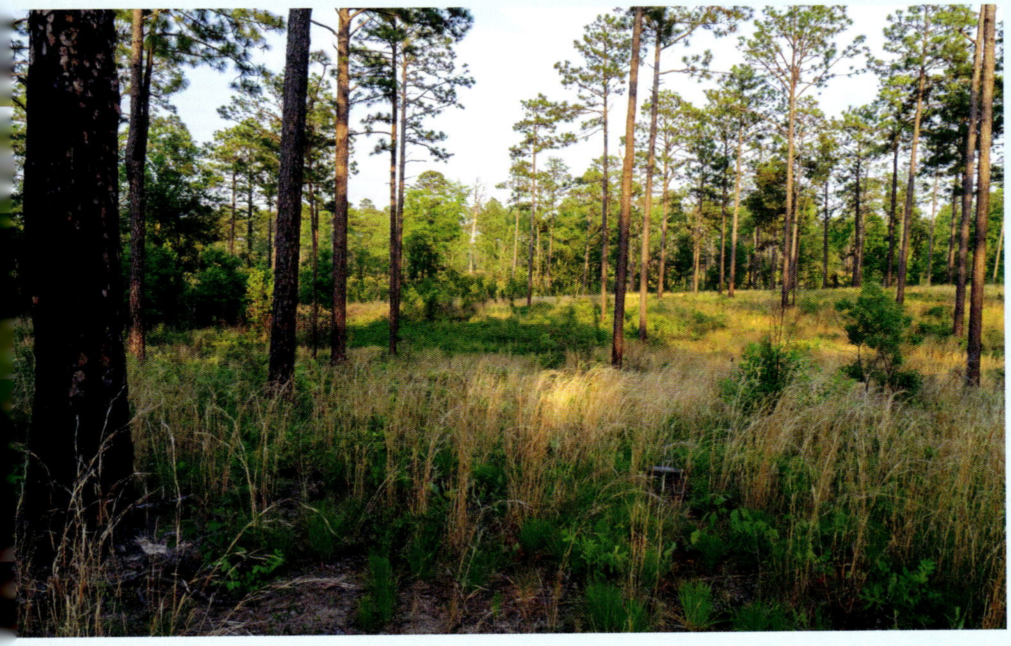

Ideal fire management facilitates natural habitat transitions by leaving areas with varied canopy cover and hardwoods versus softwood and creating an abundance of snags softened by fungus, which infects dead burned trees, allowing several woodpeckers to play keystone roles. Red-cockaded Woodpeckers create nest cavities in live pines, but Pileated, Red-headed, Hairy, and Downy Woodpeckers as well as Northern Flickers benefit from softer snags, and some of these species also enlarge Red-cockaded cavities for their own use. Yellow-bellied Sapsuckers winter in longleaf pine habitats.

The Red-cockaded Woodpecker also serves as an umbrella species for this ecosystem. Protection and restoration of its habitat benefits rare species such as the gopher frog, gopher tortoise, eastern indigo snake, flatwoods salamander, and Venus flytrap. Some populations of Red-cockaded Woodpeckers are recovering thanks to improvements in habitat management as well as population enhancement techniques, such as nest-cavity restrictors that prevent other species from enlarging and usurping their nests, cavity inserts in live trees, and translocation of young females to groups lacking females, which can establish new, more genetically diverse breeding groups and help overall survival.

Excess logging, development, and fire prevention have deteriorated and fragmented longleaf pine habitat and Red-cockaded Woodpecker populations. Maintaining the Red-cockaded and other unique life in this ecosystem requires planting seedlings, controlling invasive species while managing native ground cover, and especially conducting frequent summer prescribed burns (at one- to five-year intervals), which kill non-fire-resistant hardwoods and younger trees and keep the understory open. While the Red-cockaded Woodpecker's near-threatened status has helped implement conservation measures on public lands, little is being done for the species on private lands, furthering fragmentation of the populations.

Brooding

Once brooding has begun, an adult might be seen panting at the cavity entrance to cool itself on the warmest of days, but otherwise the birds remain hidden from view. At this stage the female is dominant and decides which of the parents occupies the cavity.

During incubation through the first several days after the eggs hatch, parents do not usually switch places at the nest until their mate is at the nest tree or otherwise close to the cavity. The arriving parent may announce their arrival with long calls or soft drumming from a distance or with *wicka* calls as they land at the nest. The brooding bird occasionally taps on the cavity wall to express its interest in changing places or to acknowledge the arrival of its mate. Cuban Green Woodpeckers make an intentional whirring sound with their wings to alert their mate as they arrive to trade places.

The duration of each parent's turn at incubating varies by species, weather, time of day, and likely the experience of the parents and can be as short as five minutes or more than ninety minutes. The goal is to keep the eggs consistently at the ideal temperature for development. On hot days later in the incubation period, the attending adult may leave before its mate arrives.

Vigilance is paramount at this stage, and adults will go to great lengths to avoid being seen arriving at or departing the nest. Sometimes they delay their arrival, regardless of whether the eggs have been left unattended, until predators, or humans, are not watching or close by. When they do arrive or depart, they often fly within the canopy or close to the ground, hidden by vegetation, before steeply climbing or dropping. On the other hand, if they know their nest has been discovered and is being approached by a nest predator, such as a snake, raptor, or human, they noisily complain, and if the invader is small, they attack to repel them. Because of this wariness and the need for the adults to be with their eggs and young, it is best to avoid being near nests, especially during this precarious period.

Woodpeckers have a short incubation period of ten to fourteen days, depending on the species. Nestlings hatch from their eggs asynchronously (not at the same time) over twelve to twenty-four hours, breaking out using a hard nodule called an egg tooth at the end of the bill. In most species the egg tooth gradually wears off before the young leave the nest, although flickers commonly show signs of it at fledging.

Nestlings

At hatching, the nestlings are featherless, and feathers come much more slowly than for open-nesting birds, making woodpeckers entirely dependent upon their parents for warmth through near constant brooding. Unusually wet or cold weather during this time is a common cause of nestling mortality.

Initially, the nestlings' eyes are shut, and they are unable to hold their heads up for long, but by the second day they can stretch their neck toward the cavity entrance for several seconds. During the first ten days, the young face one another on the cavity floor with their necks resting on the backs of their siblings. They have a disproportionately large gape (interior of the open mouth), with oral flanges (fleshy, knob-like swellings) protruding from each side. The parents must peck at the flanges to encourage the young to open their bills, but within days they respond to changes in light, vibrations, and calls of arriving parents. The first nestling to extend its bill high enough is the first to be fed. In turn, each begging bird receives food before returning to the cavity floor, allowing the next to be fed. Nestlings double their size in the first twenty-four to forty-eight hours.

Woodpecker parents share in the feeding of young, just as they did with incubation duties, although the female Lineated Woodpecker does

Here, a male Downy Woodpecker prepares to feed a male nestling. Note how the sex-specific color markings differ between the adult and nestling, which is typical in the genus *Dryobates*.

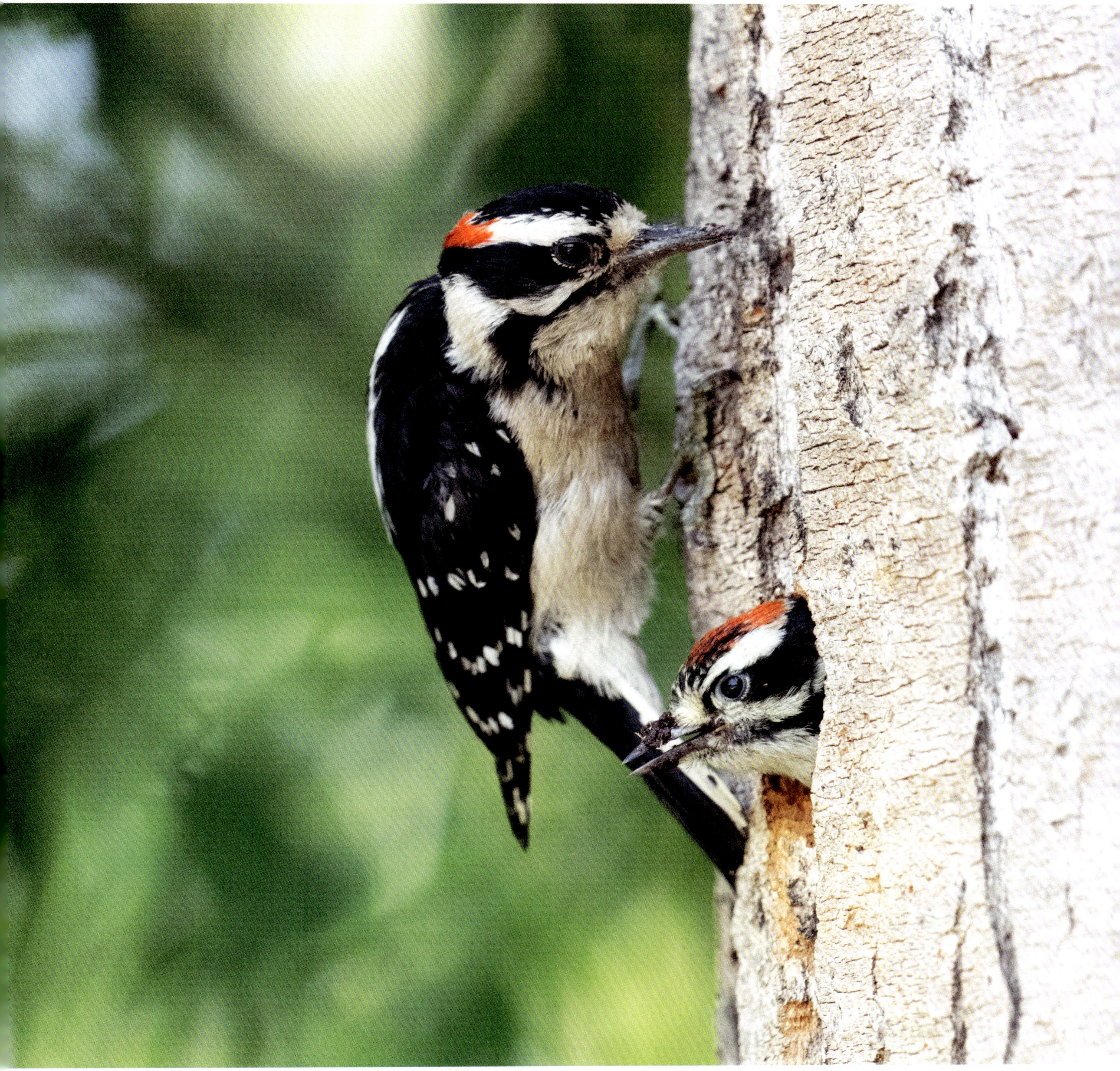

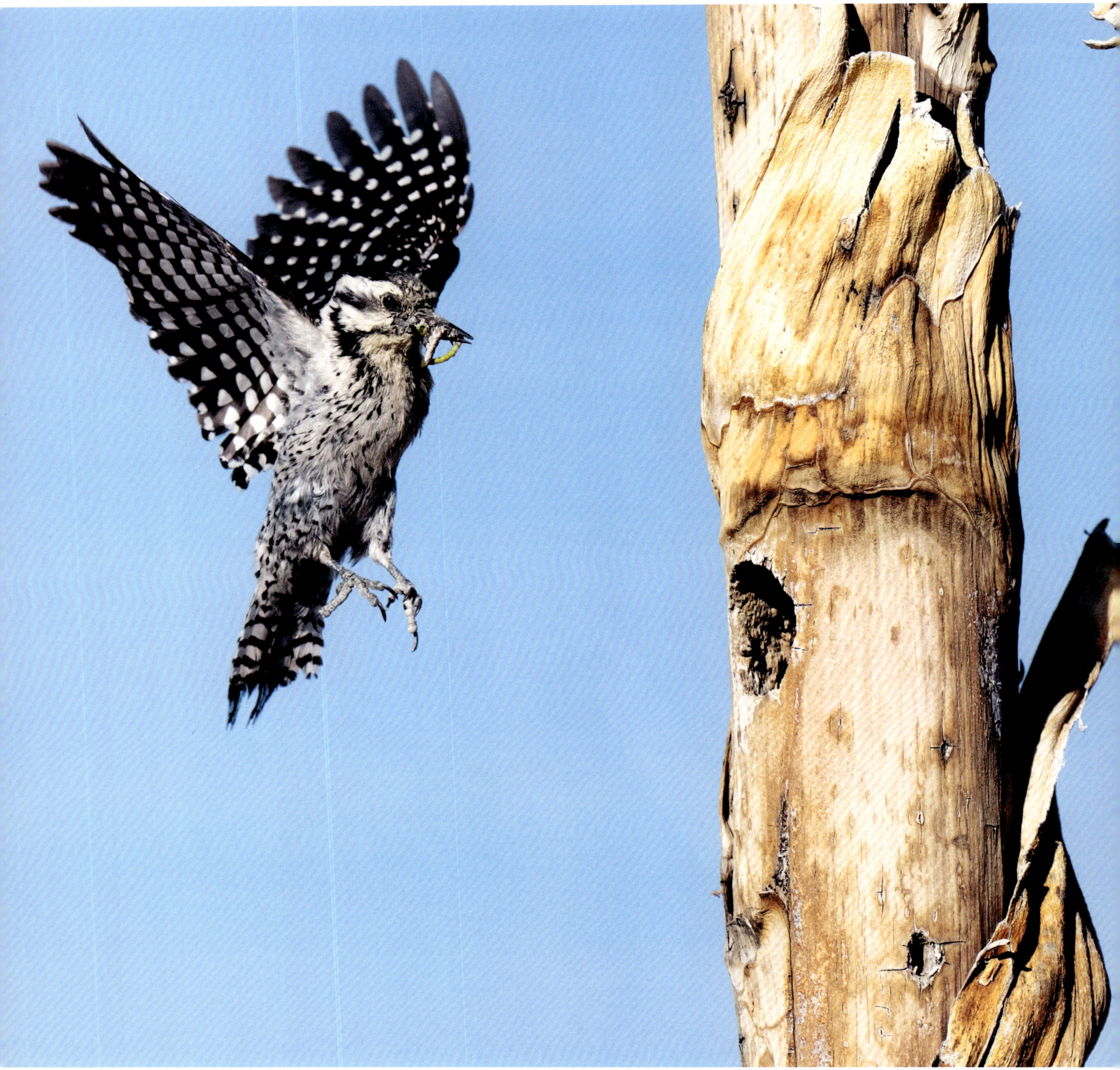

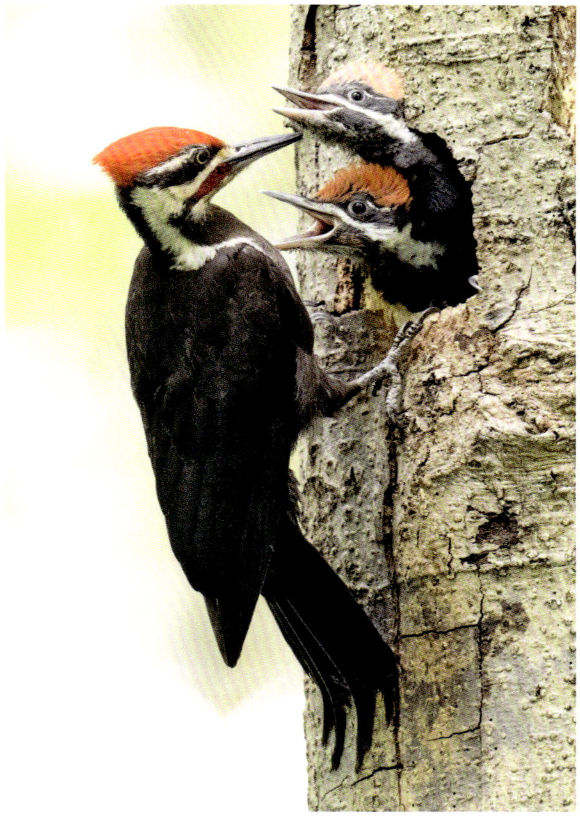

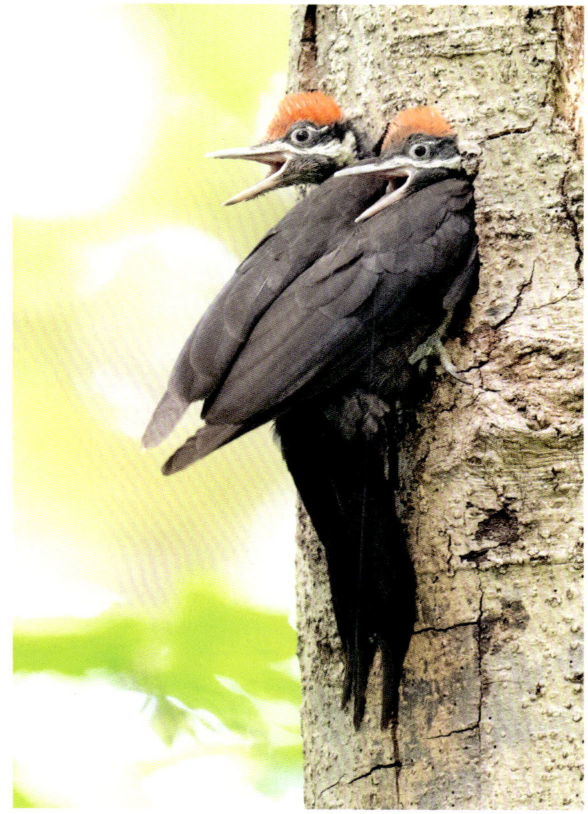

Above, left: During their last week in the nest, these Pileated Woodpecker nestlings are often peering out in anticipation of a food delivery by a parent. As an adult comes into view, they raise their crests and begin begging loudly.

Above, right: Adult woodpeckers often remove waste material up until a few days before the young leave the nest, forcing them to sometimes push their way past begging youngsters.

Opposite: A female Ladder-backed Woodpecker arrives at her nest in an agave stalk with a billfull of several species of invertebrates, including caterpillars and aphids.

more of the feeding while the male guards the nest. Adults enter the cavity completely to feed nestlings during their first seven to ten days. Nest sanitation is a priority at this stage. Most eggshells are dropped from the cavity or are carried away. The young usually defecate immediately after being fed, and adults often use wood chips from the cavity walls to soak up this material to aid in removing it from the nest. The solid waste is sometimes eaten by the parents. Males tend to play a bigger role in nest sanitation.

When nestlings are small, adults arrive with smaller prey or pieces of food that are tough to identify. With *Colaptes* species (flickers and Golden-olive and Gray-crowned Woodpeckers) and *Dryocopus* (Pileated and Lineated Woodpeckers), which regurgitate their food, there will be no obvious sign they are carrying food except for a swollen crop (expanded throat). Parents switch places at the cavity entrance at this time, with the arriving bird entering completely and brooding the young between feedings.

By roughly the end of the first week, the appetites of the growing young force adults to leave the nest unattended at times. Nest sanitation activities become more obvious at this stage as the gelatinous fecal sacks containing the excrement of the young can be fairly large. The parents carry the waste in their bills and deposit it far from the cavity.

Liquid waste is absorbed by wood chips at the bottom of the cavity, so the parents exit with bill-loads of moist wood chips as well. I have seen Williamson's Sapsuckers regularly drop this waste at a couple of specific distant snags. Nest sanitation varies by species, with some continuing to clean their nests until a few days before fledging, whereas others, such as Red-bellied Woodpeckers, may halt these efforts during the first week. Most fatalities among nestlings occur during the early part of the nest season, and their bodies are removed from the nest by the adults.

Some woodpeckers deliver different foods as their young grow. During the early nestling stage,

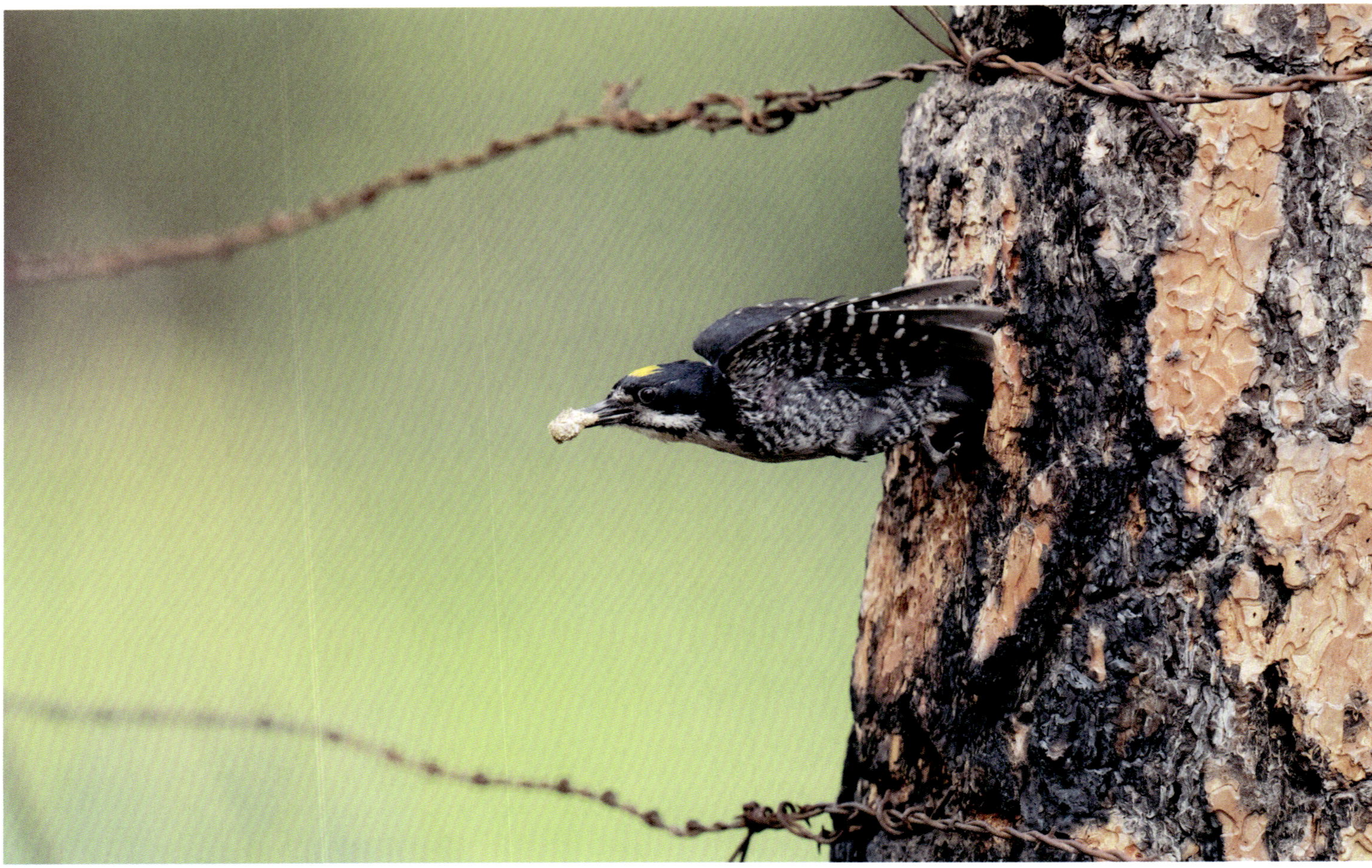

parents bring in small, soft, protein-rich prey such as ants and other small insects and larvae, including caterpillars, aphids, and moths. Sapsuckers deliver ants mixed with globs of sap during the first week, and Red-headed, Lewis's, and Gray-breasted Woodpeckers smash their food at anvils (hard, flat places on trees or snags) into indistinguishable multicolored masses from which they pull progressively larger globs as the young develop.

Woodpecker nests remain quiet the first several days of the nestlings' lives, and their weak calls are muffled by the attending adult in the nest. Sounds are limited to the subtle calls or taps adults make when trading places. During this time the only place the woodpeckers are commonly seen or heard is at valuable food sites such as fruiting trees, flowering cacti, or trees rich in grubs, termites, or ants, where they might have territorial squabbles with competitors.

As the young gain warming feathers and need more food, brooding becomes more irregular, except at night, when the male continues to roost with the young until they are ready to fledge. Usually before the end of the nestlings' second week, the parents dip into, rather than fully enter, the cavity, and the ends of their tails remain visible. Since the young can thermoregulate, both parents are able to forage at the same time. The nestlings now spread to the edges of the cavity with their bills resting against the wall.

Woodpeckers feed their young throughout the day, with peaks in the early morning and evening. The frequency varies by species and weather. Flickers and Pileated Woodpeckers, which feed their young by regurgitation, visit with prey once or twice per hour, whereas sapsuckers and Downy Woodpeckers, which feed small insects directly from their bills, may feed their young every few

Keeping a nest clean helps the young stay clean and avoids attracting parasites and betraying the location to predators through odors. Woodpeckers remove the shells and solid waste through much of the nesting period, although some waste is eaten by some species early on. Here, a male Black-backed Woodpecker removes a fecal sack and drops it far from the nest.

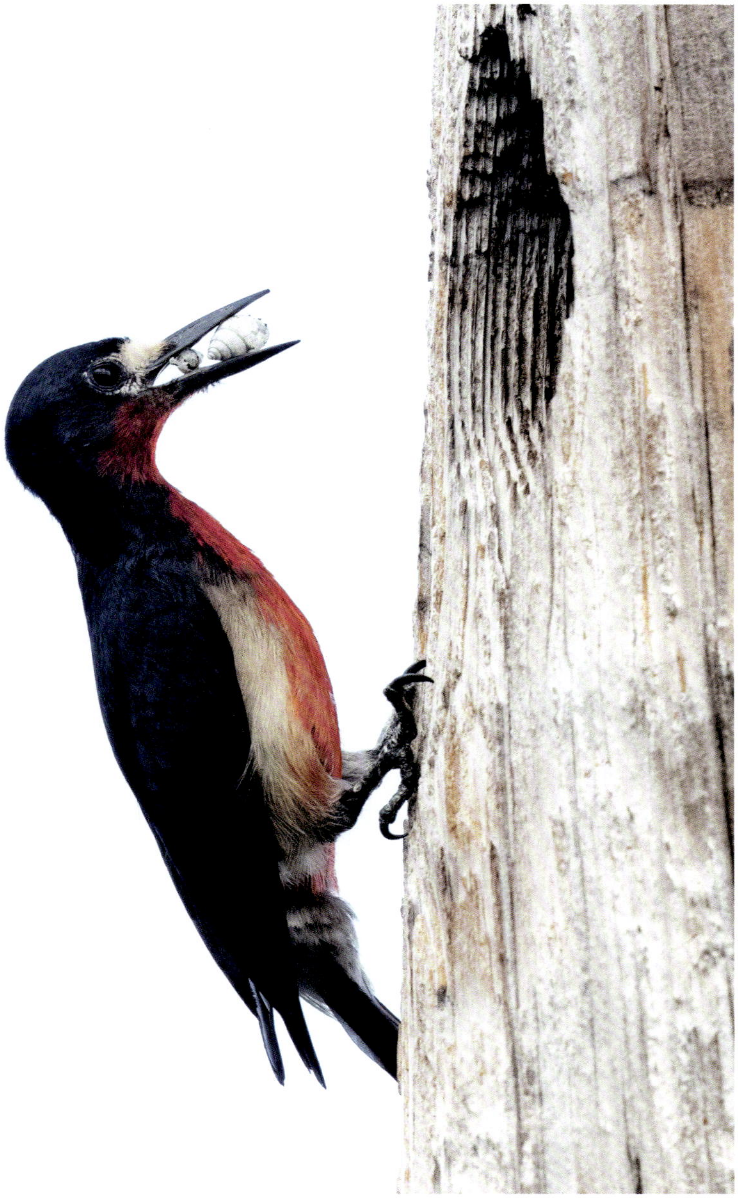

Melanerpes species like this Puerto Rican Woodpecker are incredibly omnivorous and eat a wide variety of plant and animal foods. I was surprised to see this male deliver snails, an item I had not seen listed as one of their foods.

minutes, especially when they locate a large concentration of prey. The less frequent visits by the former may help hide their nests from potential predators. Cooperative breeding woodpeckers ameliorate the risks of frequent provisioning since they have helpers on hand to recognize and repel threats.

By the beginning of the third week, parents can simply reach their head and neck in to feed the young, which are by now clinging to the vertical walls, like adults do on trees, which helps build important muscles.

Nestlings are fed most frequently and reach their maximum size as they enter their last week in the nest, which may be the third or fourth week. In the final week adults carry larger prey, which may reduce the frequency of visits. Adults may arrive with increasingly overfull bills of identifiable and unidentifiable prey items. At this stage woodpecker feeding often surprises me, such as instances when during a rainstorm Guadeloupe Woodpeckers began bringing frogs in consecutive feedings, Puerto Rican Woodpeckers delivered snails, Gila Woodpeckers fed pieces of birds, Red-headed Woodpeckers supplied bright green grasshoppers, and Lewis's Woodpeckers delivered berries of multiple colors. The young beg incessantly and lunge out to snatch food from parents, who often lean away from the sharp bills.

In some species, the waste at the bottom of the cavity is building up, attracting parasites such as lice, flies, and other insects—and potentially predators—with its odors. The combination of accumulating waste, odors, and the din of screaming young make the nest less healthy for the young and easier for predators to detect. They need to leave soon!

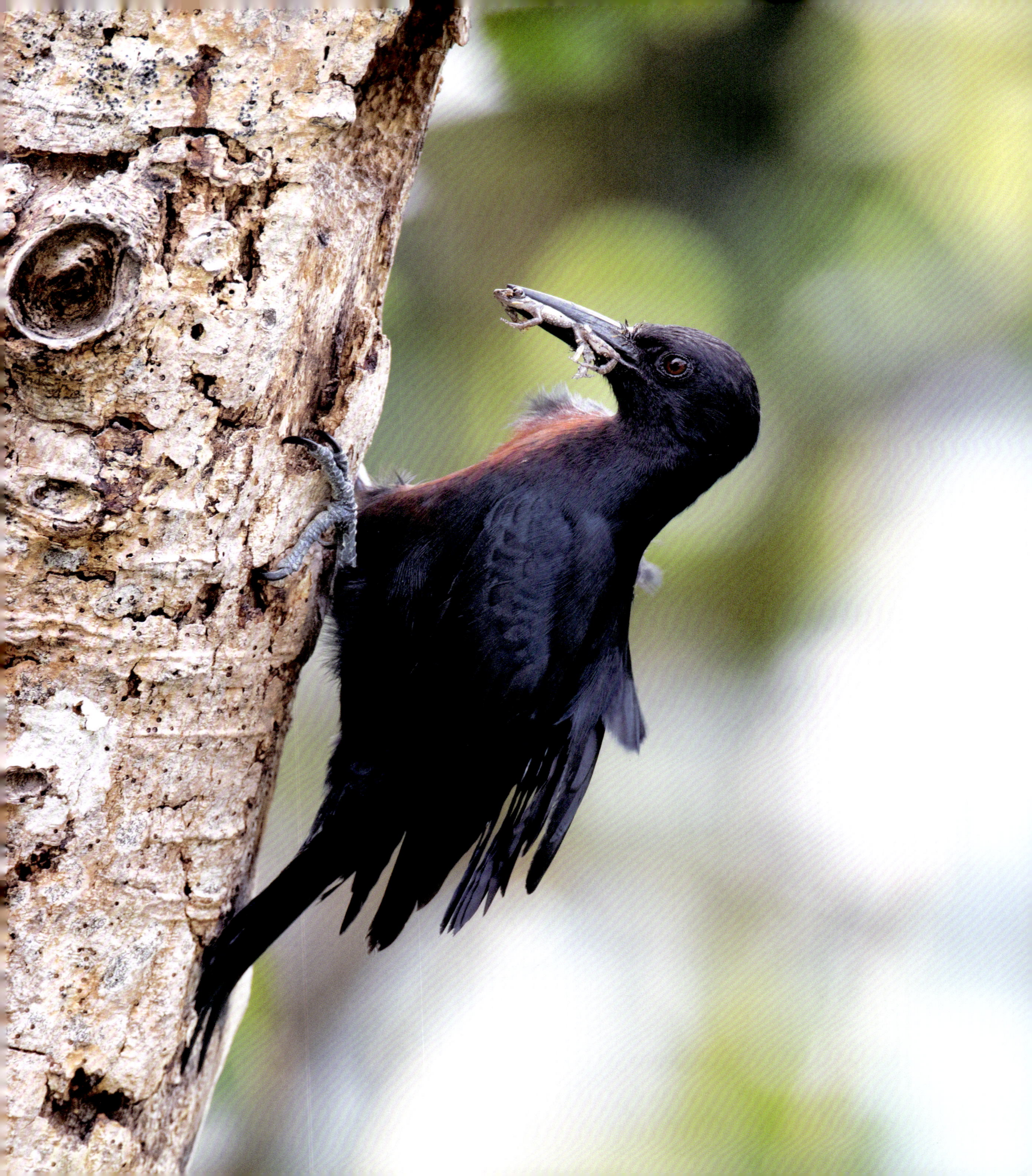

FOODS AND FORAGING TECHNIQUES

Woodpeckers have evolved to breed in response to many factors, including the amount of daylight and the temperature, among others. The number of weeks some foods remain in strong supply can impact the length of the potential windows for successful breeding. Many species shift their diets in response to seasons, weather, fire, or a closing or opening of canopy or ground cover. Some may alter their food targets or strategy, or even migrate elevationally or latitudinally.

Woodpeckers scour the surface and subsurface of the ground, trees, shrubs, and cacti in their locale as well as the sky above them for food. While most species are opportunistic, some have evolved to be specialists, and some are generalists. Some focus on thick tree trunks, others on large limbs, and still others on small twigs. Some lift or move leaves on the ground, and others look beneath lichen or bark. Many feel into holes or crevices with their bill and tongue. Some prefer open areas without cover, and others want the protection of a canopy. Some initiate excavations deep into a tree, and others find overlooked food in the excavations of other woodpeckers. In many habitats, several woodpecker species have developed or moved in to handle each of these opportunities as they arise, whereas isolated island species are generalists that exploit most niches.

Members of genus *Melanerpes* utilize a wide range of foraging approaches and foods, while others specialize in food and feeding methods to such a degree that the sighting of a specific species during a particular season can be indicative of the presence of a particular tree or prey species. For instance, in northern or high-elevation areas, the occurrence of an American Three-toed Woodpecker is indicative of spruce or lodgepole pine; in the West, an Acorn Woodpecker assures you that oaks are nearby; White-headed Woodpeckers guarantee large-coned pines; and nesting Williamson's Sapsuckers hint at either Douglas-fir or western larch being part of the forest composition.

Woodpeckers, especially those in the *Melanerpes* genus, are opportunistic feeders, so I should not have been surprised to see this Guadeloupe Woodpecker deliver frogs to its young in the nest during a rainstorm.

Foods and Prey

All woodpeckers are at least partially insectivorous, and some feed only on insects, whether eggs, larvae, pupae, or adults, with ants, beetles, and moths most frequently consumed. Other invertebrates, especially spiders, are common foods.

Ants are a dominant part of the diet for *Colaptes* woodpeckers such as flickers, *Dryocopus* like the Pileated, and *Celeus* species like the Chestnut-colored Woodpecker. Ants are also the primary prey of adult Red-cockaded Woodpeckers and nestling Williamson's Sapsuckers. The Williamson's Sapsucker eats tree ants and carpenter ants more than other prey, while the Pileated Woodpecker's diet is typically composed of more carpenter ants followed by wood-boring beetle larvae and damp-wood termites. The Fernandina's Flicker typically spends more time hunting ants on the ground but switches to ants and termites it finds in palm trees during the wet season, when the ground is inundated with water.

The *Picoides* (American Three-toed and Black-backed Woodpeckers) and the *Campephilus* (Pale-billed, Ivory-billed, and Imperial Woodpeckers) specialize(d) in beetle larvae and opportunistically switch between species of beetle larvae and other prey based on availability. Black-backed Woodpeckers and American Three-toed Woodpeckers both consume large numbers of wood-boring and bark-beetle larvae, but Black-backed Woodpeckers consume more of the former while American Three-toed Woodpeckers consume more of the latter. Beetles and their larvae are also inordinately represented in the diets of Smoky-brown, Arizona, and Strickland's Woodpeckers and are opportunistically relished by related Hairy, White-headed, Downy, and a number of other woodpeckers. Many other species will take advantage of a bonanza of beetles and their larvae in burned forests.

All woodpecker species consume some amount of plant material. *Melanerpes* woodpeckers, from Red-headed to Acorn to Gila and Puerto Rican, enjoy a large amount of fruit and nuts, and most woodpeckers will eat fruit when it is encountered. Other common plant foods include berries, seeds, sap, cambium (the single layer of cells between the bark and the woody part of the tree), buds, catkins, and nectar. Gila Woodpeckers are even known to eat honey.

Some woodpeckers also eat other vertebrates. I have seen Gila Woodpeckers dismembering birds, and although I knew that Red-headed Woodpeckers eat bird eggs, I was surprised to learn that Cuban Green Woodpeckers consume heron eggs. I have also observed Hairy and Pileated Woodpeckers feeding on roadkill. Since suet contains animal fat, it should not be surprising that most of the species that consume animal protein will consume suet at winter feeding stations: Northern Flickers; Downy, White-headed, Red-bellied, Golden-fronted, Golden-cheeked, Golden-olive, and Black-cheeked Woodpeckers; and rarely even American Three-toed Woodpeckers.

Woodpeckers consume most of their liquid in their prey, but those that eat dry food such as seeds and nuts must supplement this practice, and others will hydrate during particularly hot and dry weather. I have had the pleasure of watching Acorn Woodpeckers drink from rainwater collected in knotholes, Lewis's Woodpeckers sip from melting snow, Red-breasted Sapsuckers from overflowing street drains, White-headed Woodpeckers from streams, and Northern Flickers and Hairy Woodpeckers from ponds. Others may drink from leaking faucets, melting icicles, leaves, or pine needles.

Foraging Techniques

Excavating, probing, sapsucking, gleaning, and fly-catching are the woodpecker's main foraging techniques, but they also opportunistically feed on fruit and seeds that they find as they move through their habitat. Woodpeckers that specialize in particular foods tend to be experts in specific foraging methods and have adapted anatomically over time to take advantage of these opportunities. Generalists are equipped to deploy all woodpecker foraging methods but lack the proficiency in some approaches that are found in the specialists.

Ornithologists do not know precisely how a woodpecker knows where to find food, but they seem to show up when there is an unusual abundance, such as when Black-backed Woodpeckers arrive just after a fire to feast on wood-boring beetles and their larvae inside burned trees. Can they somehow detect a prey spike from a distance, or do they simply wander when food runs out or during dispersal and stay where food is plentiful?

Once in the right area, woodpeckers likely listen for high-frequency sounds made by insects. It is common for them to peck on a tree before exca-

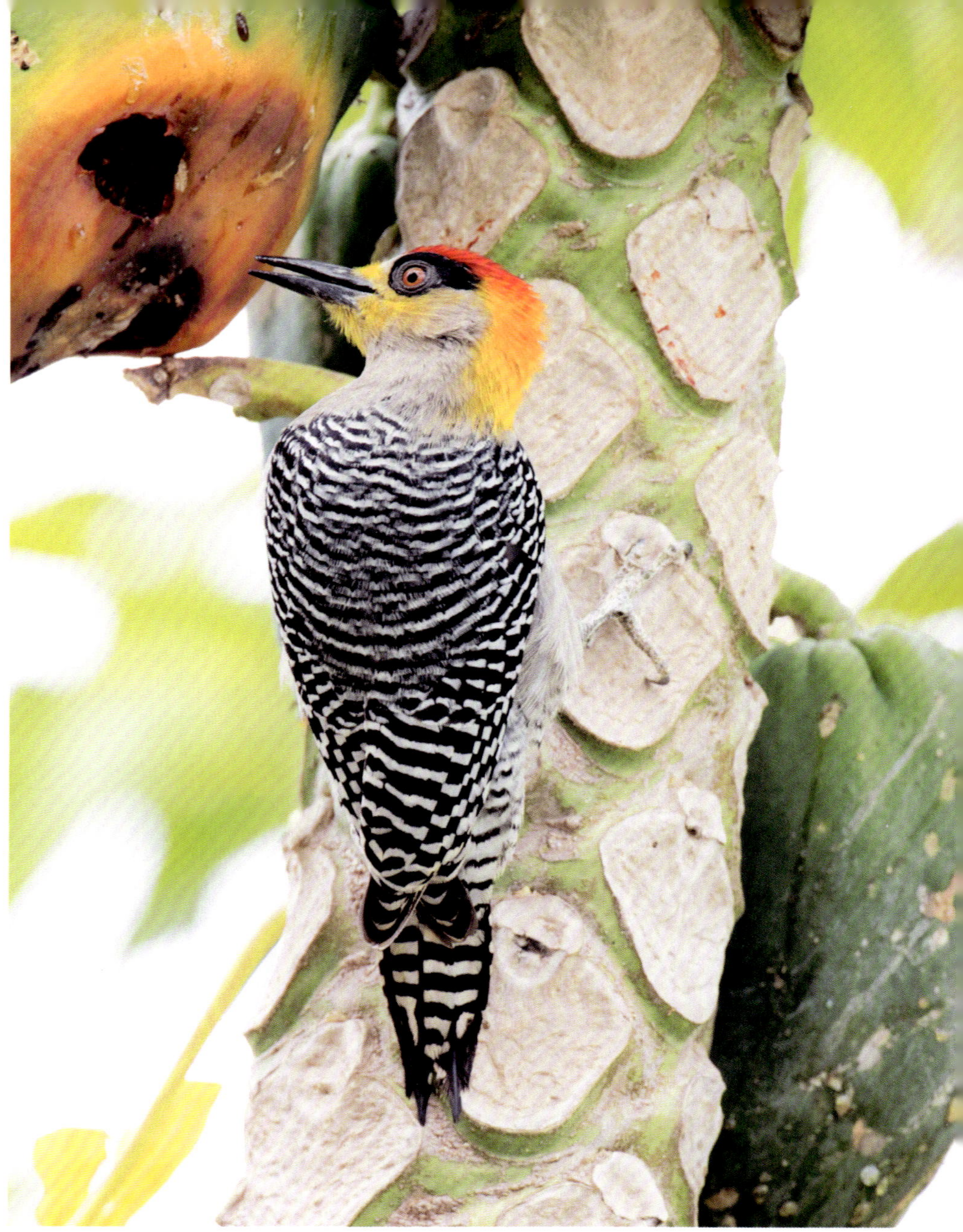

In parts of western Mexico, from the Pacific lowlands to 4,900 feet, endemic Golden-cheeked Woodpeckers forage at middle and upper levels of the forest, where they pursue everything from beetles, ants, and seeds to fruits like this papaya.

vating and then turn their head to listen for the sound of retreating prey. They also likely watch and listen for concentrations of other birds with similar food preferences, a behavior that can be witnessed at feeding stations when Hairy, Downy, or Red-bellied Woodpeckers appear after chickadees or nuthatches find these rich food sources.

Woodpeckers also look for visual cues. The White-headed Woodpecker spends a lot of time looking for prey on branches and needles and is thought to notice the woody debris at the edges of newly bored holes, where it proceeds to widen the hole with its bill before using its barbed tongue to extract the larvae.

Once woodpeckers pinpoint a location with insects, they deploy a range of techniques, from subsurface excavating to the scaling or stripping of bark, to probing with their bill and tongue in wood, in soil, or under leaves or debris, to gleaning their prey from the surfaces of leaves, bark, needles, or the ground.

EXCAVATING

Excavating is the epitome of the woodpecker's specialized foraging. Woodpeckers examine a tree the way a doctor might use a stethoscope, listening for clues of what is happening beneath the surface. They frequently tap on a tree, listening for a difference in resonance that might indicate a hollow insect chamber or induce hidden prey to move and betray their exact location. The sounds of the scratching or scampering of insects, spiders, or mites betray their travel direction. Armed with this information, woodpeckers excavate the surface of a tree through repetitive pecks or taps to access insects, larvae, or eggs. The precise method varies by species and the prey they are pursuing as well as its location within the substrate.

Excavators have the straightest, most chisel-shaped bills, and in the most extreme adaptation, in the Black-backed and American Three-toed Woodpeckers, only three toes, which allows them to pull their body farther back from the tree to exert more force with each blow. Woodpeckers typically deliver alternating blows or a series of blows from their left and right, to create an opening from which they access their prey. The hole can be shallow, deep, wide, or narrow depending upon the species and the prey being pursued. North of Mexico the most noticeable of these are the long and deep excavations the Pileated Woodpecker leaves behind as it exposes the tunnels and hiding places of carpenter ants inside snags. Such deep excavations provide rare access to a bounty out of the reach of smaller-billed woodpeckers.

Pileated Woodpecker excavations are often so extensive as to enhance habitat for several other species, including Williamson's Sapsuckers; Downy, Red-bellied, and Hairy Woodpeckers; and Northern Flickers. Flickers and sapsuckers are even known to linger in breeding areas longer to feed in the wake of Pileateds. These excavations also create shelter and even homes for other creatures as evidenced by a Townsend's Solitaire I once photographed brooding its young in a woven nest placed inside a Pileated Woodpecker excavation. Acorn Woodpeckers are often trailed at a safe distance by Nuttall's Woodpeckers, which will sometimes capture food dislodged by the intolerant relative.

Bark is the first and sometimes the only obstacle to accessing prey and is removed in several ways. The *Campephilus* species such as the Pale-billed Woodpecker will strike bark with slanting blows before prying strips off with their chisel-shaped bills. The Black-backed and American Three-toed are experts at scaling, which is the removal of patches of bark with sideways strikes and flicks of their bills or pulling the plate off with their bill after loosening. Some *Dryobates* species like the Arizona Woodpecker and, less commonly, Red-cockaded and White-headed Woodpeckers sometimes scale the bark by flaking it off with a quick scratching movement of both feet simultaneously.

Woodpeckers that specialize in subsurface excavations can access subsurface insects and their larvae that are active year-round, and are thus able to remain at the highest elevations and latitudes during the harshest weather, so it is no accident that woodpeckers that spend the entire year farthest north and at the highest elevations are adept at this strategy.

Since woodpeckers are often excavating on a vertical surface, the prey may fall or bounce out of reach when they strike. To counteract this, several species keep their stomach or breast pressed to the tree as they work, and when the food falls against their body, they simply reach down with their bill to retrieve it.

I have witnessed Lewis's and Red-headed Woodpeckers execute this maneuver quite neatly but was impressed when I observed a Pileated Woodpecker adding a particularly effective refinement. A female was excavating wood-boring beetle larvae from a red alder with her belly pressed against the trunk, when suddenly she hunched forward and swung her wing against the tree before quickly dipping her bill into the space between wing and breast and retrieved the larvae. Apparently, Downy, Hairy, Red-bellied, Acorn, and Red-cockaded Woodpeckers also use this technique.

PROBING

Visually inspecting substrates, combined with probing holes, cracks, and crevices with their bill while using their tongue to feel for a hidden bounty, is a common foraging technique for woodpeckers. Each species or even gender within a species prefers to probe in different places. I have many times watched White-headed Woodpeckers alternate between probing and looking as they pursue insects within the deep furrows in the bark of ancient pines and firs. I once watched a Jamaican Woodpecker for thirty minutes as it seemed to probe every hole in an old tree with its bill and tongue before finally finding food in the hollow at the end of a broken branch. On another occasion I observed a Golden-olive Woodpecker probe within a tangle of vines as it scrambled up them. Watching a colorful Black-cheeked Woodpecker probe the depths of an equally beautiful heliconia flower was a particular delight.

All woodpeckers have specialized tongues that wrap around the jaw and over the top of the skull, allowing them to extend it into holes. The length, number, and type of barbs, especially at the pointed but sensitive tip of the tongue, vary according to the preferred foraging method of the species and allow it to feel, collect, and retract the food into the woodpecker's mouth.

Colaptes species, such as flickers, which probe more than peck, have a longer but less barbed and pointed tongue that can more easily find

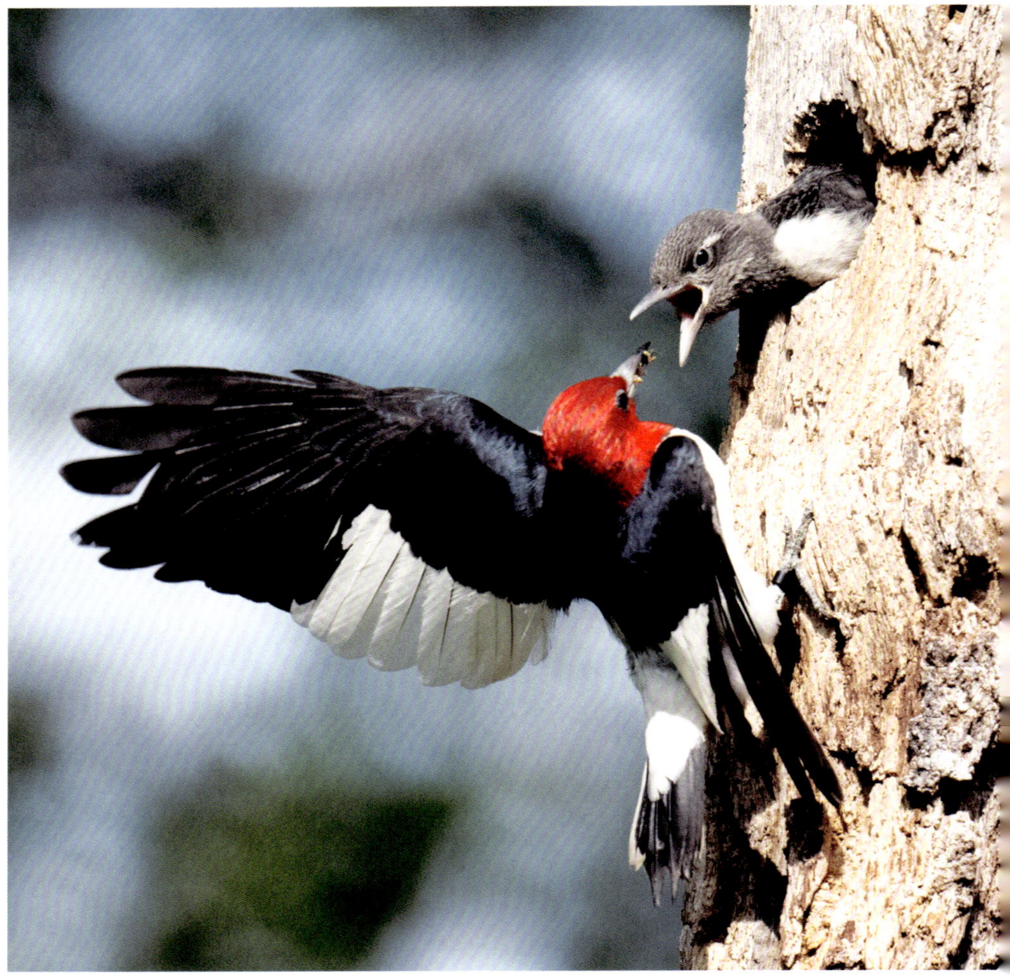

Food deliveries are more precarious when the young are in their final days in the nest. Nestlings, like this Red-headed Woodpecker, aggressively anticipate each delivery with lunging heads and snapping bills.

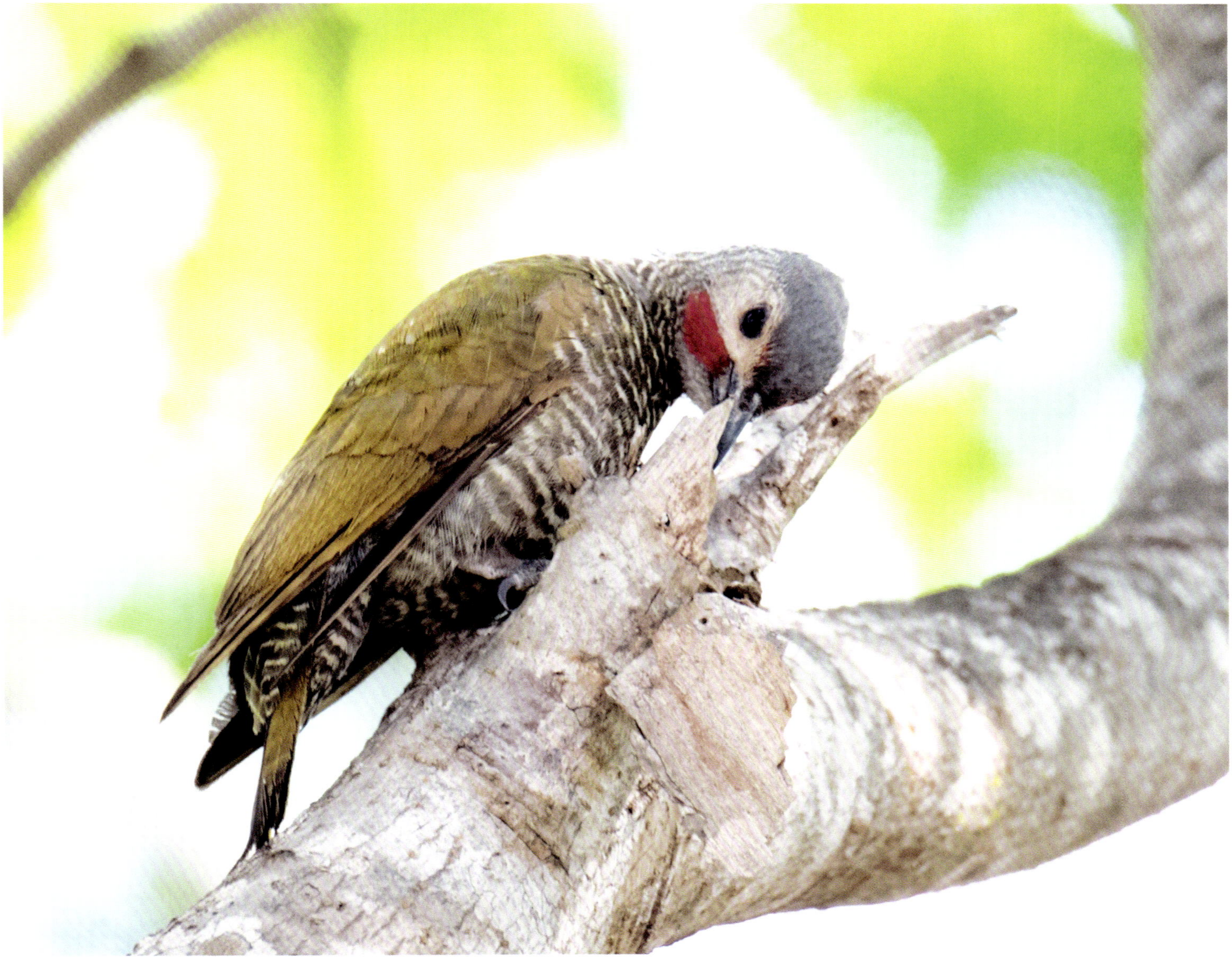

Most woodpeckers, like this Gray-crowned Woodpecker, spend some time probing in insect tunnels, hollow branches, furrowed bark, or even in flowers in pursuit of prey. Termites, ants, and other invertebrates comprise the majority of their diet, making probing a particularly important foraging method.

and retract ants and other prey from deep tunnels into the woodpecker's mouth. The Northern Flicker's tongue can extend roughly two inches beyond the tip of the bill, longer than that of any other woodpecker. These birds put this feature to good use when they probe the ground or under stones for ant colonies. In suburban areas, flickers will probe the gaps between sidewalks and grass or rocks and soil. I have enjoyed watching flickers hop across the landscape, focusing on sunny areas during cool weather and shady areas during hot weather to take advantage of the responding concentrations of ants. They are also known to break up chunks of dung to withdraw hidden ants and beetles.

Probers like flickers tend to have a longer, more pointed, and curved bill to reach deeper in contrast to excavators, which tend to have shorter, sturdier, chisel-shaped bills for breaking wood. Excavators also tend to have a medium-length tongue with hairlike, rather than barbed, bristles at the tip. That said, one of the strongest excavators, the Pileated Woodpecker, utilizes a long, extensible, pointed tongue with barbs and sticky saliva to snare ants and pull them from their subsurface tunnels.

PONDEROSA PINE HABITAT

Yellow-plated ponderosa pines rise abruptly one hundred feet or more like sentinels as you move upslope from grasslands, deserts, sagelands, and oak-savannas. The deep taproot of these heat-loving trees is part of what allows them to survive drought, and their thick bark and self-pruning lower branches enable them to live through the periodic fire required to maintain healthy stands of this species. Ponderosa pine can be found from British Columbia into Arizona and New Mexico and from the Pacific to Nebraska in areas with twenty-five inches or less of precipitation annually. As a rule, tree species and their required habitats occur at lower elevations in northern locations, and ponderosa pine is a great example, growing from sea level at its northern limits to 9,515 feet in the Southwest.

Ponderosa pine is the most important tree in several woodpecker habitats of the West, including savannas (<30 percent canopy cover), woodlands (<60 percent canopy cover), and ponderosa pine or mixed-conifer forest. Canopy cover increases in areas with less frequent fire, more moisture, lower temperatures, and poor forest management.

Quaking aspen is one of the first trees to join ponderosa pine stands in lower-lying, moist areas such as in seeps near streams, ponds, and lakes. Moisture and shade encourage a drought- and fire-resistant variety of Douglas-fir to germinate. With longer fire intervals, more flammable firs and spruce join the mix. Western larch is an important fire-resistant addition and replacement in colder and wetter locations at higher elevations in the mountains of the Pacific Northwest.

Fourteen woodpecker species utilize ponderosa pine habitats, including four "species of concern"—Lewis's, White-headed, and Red-headed Woodpeckers (in the far western part of their range) and Williamson's Sapsucker. These four can be considered keystone species for the nest cavities each creates, which house many species, including Flammulated Owls and Northern Pygmy-Owls, Mountain Chickadees, Mountain Bluebirds, and flying and western gray squirrels. Sapsuckers create sapwells critical to hummingbirds.

Lewis's and Red-headed Woodpeckers favor savannas, woodlands, and, especially, burned stands where they can hunt plenty of insects on the wing. Lewis's may require as many as one hundred well-decayed snags (twelve to twenty inches wide and thirty feet tall) per acre to attain population densities as high as several adults per acre. Both species benefit from clumps of snags to meet nesting and caching needs.

White-headed Woodpeckers have historically been associated with connected stands of old-growth ponderosa pine woodlands and forests, although recent studies show that they can survive in younger stands provided there are plenty (twenty to fifty-plus per acre) of six-plus-inch diameter snags with a hard exterior and soft interior. Predation appears to be positively correlated with the amount of ground cover. The most stable populations may occur where ponderosa are joined by another large-coned pine species such as Jeffrey or Coulter. They depend on fire as long as snags are retained and adjacent unburned pines remain for foraging.

Williamson's Sapsuckers breed in middle- to high-elevation mature conifer and mixed-conifer forests, commonly with mature ponderosa pine, western larch, Douglas-fir, quaking aspen, a relatively open canopy, and some small to medium-size live pines for harvesting sap. They may nest in burned snags, but they predominantly forage in unburned areas. They require heartrot in aspen, pine, or larch to allow for cavity excavation.

Large ponderosa pines are often targeted for commercial harvest and in fuel reduction treatments to the detriment of resident woodpeckers. Retaining large, decayed snags and older trees, which will become snags in the future, is critical. Additionally, it's important that fire burns in a pyrodiverse way, resulting in stands of trees of different sizes, ages, and canopy coverage. This creates the broadest range of habitat types to meet the

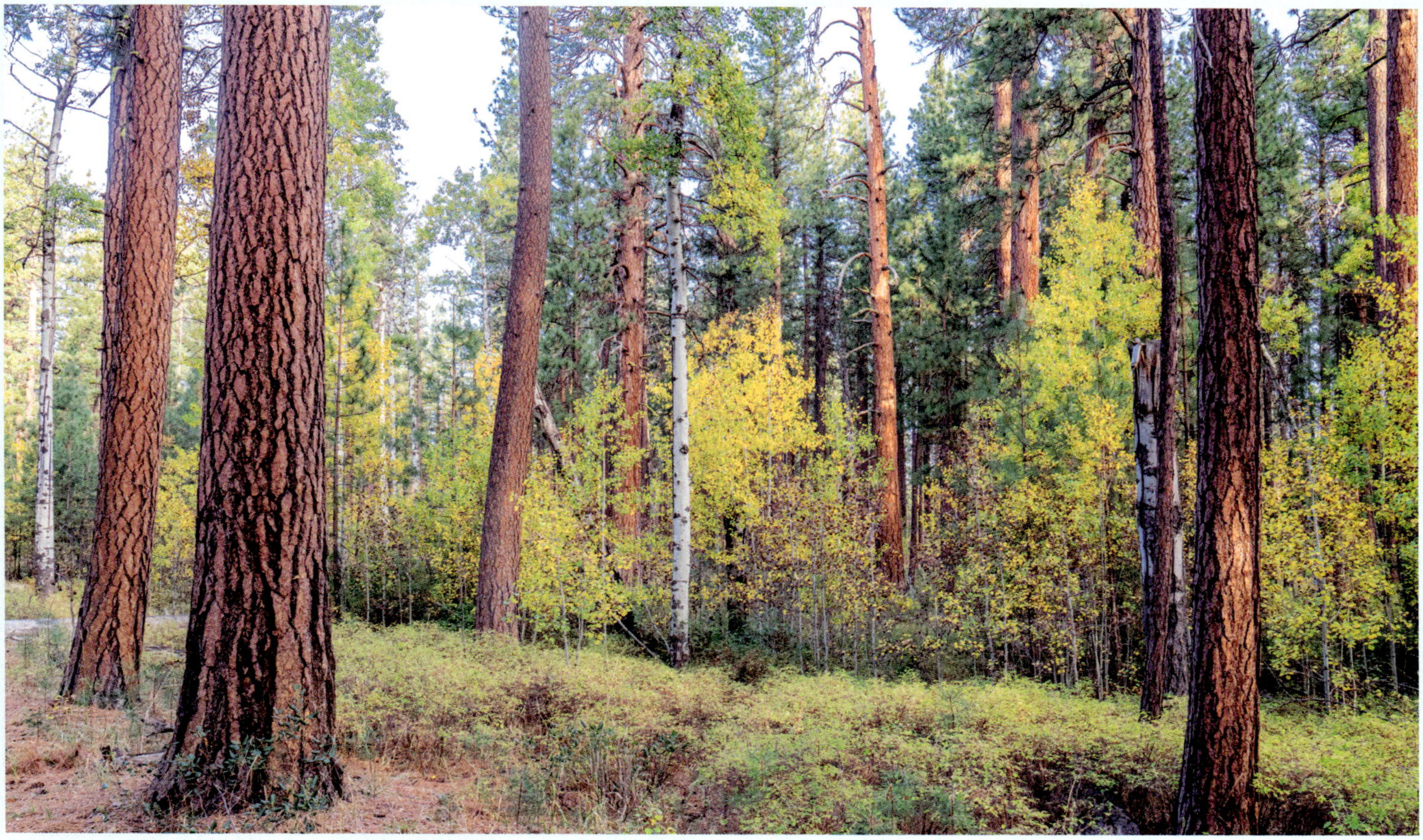

Fire-maintained, open ponderosa pine habitats with plenty of soft snags (standing dead trees) and quaking aspen for nests are the most species-rich habitat in the United States and can host fourteen woodpecker species, including several species in a relatively small area.

needs of most species and generates the savannas preferred by Lewis's Woodpeckers, woodlands profiting White-headeds, and mature forests required by Williamson's Sapsuckers.

Fire in the form of prescribed burns can help maintain the open stands of pine by removing fuel—flammable firs and young pines that would also become ladders to the vulnerable crowns of otherwise fire-resistant trees. In response to the warming climate, burns must occur more frequently than the natural five- to fifteen-year cycle.

Unrotted live pines are impossible to excavate, so retaining mature aspen and cottonwood stands with large snags is critical. These species are susceptible to heartrot, making them easier to excavate. Williamson's Sapsuckers are predicted to lose a net 30 percent of their breeding range with a warming of 2.7 degrees Fahrenheit. Douglas-fir and ponderosa pine have different rot rates and therefore, after a fire, offer snags of varying sizes, softness, and stage, benefitting woodpeckers searching for the best nesting options.

Management has resulted in large tracts of small trees, which lack resistance to fire and insect outbreaks. This makes them more vulnerable to climate-fueled, often human-ignited uncharacteristic fires—major threats to sensitive woodpecker species in ponderosa pine habitats. Cattle grazing can deter germination of both aspen and cottonwood and increase the rate of conifer germination, both by selecting deciduous over conifer and by constantly stirring up the ground to expose bare mineral soil. Maintaining large, mature connected stands that meet the requirements of sensitive woodpecker species requires retention of snags and fire management that maintains, and in some cases, reintroduces habitat diversity while protecting the oldest trees and stands. Mature, fire-maintained landscapes with stands of various ages and plentiful large trees and snags are most valuable for all woodpecker species and provide more nest sites.

SAPSUCKING

Williamson's, Yellow-bellied, Red-breasted, and Red-naped Sapsuckers all have a shorter, less extensible tongue tipped with stiff hairs to which sap adheres, allowing them to mop or soak it up. Sapsuckers create and feed from two kinds of tree wells.

Xylem wells are made in the winter or early spring before deciduous trees develop new leaves and are small, round holes drilled into the tree's sapwood. The targeted xylem tissue that carries nutrients up from the roots to the branches to promote leaf production is about 2–3 percent sucrose. These holes look like small nail holes and sometimes encircle the tree in parallel rows.

Phloem wells are shallower, rectangular holes that start as grid-like patterns. These wells are drilled year-round in conifers and in deciduous trees after new leaves begin growing, and produce sap that is usually between 10 and 25 percent sucrose (maybe as high as 48.5 percent at times in some willows) and flows through the outer phloem tissue down the trunk to fuel the tree's growth. Sapsuckers chisel the outline of each cell in the grid before picking out the interior patch with their bill, often eating a portion of the material, then flicking the remaining core aside. They maintain many nearby wells, periodically visiting each as it fills with sap, sometimes extending the edges of each cell to increase sap flow, with the resulting wells looking less like a grid and more like a puzzle. Yellow-bellied Sapsuckers tend to create phloem wells in the most sucrose-rich deciduous trees, Williamson's prefer conifers, and Red-naped and Red-breasted Sapsuckers utilize both deciduous and conifer trees.

Like other woodpeckers, sapsuckers guard food stores, and these wells are no exception, but many other animals manage to sneak their way in for a few sips now and then. Dozens of insect species such as wasps, bees, and butterflies visit these wells and sometimes fall prey to the sapsuckers in the process. Many mammals, including squirrels, mice, and even porcupines, are also known to visit the grids. The wells of sapsuckers are thus a boon for ecosystems and another way that woodpeckers serve as keystone species. The wells feed their architects but also provide a rich source of energy for a wide variety of animals in the early spring, when insects and pollen are scarce.

The sap from phloem wells features a similar liquidity and sucrose content to the flowers that hummingbirds feed on and offer great feeding opportunities for these nectar-loving birds. Rufous and Ruby-throated Hummingbirds have long migrations and breed farther north than any

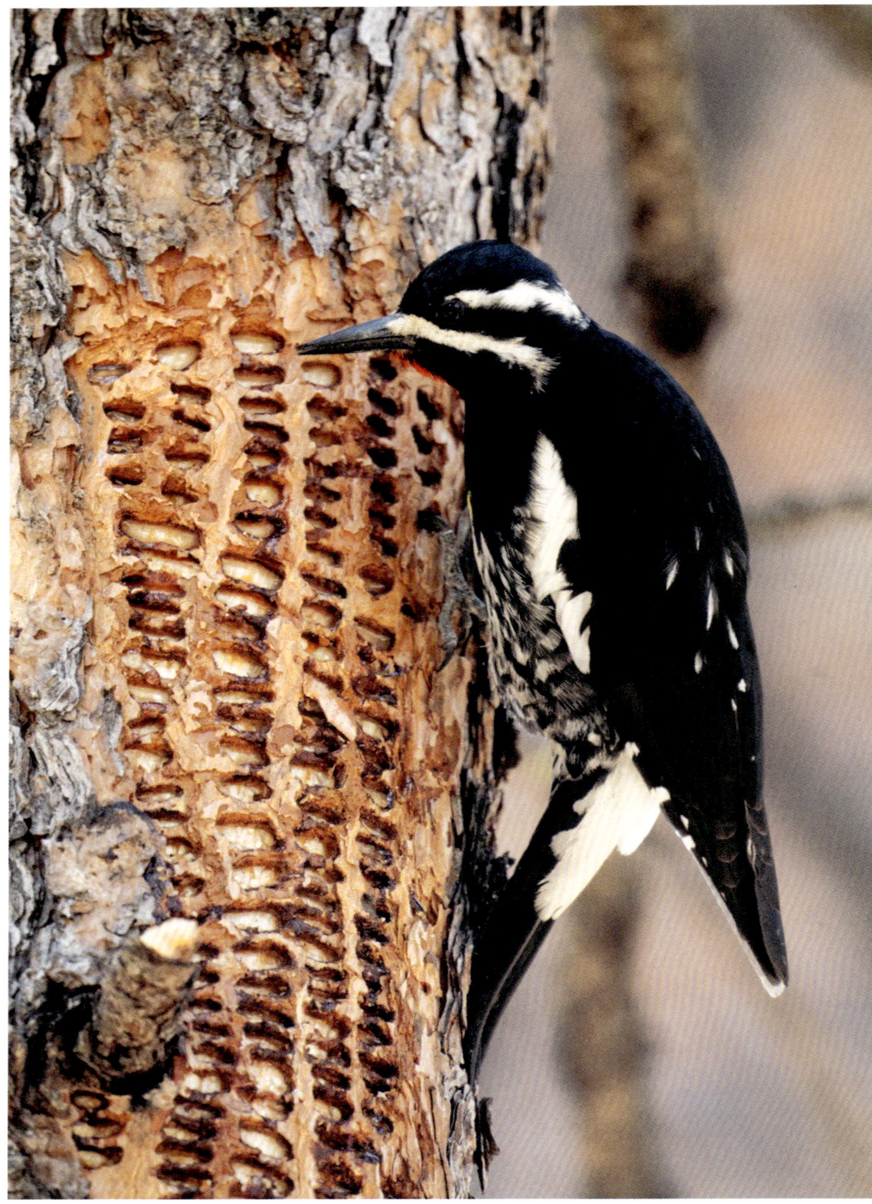

Sapsuckers create and manage wells in trees. Phloem wells, like the one this Williamson's Sapsucker is feeding from, are shaped like cells. The birds continually enlarge them to encourage sap to flow, eventually connecting cells.

This female White-headed Woodpecker is gleaning ants and aphids from an aspen, a process of plucking insects from the surface of branches and leaves.

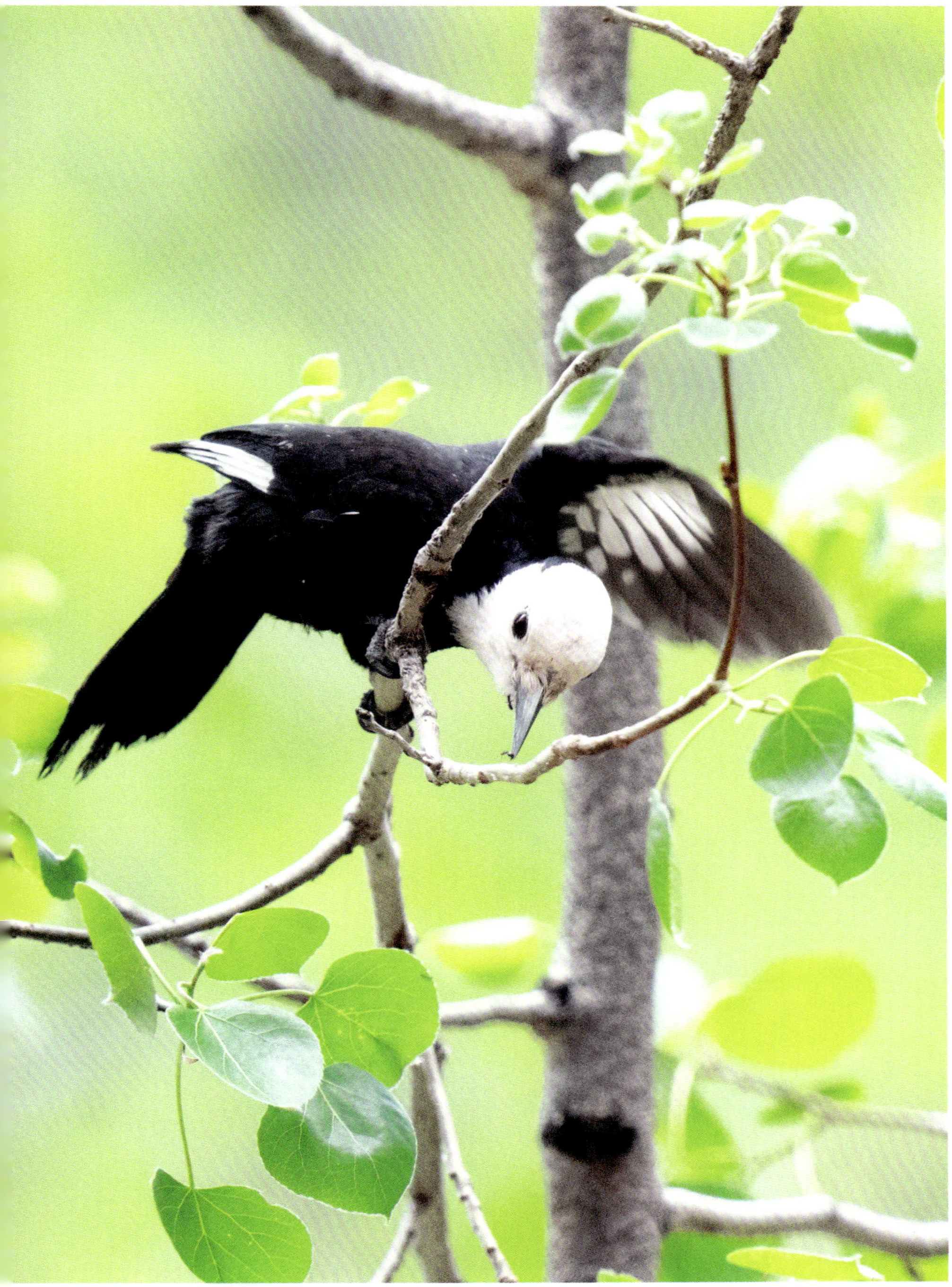

other hummingbird, and it is believed that their northern limit is determined in part by the presence of sapsuckers. The Ruby-throated seems to rely upon Yellow-bellied Sapsuckers, and the Rufous Hummingbird on Red-breasted.

Rufous and Ruby-throated Hummingbirds seem to time their arrival on northern breeding grounds with the availability of the phloem wells of these sapsucker species, arriving from their winter homes a week or two after the sapsuckers. The hummingbirds even follow the well-makers to discover the whereabouts of their wells and set up their nests nearby. I have also seen Calliope and Broad-tailed Hummingbirds visit Red-naped and Williamson's Sapsucker wells, and there is likely a similar reliance.

All hummingbird species visit sapwells, as do Black-throated Blue, Cape May, Orange-crowned, Nashville, Wilson's, and Yellow-rumped Warblers. Nuthatches, kinglets, titmice, sparrows, siskins, goldfinches, chickadees, wrens, and at least six woodpecker species, including the Red-bellied, Downy, Hairy, American Three-toed, Nuttall's, White-headed, and Acorn, gain calories from this sap.

White-headed, American Three-toed, and Acorn Woodpeckers are known to create their own sapwells. The American Three-toed's appear as loose horizontal rings around the trees, and the Acorn's are a bit more random. I have twice witnessed White-headed Woodpeckers create sapwells by removing patches of bark on young pines and drilling holes to access the sap.

GLEANING

All woodpeckers, no matter how specialized, will pluck prey from the surface of trees, leaves, the ground, or their sapwells in a technique called gleaning. When woodpeckers move quickly up a tree in a spiral or zig-zag fashion, they are usually gleaning insects or spiders. Williamson's Sapsuckers spend much of their time gleaning wood ants from tree trunks as the ants move between aphid colonies.

GROUND FEEDING

Most woodpeckers feed on trees, snags, cacti, or even human-made structures, but some frequently feed on the ground: the Northern, Fernandina's, and Gilded Flickers, along with Lewis's, Red-headed, Pileated, Golden-fronted, Gila, Nuttall's, Puerto Rican, Guadeloupe, Gray-breasted, and West Indian Woodpeckers. I have also seen Red-breasted Sapsuckers and White-headed Woodpeckers glean ants from the ground and would not be surprised to see Williamson's Sapsuckers do the same.

FLY-CATCHING

Insect food is not limited to the ground or trees, and several species are capable of accessing the richness in the air. The most common fly-catching woodpeckers are all *Melanerpes*, including the Acorn, Red-headed, Gila, Gray-breasted, Jamaican, and especially the Lewis's Woodpecker, although I have also seen all four sapsuckers sallying for insects. I have also been surprised to see Downy, White-headed, and American Three-toed Woodpeckers opportunistically catch insects in the air, although this is less common. Strickland's and Nuttall's Woodpeckers have also been reported to fly-catch, and it is quite likely that most species possess enough plasticity to pursue flying insects with varying degrees of success.

Fly-catching for Acorn, Red-headed, Lewis's, and Gray-breasted Woodpeckers usually begins the same way, with the woodpecker finding a high, exposed perch over an open area. The bird then scans the skies, and when it spots a target, it takes flight straight up into the air or in a dive. In either case, it returns to the same perch from which it launched or ends up at another exposed landing spot. Lewis's Woodpeckers lean on this strategy more than other species, and I have seen them utilize their surprisingly long and broad wings to fly-catch for nearly thirty minutes at a time, especially during insect hatches.

Stooping is executed opportunistically by many woodpeckers when they drop from prominent perches to pluck prey from the ground or a log.

Fly-catching is an important way for Red-headed Woodpeckers to acquire food and is one reason open habitats are important for this species. During the breeding season, they have been observed spending the majority of their foraging time either fly-catching or flying to the ground from elevated perches to capture prey.

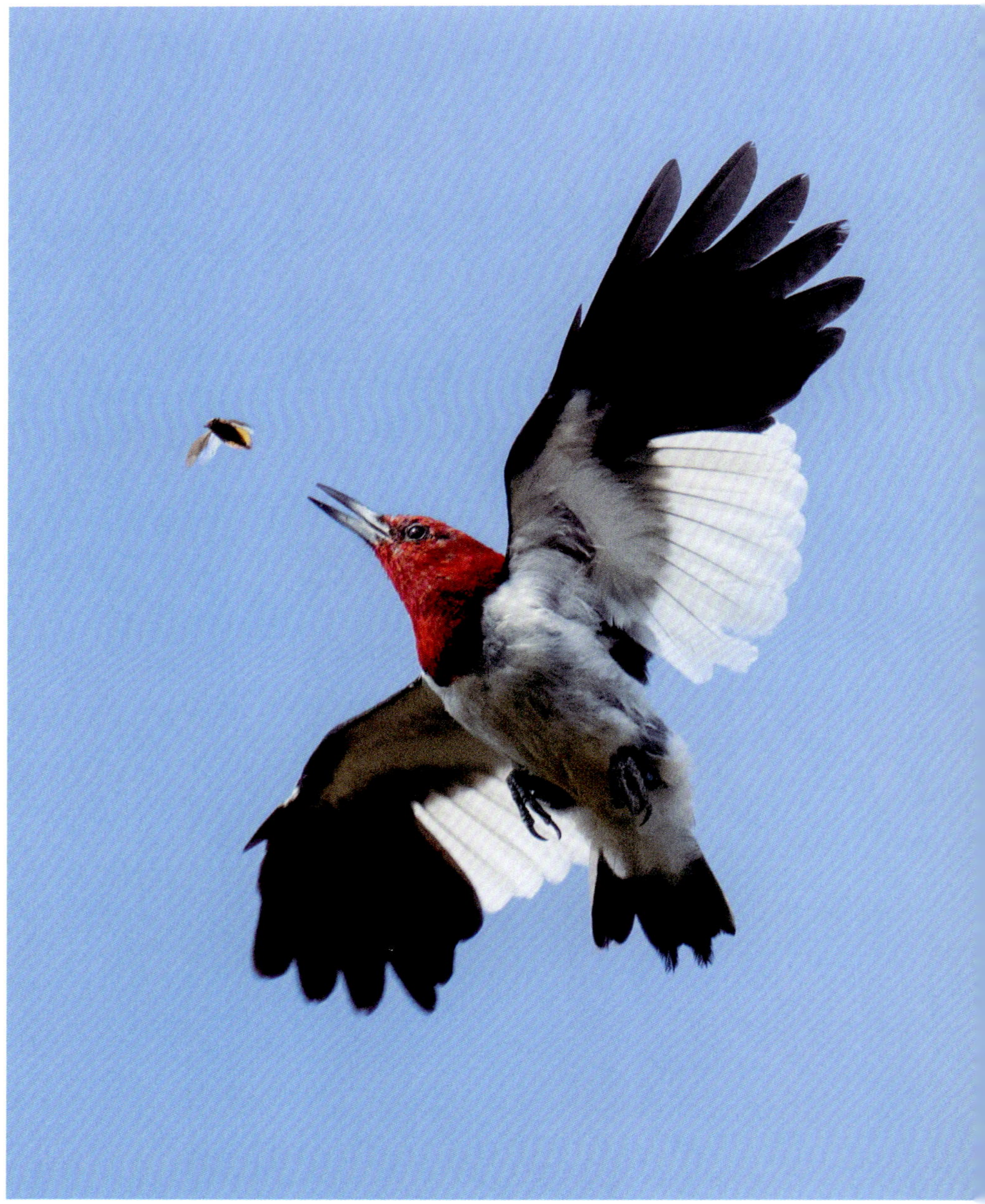

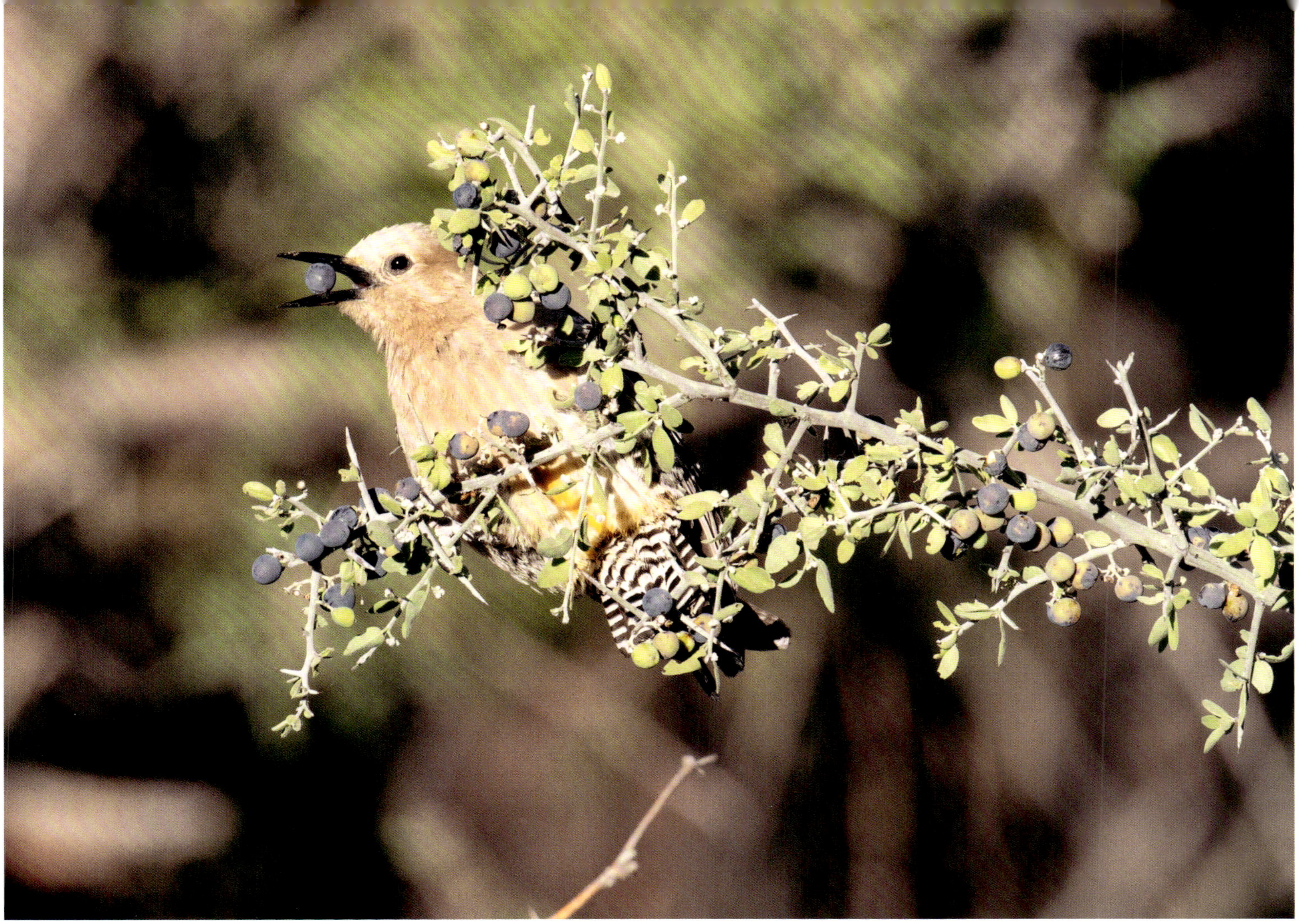

The berries of the bitter condalia make a tasty meal for this female Gila Woodpecker to deliver to her young, who are waiting in a cavity in a saguaro. The Gila, like other *Melanerpes* woodpeckers, will consume an impressive variety of food, from plant nectar to baby birds to suet and berries to insects.

Seasonal Variations

The food consumed by each woodpecker species varies by season, habitat, geography, and year. Sapsuckers may consume xylem sap in winter, switch to phloem wells in deciduous trees in spring as trees gain leaves, and rely more upon ants during nesting season. American Three-toeds may utilize sapwells in the early spring or fall, but where there is a bark beetle outbreak, will rely upon that species. Lewis's and Red-headed Woodpeckers' main diet is insects in spring and summer, fruit and berries in late summer and early fall, and conifer seeds and acorns in winter. Gila Woodpeckers alternate between insects, pollen, the eggs and young of birds, cactus fruit, and other food as the opportunities arise. Pileated Woodpeckers always favor ants but utilize other arthropods during the breeding season and fruit and berries in fall.

Woodpeckers show problem-solving skills in pursuing nourishment. Red-headed Woodpeckers have been observed dropping nuts and pine cones on the road to be run over by passing cars, which breaks them open and separates the food from the inedible parts. Gila Woodpeckers have been observed using bark, grain, and sunflower seeds to scoop honey from a saucer to feed their young. Sapsuckers regularly dip ants in sap to feed their young. Juvenile Nuttall's Woodpeckers have been observed feeding on sap from a willow by pecking at the natural bark break to encourage sap flow. Acorn, Lewis's, and Red-headed Woodpeckers utilize anvils to break food into the exact right sizes and shapes to be cached in available holes or fed to young.

Woodpeckers' prowess in finding abundant food allows them to keep the young well fed during their first three weeks in the nest, but soon they will begin the most dangerous half of their year and learn to find their own food.

Below: Lewis's Woodpeckers exhibit the greatest agility in flight displays and fly-catching. With more aerial skill than any other woodpecker, they soar, sally, and dive through the sky in pursuit of winged insects.

Right: Once the young in the nest are fully feathered, the adults, like this male White-headed Woodpecker, leave the nest as soon as they deliver food, without waiting for their mate.

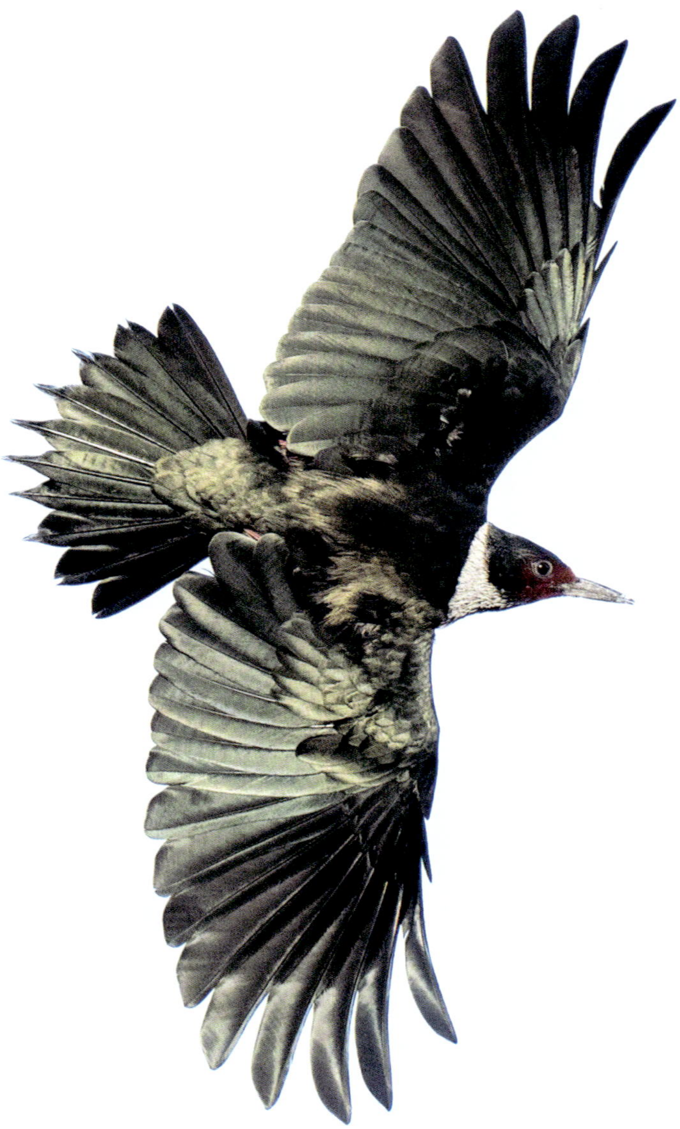

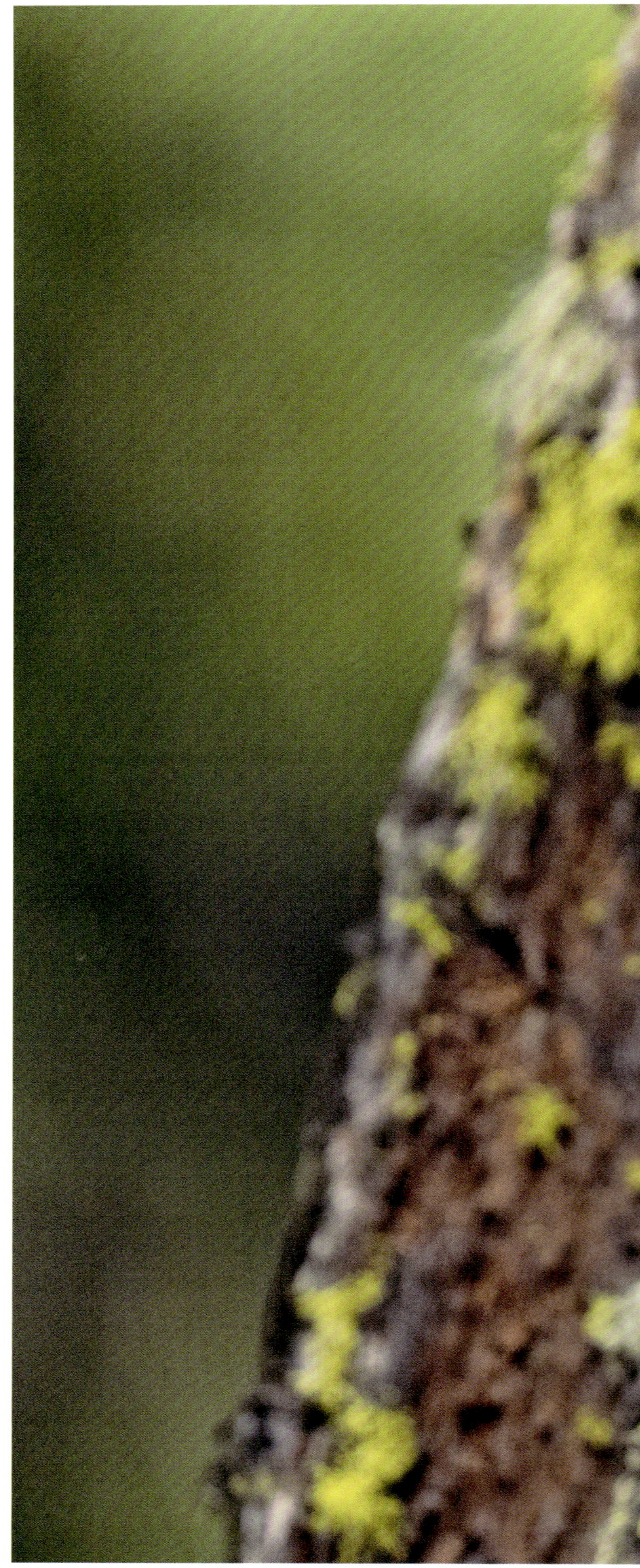

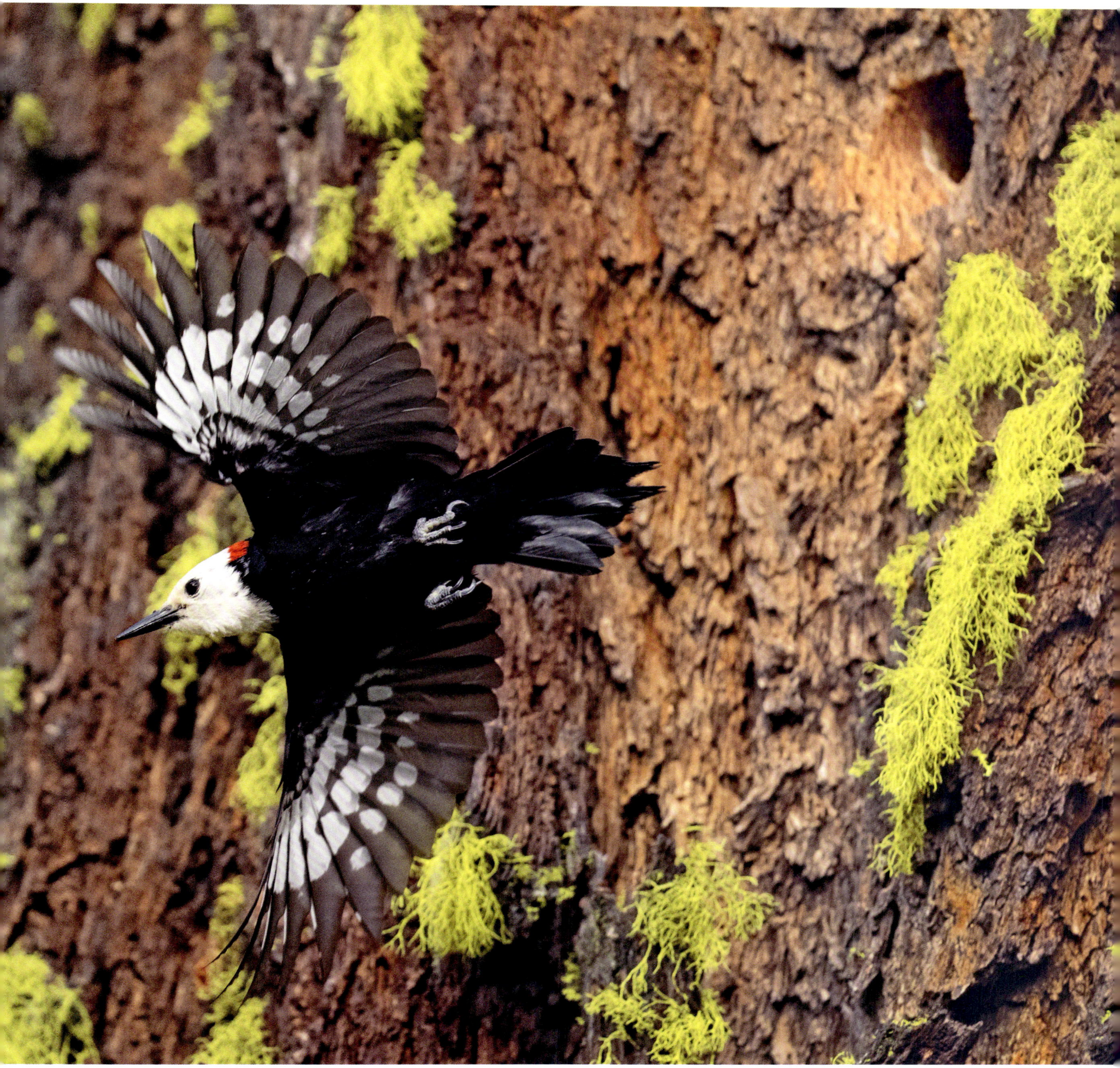

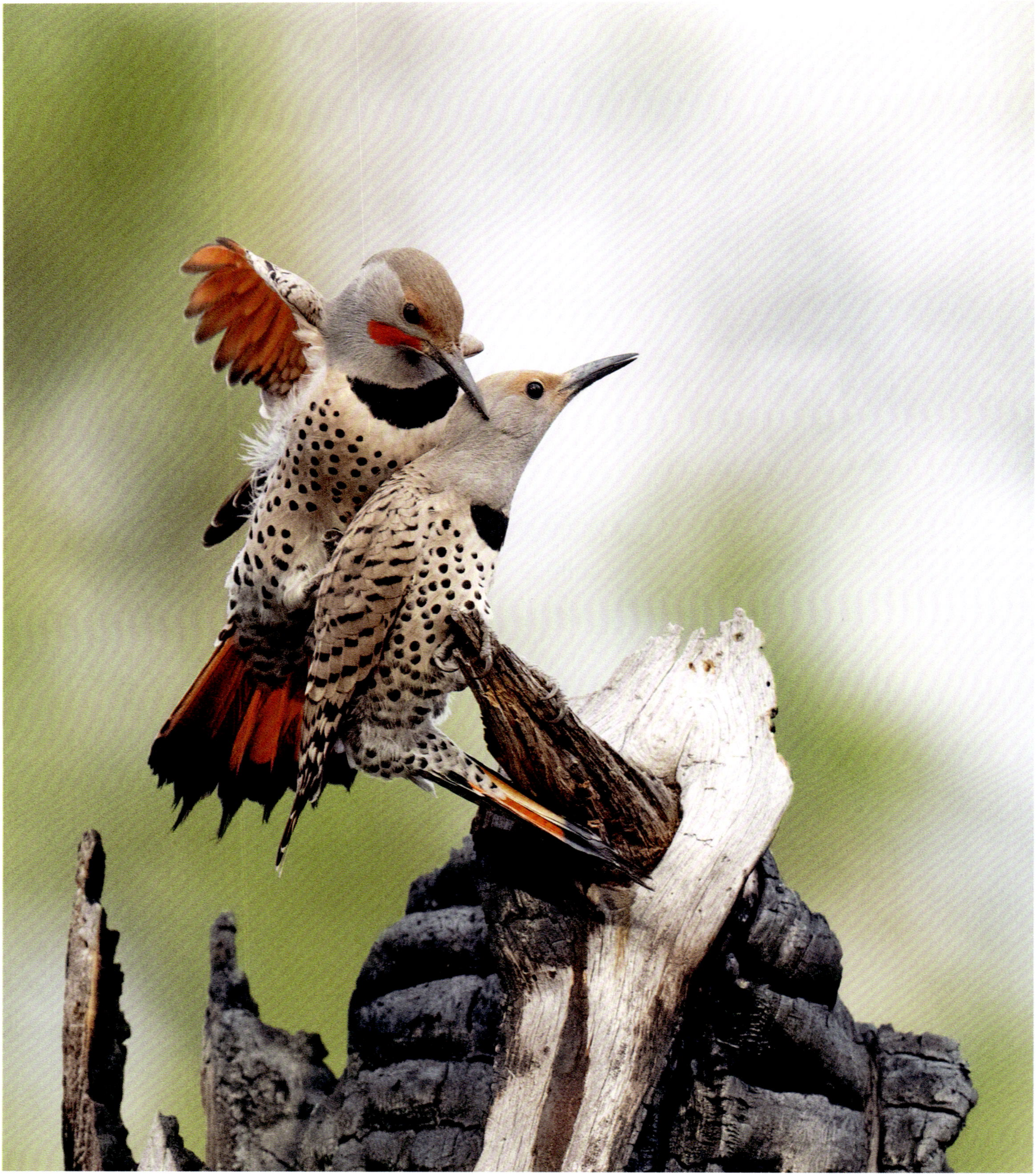

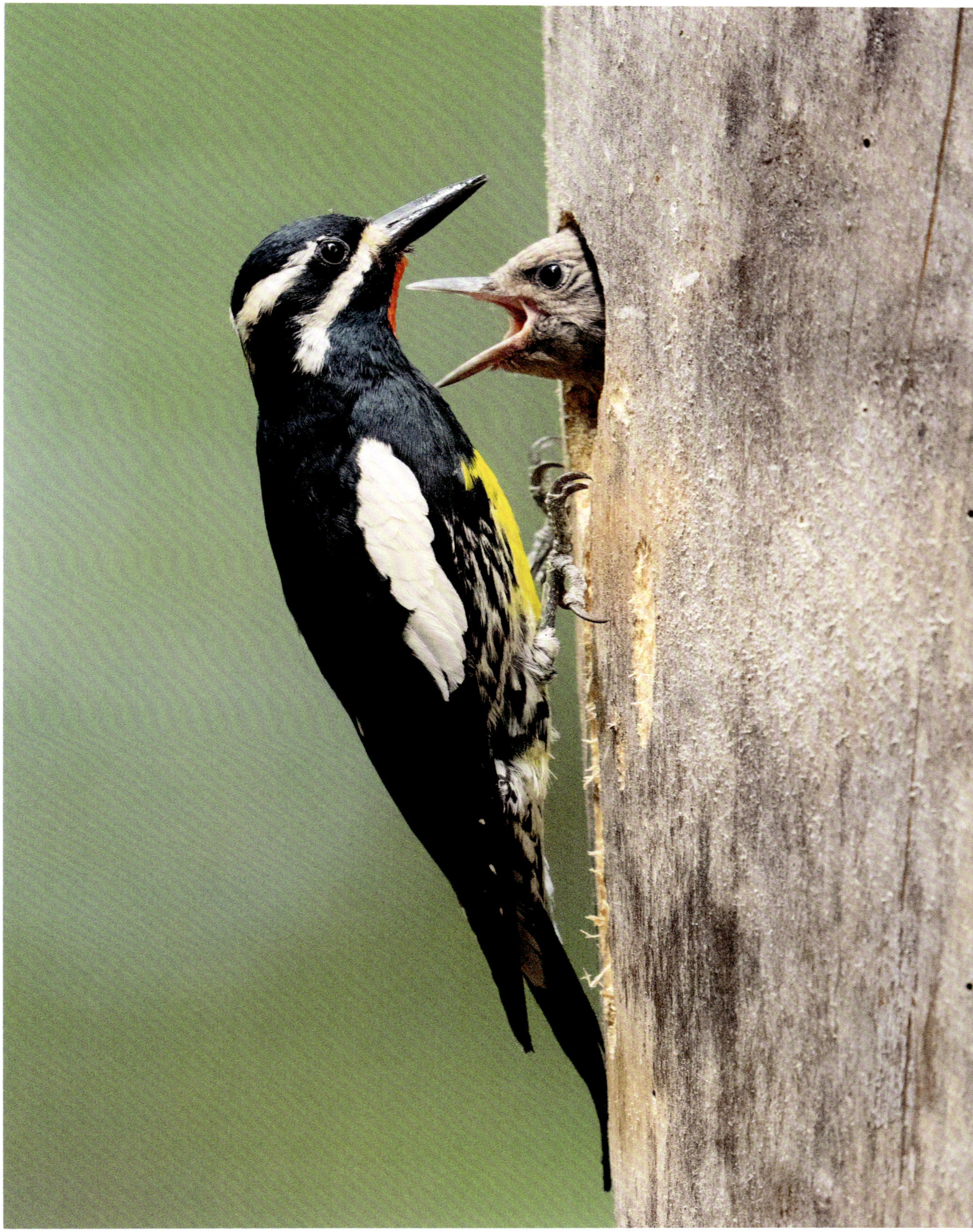

A female Williamson's Sapsucker nestling fills the cavity entrance immediately after the adult male delivers ants to her sibling. During the last week in the nest, it can seem as though there is always a head occupying the cavity entrance.

Opposite: Northern Red-shafted Flickers copulate on a snag. Once they have chosen a nest and as the female begins laying eggs, woodpeckers copulate frequently, oftentimes near the nest when the female arrives to relieve the male of the nest-guarding role.

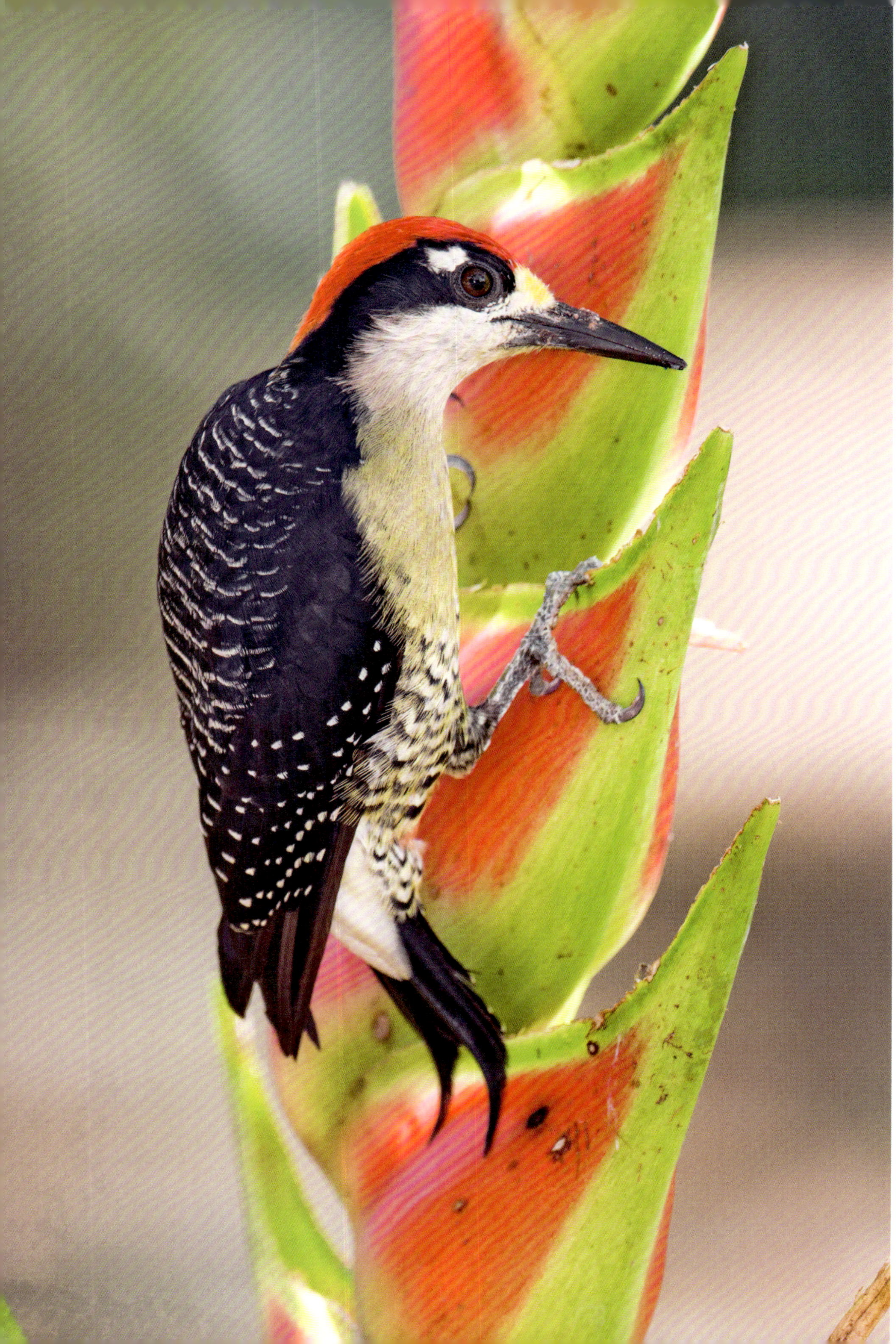

Probing flowers such as heleconias can be profitable for the Black-cheeked Woodpecker, who may find seeds, nectar, or insects inside.

Opposite: The acrobatic Antillean Piculet, the only member of the *Nesoctites* genus, can easily hang from the underside of small branches as it does when feeding from the berries of wild mahogany (*Trichilia hirta L*). Unlike other woodpeckers, they often utilize their large feet rather than a long tail to maintain their balance.

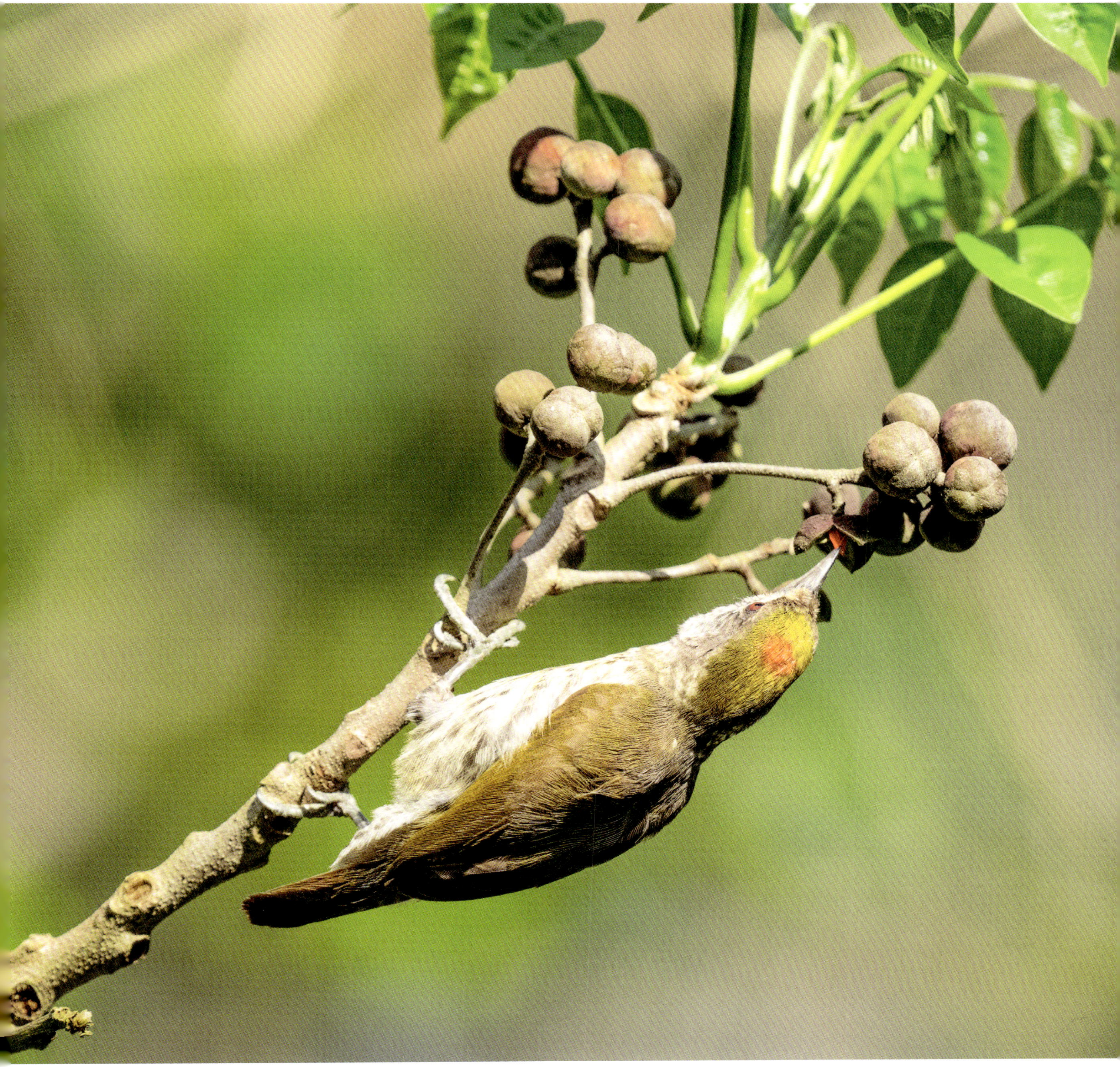

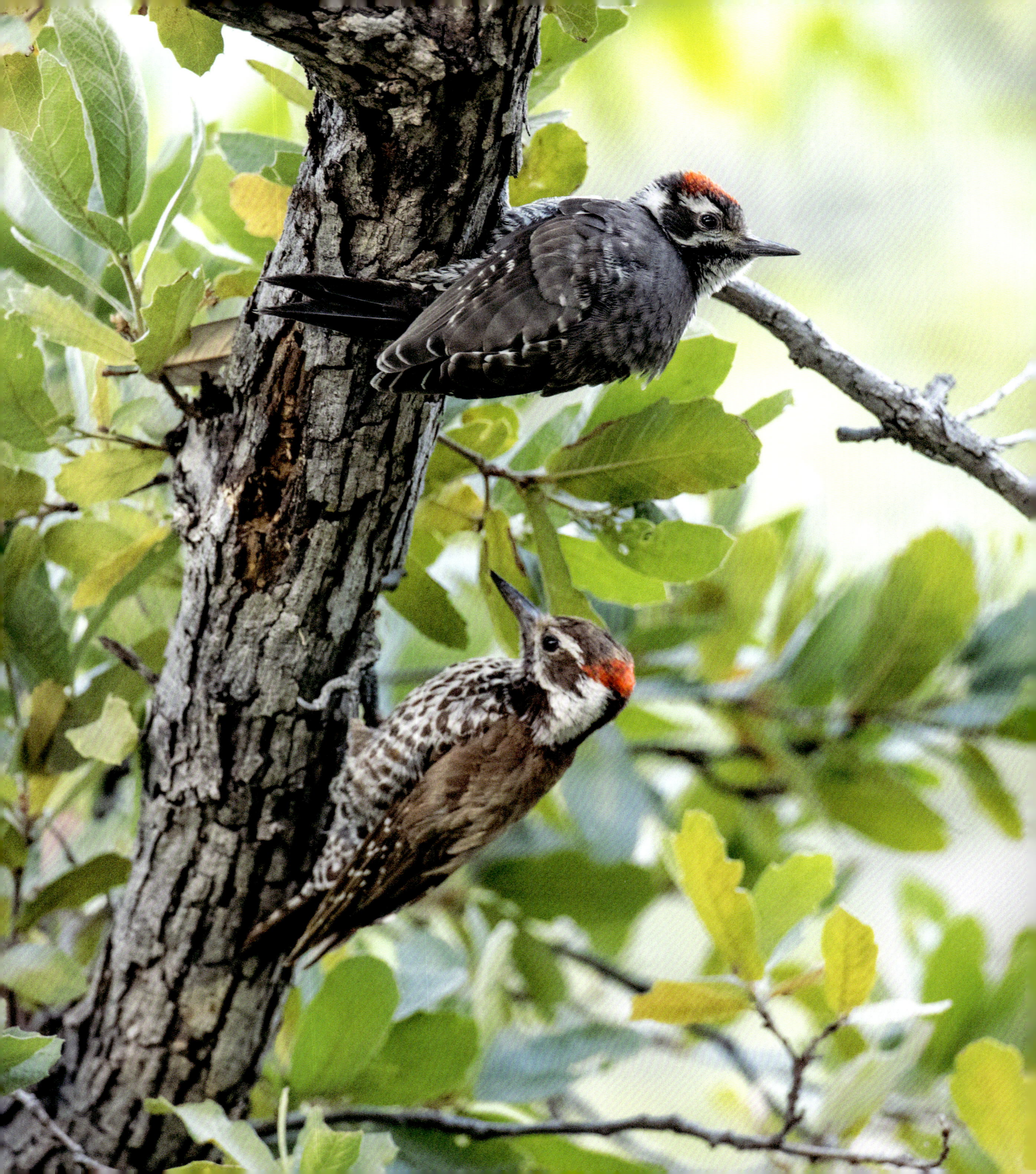

4

Gaining Independence

The dark crowns of oaks seemed to float like islands atop the ground fog early in July in Minnesota, and I could feel tension in the air. After a month of relative silence followed by a week of constant clamor, an hour could now pass without sound before a single woodpecker call might initiate an eruption of begging and drumming. The bridge between dawn and sunrise was the most predictable time as adults left evening roosts and the young waited to unleash their grievances to their parents after a long, hungry night.

The Red-headed Woodpecker nestlings could fledge at any moment. After days of infrequent feeding, it seemed they could be lured out with the promise of food at a time of their parents' choosing. The initial hours of light offered the best chance to witness this singular moment and confirm how many young had survived to this stage.

The past few mornings it sometimes took thirty minutes before I knew if the young remained in the nest. The morning chill anchored them until they knew food was near. The instant an adult began churring, a fluffy, ashen face filled the cavity and lifted my spirits. As if solar powered, the young woodpecker animated with the rising sun and stretched out of the entrance up to its shoulders. It halted the precarious lean by spreading its wings inside the cavity and bobbed with its neck extended for over ten minutes, then exploded into flight without any hint it was about to leave. After landing awkwardly on an oak thirty yards away, it hitched its way to the top of the twenty-foot-tall tree, then made a wobbly flight to join the adult. A few minutes later, distant drumming lured another bird to the entrance. The second quickly extended its neck like the first, gripped the lip of the cavity with its claws, froze for several moments, and without tell was on its way to join the adult and the first fledgling in a pin oak fifty yards away.

Within an hour, six weeks in the safety of the nest ended, and the most dangerous six months were just beginning for both fledglings.

Recently fledged woodpeckers are capable of competent flight within a few days of leaving the nest. A juvenile male Arizona Woodpecker (top) stays close enough to his father to receive food as the adult finds it.

The predictable becomes perilous for the large-eyed young peering from cavity entrances as they confront the first of several dangerous thresholds. Over the next few months, they will leap from the safe confines of their nests, learn where to find, seize, and collect food with limited guidance from their parents, and be driven from newly familiar landscapes to find their own territories before winter or live on their parents' territory as floaters.

FINAL DAYS IN THE NEST

Nestlings cling to and in some cases sleep on the walls of their cavities during their final days in the nest, which in some species helps them avoid any waste that has accumulated. A young but surprisingly mature face sometimes occupies the cavity entrance.

Nestlings begin to look like adults at this stage except for sometimes noticeably shorter bills, and present their final juvenile plumage, which is often very close to that of adults. Depending upon species and individuals, color badges may be closer to the bill, brighter or duller, and are often absent when compared to adults. In some cases, female nestlings may display evidence of a smaller amount of color than the males, which will disappear as they mature.

Parents of some species deliver larger prey during the last few days, which can result in fewer visits to the nest. Dominant nestlings hang their heads out of the cavity or are ready to do so at any sign of their parents, whether that be a shadow, a call, or a vibration on the face of the tree. Less frequent food deliveries motivate nestlings to jostle for the prime position in desperation to be fed. The climbing and jostling help develop and strengthen their tails and legs, which they'll need to securely brace themselves and climb on precarious surfaces and find food upon leaving the nest. Strengthening their immature flight feathers is more challenging, however, and delays their ability to control their flight.

At the slightest sound, the belligerent young jab their bills at one another and at arriving parents. Nervous adults approach the cavity carefully during these last few days to avoid being poked. They alight on the nest tree and cautiously hitch up, down, or sideways, carefully choosing the safest moment to fill the snapping bills.

The dominant nestling will be fed first and more frequently, so the young now clinging to the cavity walls struggle, sometimes dragging siblings from the prime spot at the opening. Most cavities allow only one nestling at the entrance at a time, veiling much of this competition. The larger holes of Pileated Woodpecker and flicker nests allow observers to witness how aggressively the nestlings fight, sometimes even grabbing one another by the neck.

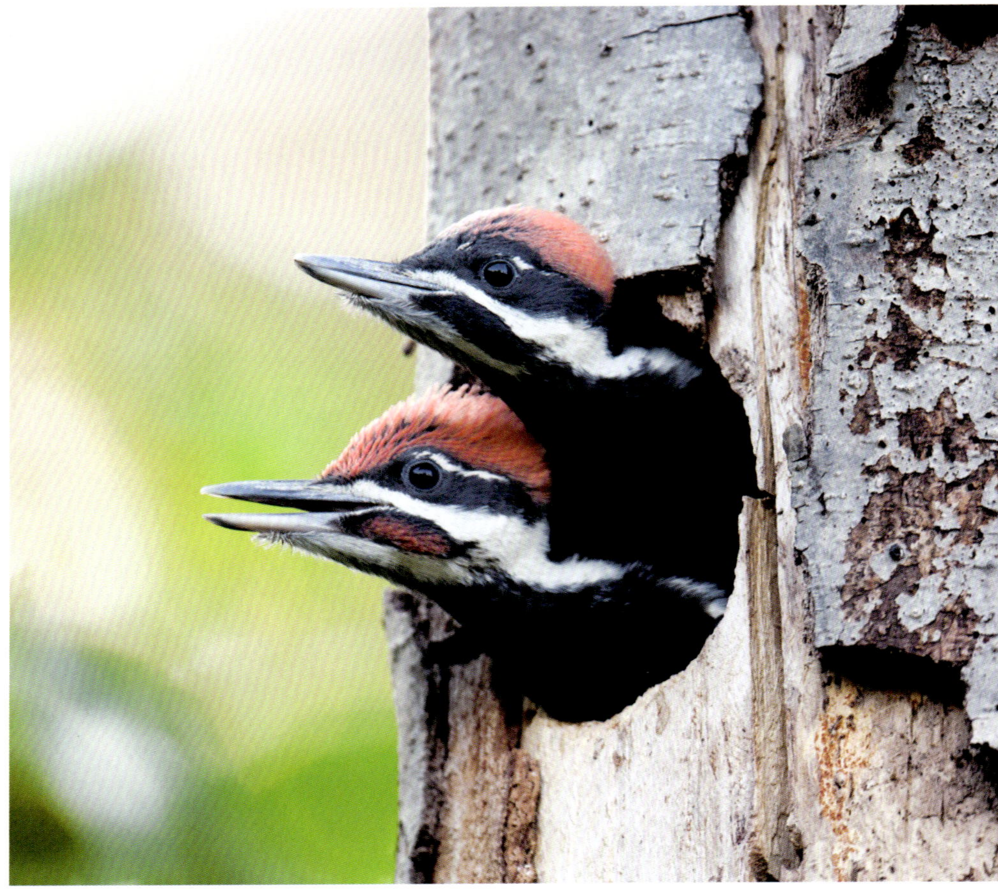

Pileated Woodpecker nestlings peer from the nest cavity during their last several days in the nest. At this time it is possible to establish the sex of the young with some degree of confidence in most species. The male nestling can be distinguished by his red crown, which reaches the bill, and the red at the front of the malar stripe.

Opposite: As nestlings edge closer to fledging, they often rest with their bodies protruding farther through the entrance. This Red-cockaded Woodpecker nestling can be distinguished as a male by the red on his crown. As he becomes an adult, the red will be limited to a few red feathers that remain hidden unless the bird is excited.

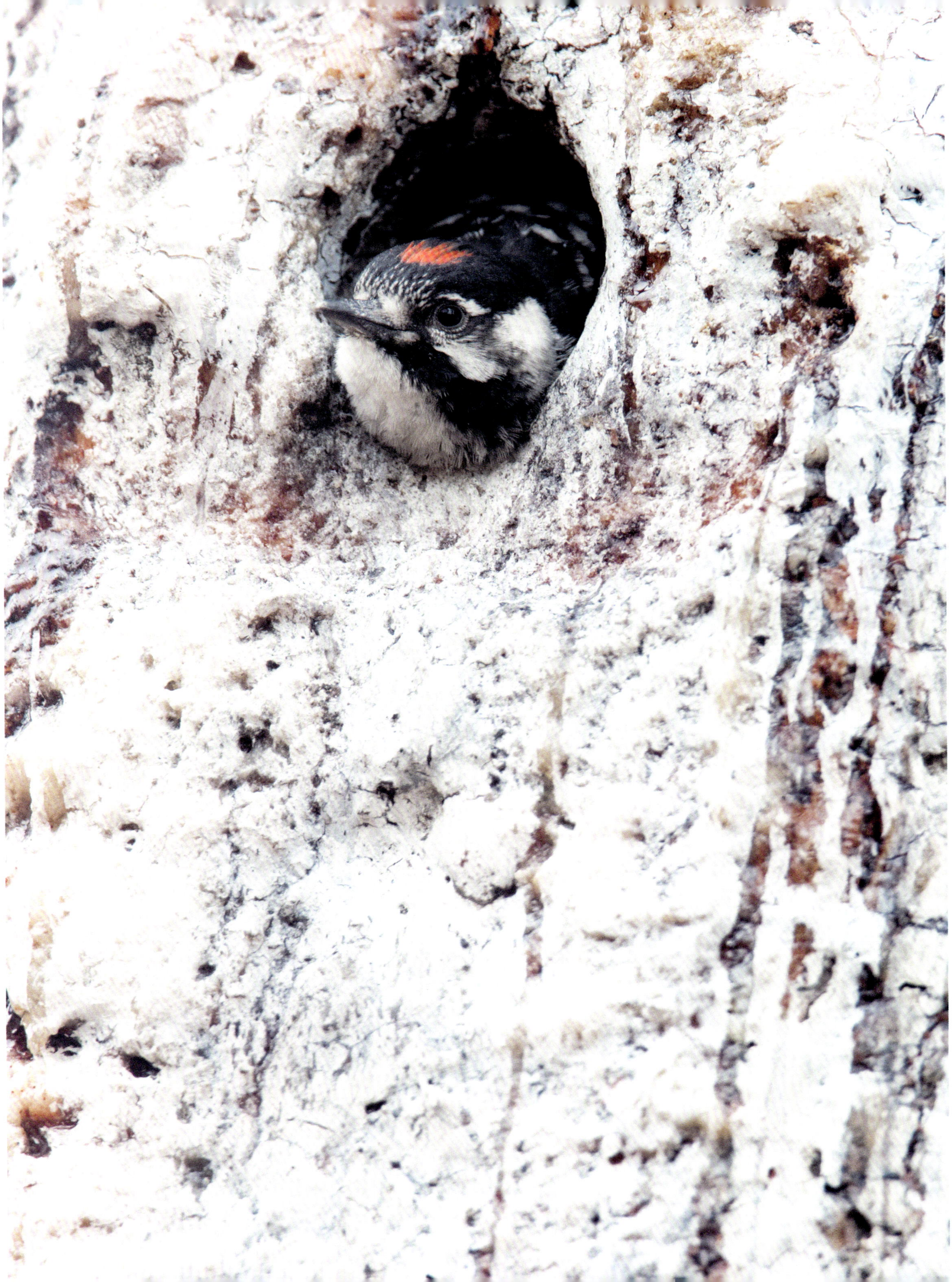

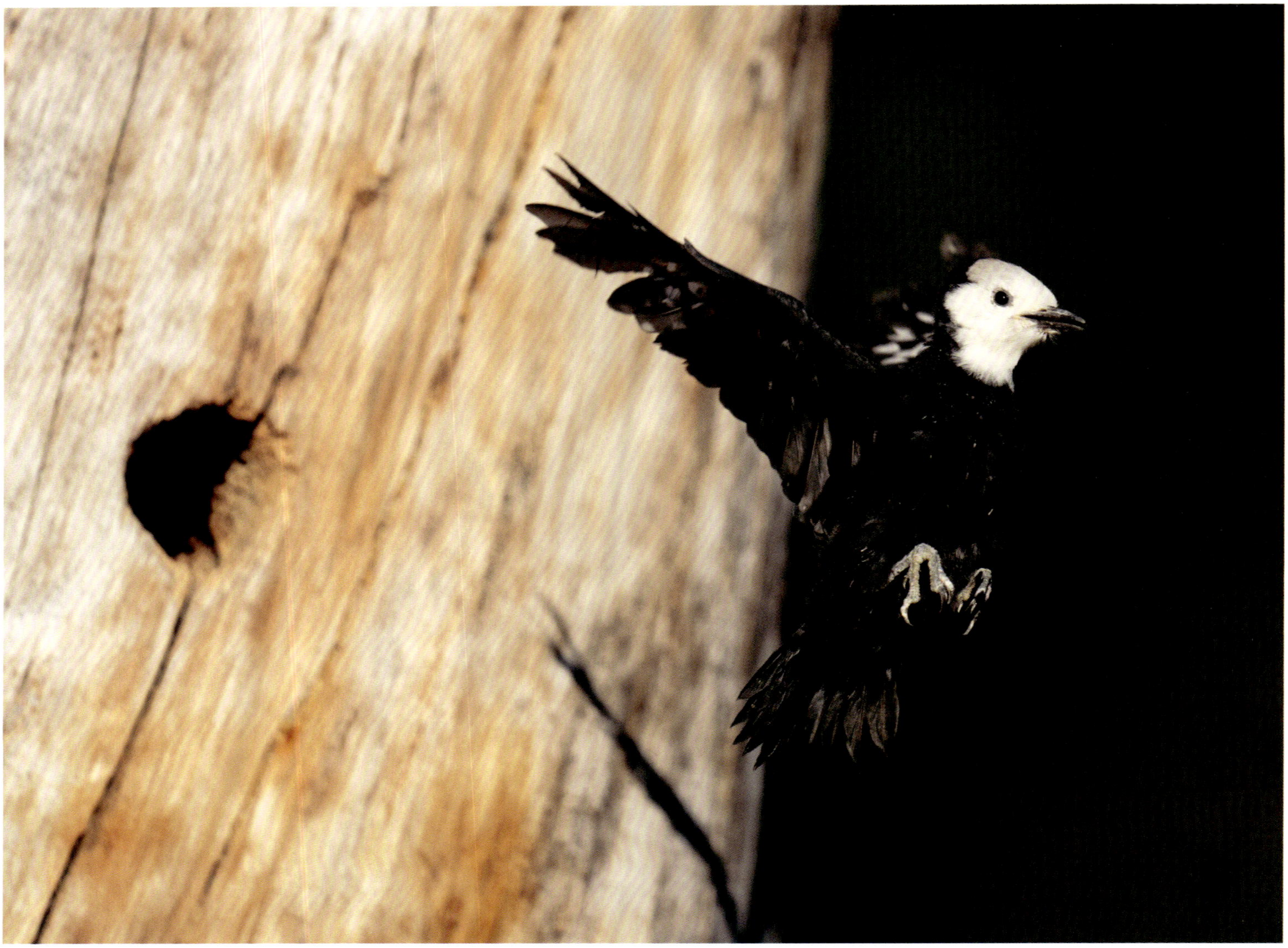

FLEDGING

In the last hours before fledging, larger nestlings may consume the entire delivery that had previously been shared, increasing the clamor in the nest. Nestlings fledge roughly three to four weeks after they hatch, with the extremes represented by Downy Woodpeckers leaving as early as eighteen to twenty-one days, and Lewis's Woodpeckers staying as long as twenty-eight to thirty-four days.

The young begin to lean out of their familiar confines, claws overhanging the lip, sometimes motionless for several minutes. Watching this process is at once awe-inspiring and attention-challenging, as very little changes in the final twenty-four to seventy-two hours in the nest.

Sometimes the cavity entrance appears empty until a parent vocalizes or drums. Quickly, at least one nestling fills the entrance. When it does appear, it displays increased curiosity about the world outside, sometimes snapping at passing insects and occasionally catching them. Flicker nestlings often extend their tongue and explore the cracks or crevices just beyond the cavity entrance.

Between feedings, some nestlings may lean out and stretch their necks, then grip the cavity lip with their claws and extend themselves so that much of their torso and wings are visible. Just when you think the young are going to leave, they withdraw and are fed by the parents, which seems to start the process anew.

On dozens of occasions, I have had the privilege of watching woodpeckers of several species during their multiday process of leaving the nest. These included American Three-toed,

A leap from the lip of the cavity, a raising of wings, followed by one solid downstroke, propels a female White-headed Woodpecker nestling into the outside world. A male fledgling would have some red on his crown.

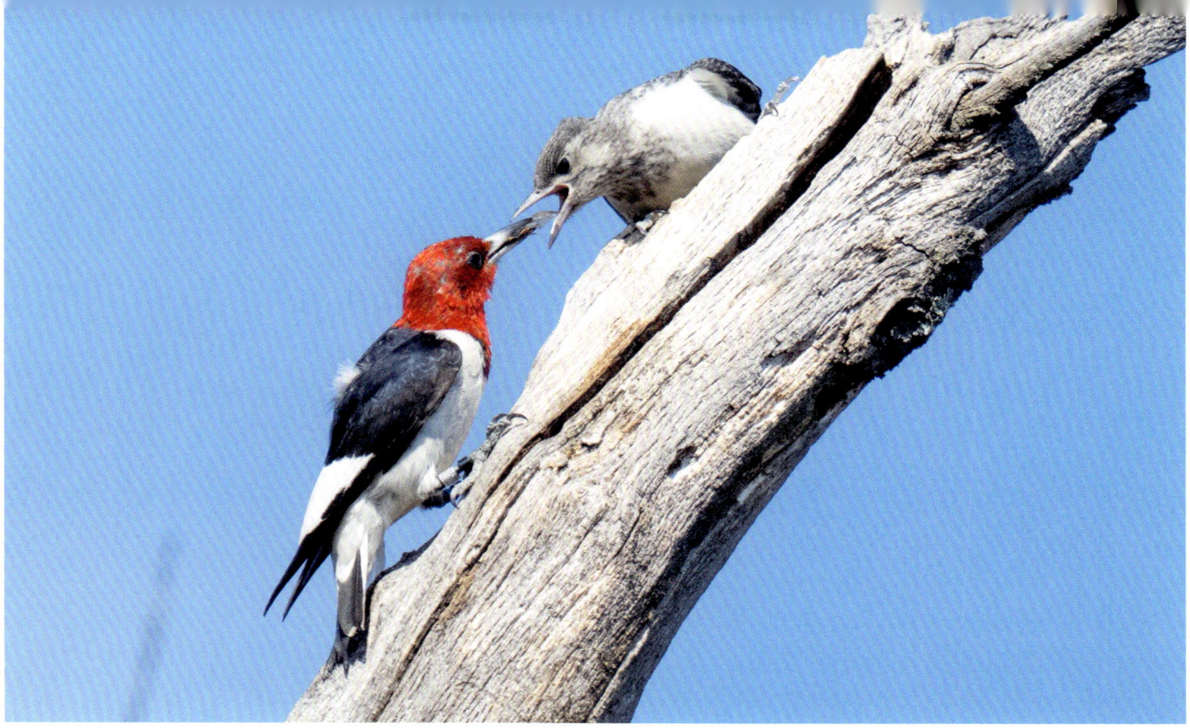

When an adult woodpecker, like this Red-headed, is actively fly-catching, dependent juveniles will situate themselves higher in trees so they are well positioned to receive food after a successful aerial capture.

Black-backed, Downy, Hairy, Puerto Rican, Red-headed, and White-headed Woodpeckers, as well as Northern Flickers and Williamson's Sapsuckers. I watched and photographed most of these nests from blinds a great distance away, but observed two flicker nests through the window of my home. Some behaviors were virtually identical, but others differed greatly, which could have been impacted by many factors including but not limited to the species, individual personalities, or experience of the parents, as well as height of the nest.

The greatest variations occurred in the parental involvement in the process and the predictability of fledging. At one low Williamson's Sapsucker nest, the parents seemed to lure the young from the cavity with food nearby, while at another nest the feeding continued in a consistent way until the young fledged without fanfare between feedings, giving me no warning of what was about to happen.

This unpredictable process was very similar at the White-headed, Red-headed, Black-backed, and Pileated Woodpecker nests that I witnessed. At the Black-backed nest and also at a White-headed Woodpecker nest, seemingly without encouragement a nestling climbed, instead of flying from the nest. In both cases the young hitched up the face of the snag. The young Black-backed was quickly joined by a parent that fed it before both flew in the same direction. The White-headed Woodpecker climbed to the top of the snag before flying into the dense canopy of a nearby pine.

The Northern Flicker nests that I watched, including the two witnessed through the windows of my home, showed very similar behavior. On two occasions, one in North Carolina with Yellow-shafted Northern Flickers and once in Oregon with Red-shafted, the clamor of calls and drumming led me to nests previously unknown to me. The adult flickers were drumming and utilizing both contact and long calls from a distance and would fly into view of the young before resuming their signals from farther away. Only when the young stopped begging did the parents fly in with food, but delivered only a portion, leaving the youngsters protesting and lunging for more. The rotating heads of the young told me that they were tracking the movements of the adults.

On one occasion I witnessed a flicker nestling stretch a wing at the cavity entrance, and to its surprise the wing popped out of the hole and accidentally propelled most of the youngster's body out. While one wing flapped outside and one inside, it frantically attempted to back into the nest but was repelled by the jabs of a sibling, which forced it into flight.

Towering white fir, ponderosa pine, and incense cedar shaded the edges of a bowl of mature quaking aspen high in the mountains of southern Oregon. The surrounding conifers muffled the cricket-like begging of young Williamson's Sapsuckers inside a large, barkless aspen snag. As the sun rose, parents scuttled larger billfuls of ants into the noisy mouths more erratically, prompting shrill cries as they landed farther from the cavity opening, often sliding on unfamiliar parts of the slick surface. Each parent took more time shuffling up the face of the tree toward the necks stretching from the hole. The male favored creeping up from below with his neck and torso leaning back and just out of reach of the lunging bills. The female seemed to prefer approaching from the side, encouraging the occupant of the hole to stretch in that direction.

With each delivery, the parents seemed to entice the young out, little by little. The male flew toward the cavity, head and belly pulled back, touching one foot for the briefest of moments, then dropping, causing the nestling to reach too far and find one wing out of the cavity. It quickly but awkwardly flapped with the grace of a large leaf in the wind to the tree in front of it. Over the next hour, the parents seemed to challenge the second youngster to reach farther for food until it too fell and feebly flew. The third nestling lingered into the next day and abruptly dropped from the nest after several feedings, surprising its parents, who arrived with bills full of ants moments after it left. They seemed perplexed as they lingered on the face of the tree, returning three times over the next several minutes and dipping their heads in to confirm that the cavity was empty.

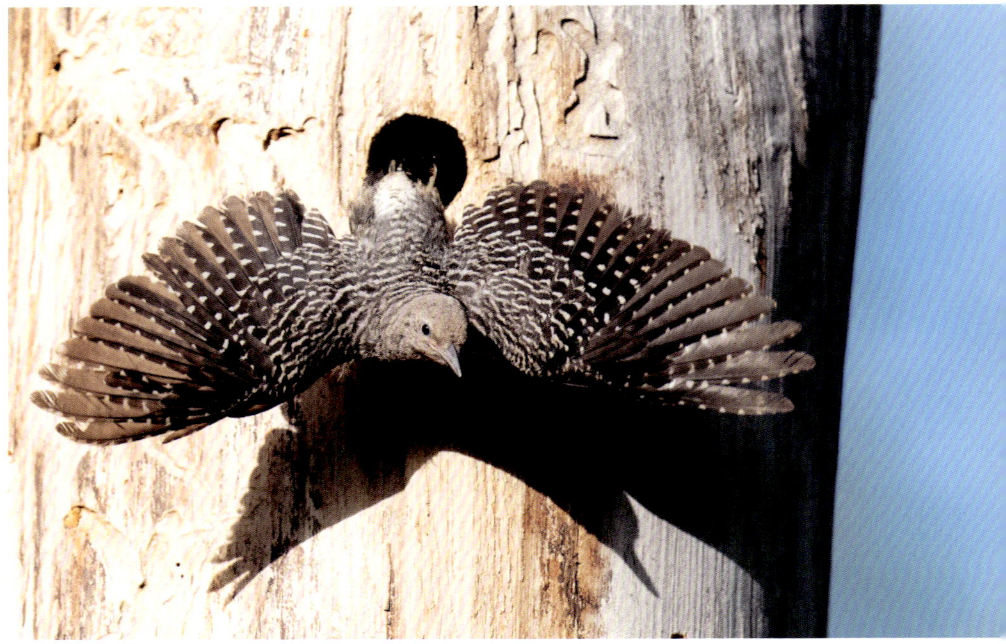

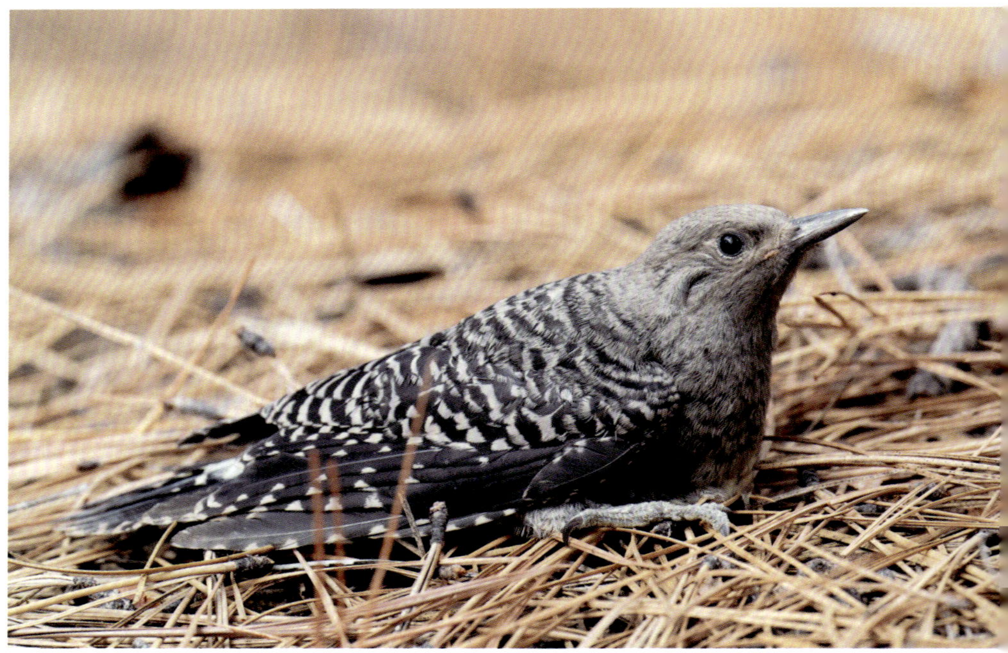

Top: As woodpeckers prepare to leave the nest, they sometimes lean too far forward and are forced to extend their wings and fly out. I have witnessed this a few times with Williamson's Sapsuckers, including this female (distinguishable by her brownish coloration).

Bottom: Most fledglings end up in nearby trees after flying from the nest, but occasionally young, like this female Williamson's Sapsucker, fall to the ground. At this point, steep flights are difficult for fledglings but they must either make a low-angled flight to a nearby tree or climb one to be safe from predators.

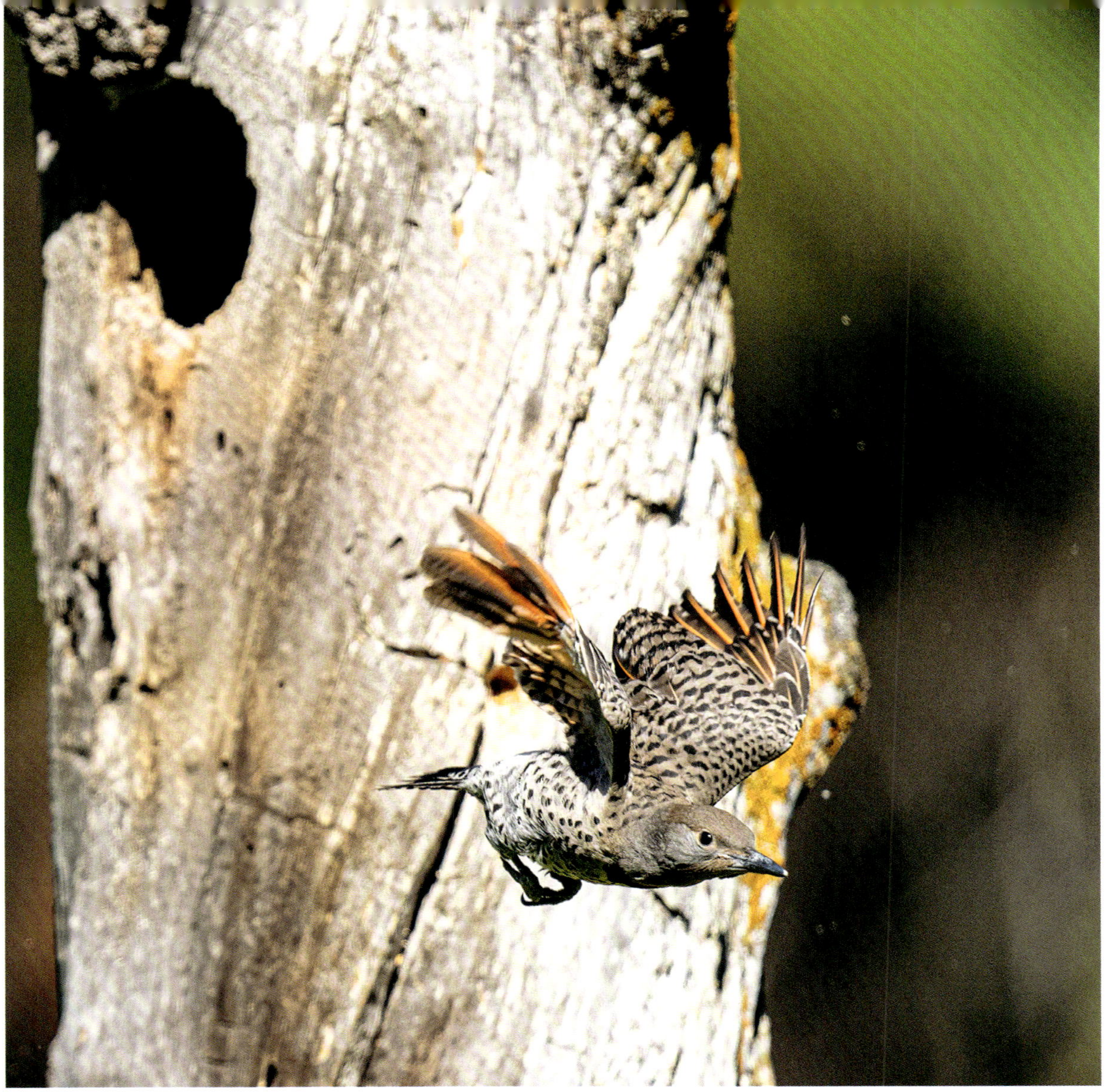

Flickers, like this female Northern Red-shafted fledgling, are often surprisingly strong fliers when they leave the nest and seem to have more control over their direction of flight and where they land, often joining parents and siblings several dozen feet from the nest.

The Northern Flickers nesting in a snag in the adjacent meadow provided the greatest contrast to the Williamson's Sapsuckers in how abruptly they fledged and by the strength of their initial flights. The parental coaxing came from farther away, a combination of single *peah* calls and repeated *kik-kik kik-kik* calls. After each visit, they flew several dozen yards away and called back to their young. The nestlings appeared ready, claws at the entrance lip and eyes tracking the distant adults, but instead of leaning and falling out, or giving any sign of imminent action, they flew. Some seemed to fly in response to calls from the parents, but others departed when all was quiet, with no adult in sight. At the flicker nests I have watched fledge, the young were surprisingly controlled in flight and landed on a tree anywhere from 30 to 150 feet away. They often landed awkwardly but seemed to be guided to a general location by the calls of parents.

Pileated Woodpeckers a few miles away also fledged less predictably, usually when parents were silent, and while the strength of their initial flights varied, they did not appear as strong or fly as far as the flickers. I watched them fledge with wobbly flights of thirty yards or so, each to a separate tree on which they slowly climbed.

Some woodpecker nestlings leave the nest within an hour of the first, but more frequently they leave over two to three days, particularly with larger broods. With each successive fledging, the parents focus less on the nest, perhaps making young hungrier and more likely to fledge without encouragement.

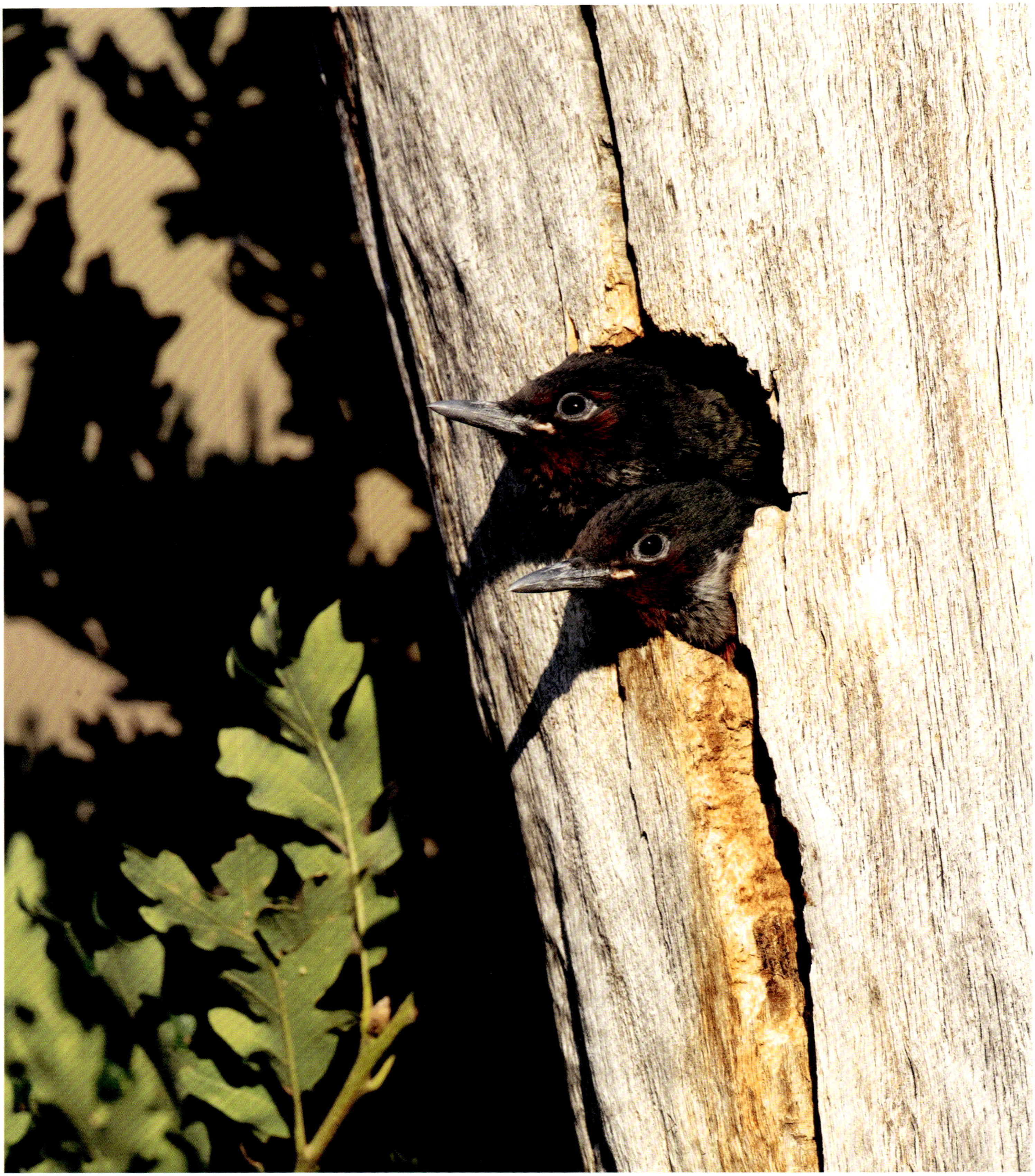

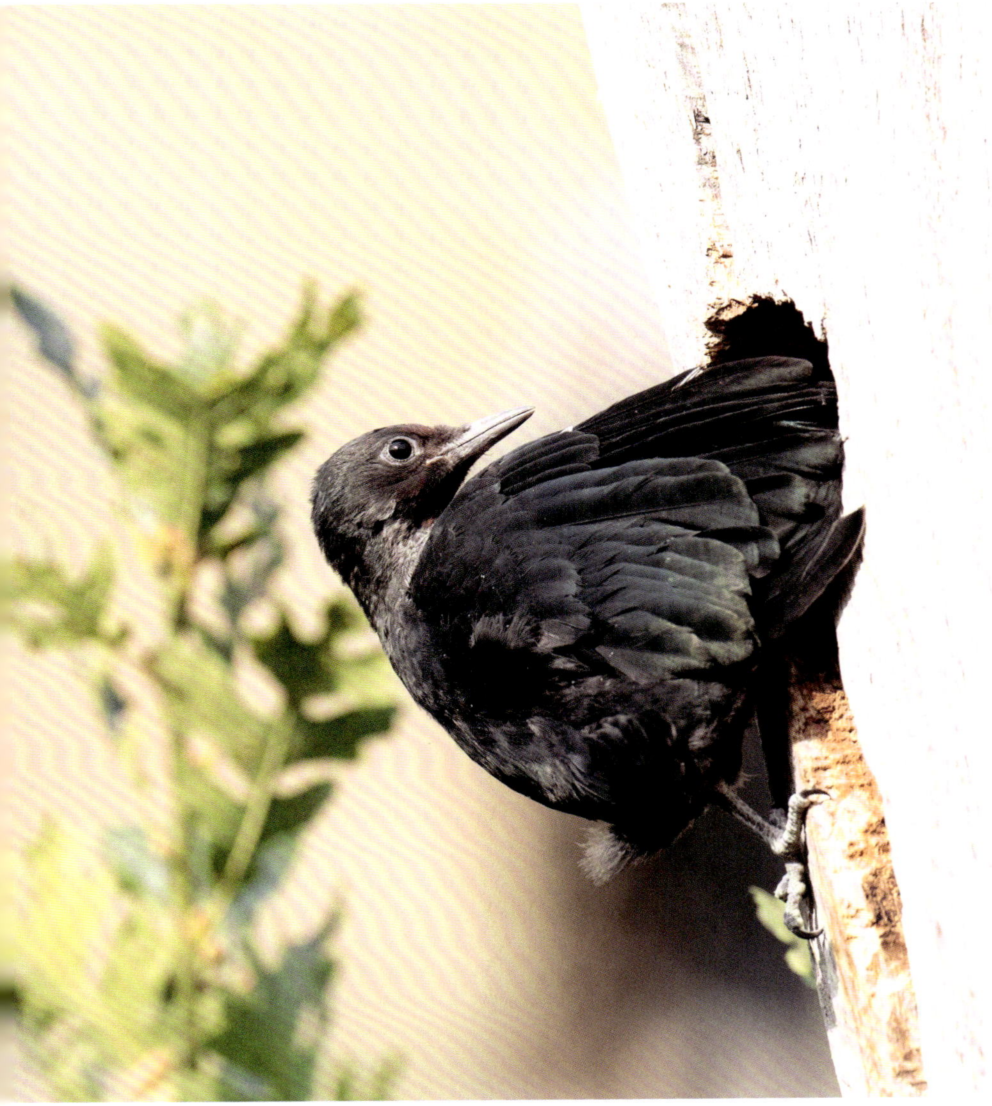

Some woodpeckers, like this Lewis's, actually climb or crawl from the nest when they first depart. *Melanerpes* species such as Lewis's, Red-headed, and Acorn Woodpeckers are among the minority that sometimes return to the nest cavity after leaving.

Opposite: A single nestling often fills the cavity entrance as they near fledging. However, Northern Flickers, Pileated Woodpeckers, and Lewis's Woodpeckers (shown here) utilize larger cavities that allow two or more to share the prime position.

One of the benefits of experiences with wildlife is the foundational knowledge you build to which you add study and research. This base allows you to focus on behavioral inflection points, anomalies, or undescribed behaviors. For instance, I believed that when woodpeckers leave the nest with one fateful thrust forward, followed by an awkward, shaky flight, they never return. I had seen this with Red-headed Woodpeckers and assumed I would see the same with the closely related Lewis's Woodpeckers, but I did not think I would get to witness a far less common way of leaving the nest.

Years ago in an oak woodland in southern Washington, I watched a gray snag figuratively shaking and coughing debris from its cracks as several young Lewis's fought for the prime positions inside. One chunky purplish youngster pushed its head past its siblings and into the bright sun as it had done on several other occasions. In a move indicative of imminent fledging, it gripped the entrance lip with its claws and leaned so half of its body was outside the cavity. My finger was on the shutter button, so hopeful of finally capturing the forthcoming critical moment of fledging. Then the youngster was three-quarters out, but its head dropped. I thought it was falling until it simply pulled its tail from the inside and rotated its body 180 degrees, and just like that, it was out and clinging to the tree. I had not expected this. The youngster's movements seemed nonchalant, even practiced. Slowly and awkwardly, it hitched its way up the snag and was replaced in the nest entrance by a sibling.

My second surprise came when this youngster flew feebly to a snag nearby, only to fly back to the nest tree, hitch down to the cavity entrance, and push its way back in! I have watched several Lewis's Woodpecker nests and seen this reentry behavior only at this one cavity, but the young at this site did this on several occasions over three days until all had left the nest.

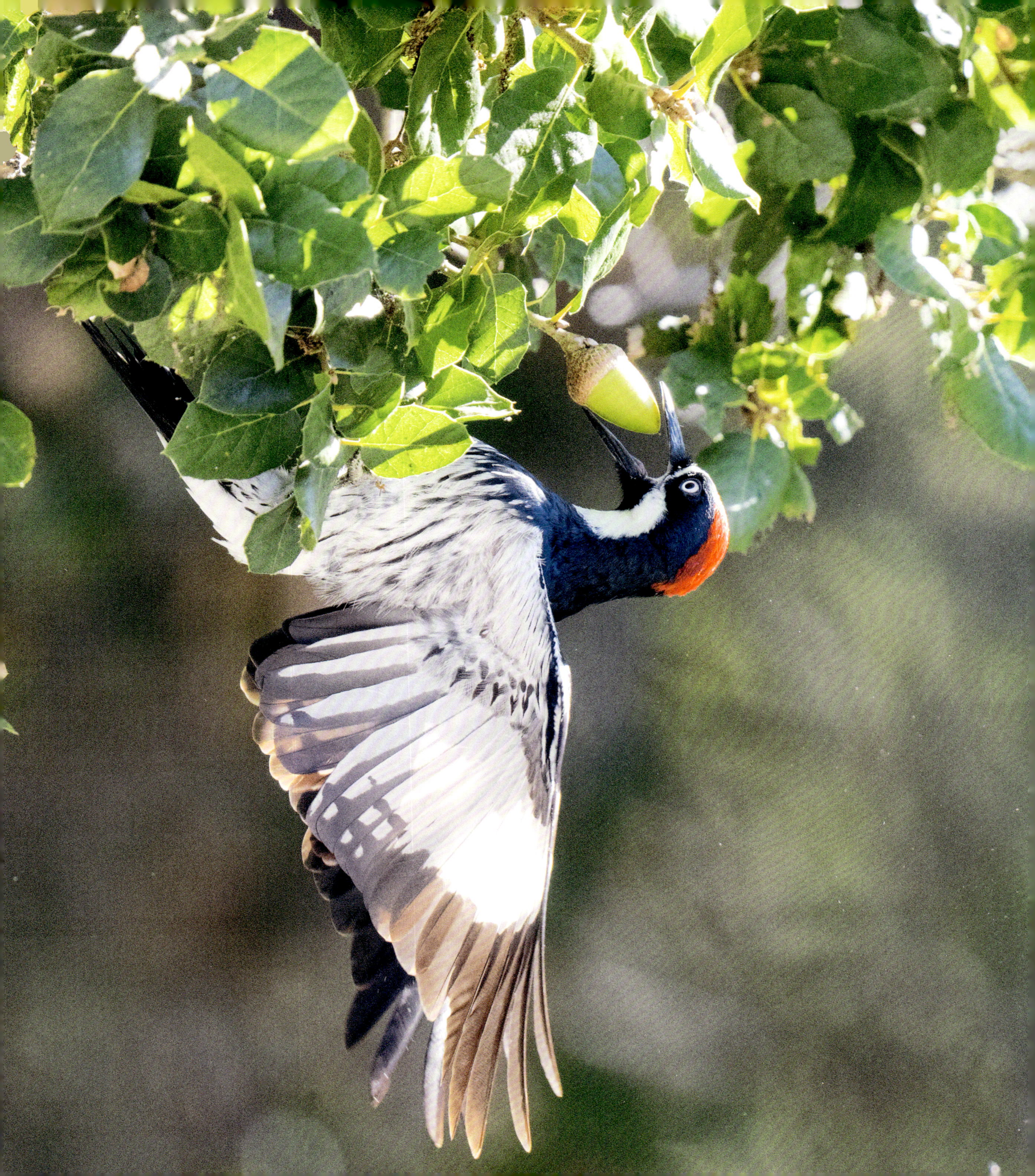

LEARNING TO FIND FOOD AND SHELTER

Fledglings are especially vulnerable in their first several weeks out of the nest. Juvenile flight feathers prevent strong and controlled flight, making it more difficult for fledglings to follow their parents and making them more vulnerable to predation, particularly from diurnal raptors such as kestrels, goshawks, and Sharp-shinned and Cooper's Hawks. Most newly fledged young hide quietly in dense tangles of branches or vegetation.

Fledglings can suffer heavy rates of predation during their first week out of the nest, as in the case of the Lewis's Woodpecker because the young sometimes climb out of the cavity and explore the nest snag for a few days before finally taking flight. Some even continue to return to the nest tree for a few days to compete with nestlings for food from adults. Research on Red-headed Woodpeckers in Minnesota indicates that most young that make it past that the first week out of the nest survive at least until the winter.

During the first few days, fledglings of many species hide quietly near their nest, until a parent approaches with food. Later, they must follow their parents to learn where and how to secure food. All this must happen before the adults lose interest in feeding or even sharing a territory with them. Finally, it may take the young of some species several weeks to several months to find a cavity to roost in at night, and until that time they must sleep outside.

The young scatter initially, but in some species they later break into groups of one or two that follow one parent at a time. When woodpeckers have more than one brood, one parent may be excavating a new nest cavity in preparation for a second brood, which will limit the attention given to existing young. Young woodpeckers must learn what they can about securing and storing food before they are driven from their natal territories or prepare to migrate.

The life paths of young woodpeckers of different species diverge after they climb, flutter, or fly from their nests. The period between leaving the nest and establishing their own winter territory is one of the least studied and documented periods of a woodpecker's life. The studies that have been conducted were typically narrow in scope and geography and may or may not apply to broader populations.

Smaller woodpeckers that glean ubiquitous food like ants from the surface of a tree's bark, twigs, or leaves may more quickly learn to feed themselves. More specialized knowledge and skills, such as which trees provide the most sugar-rich sap and how to access it, likely take more time to acquire. Species that dig out larval insects from deep inside a tree may need experience to understand where and how to listen for their preferred prey and how to seize it. In some cases a young bird must rely on further development of its chisel-like bill. Some species need practice to effectively capture their food in the air.

Williamson's Sapsuckers appear to become independent more quickly than other North American woodpeckers. This speedy self-reliance may be a function of how easy it is to learn to glean their primary food, the abundant ants moving up and down trees as they tend aphid colonies. Adult males sometimes leave the nest a day or two before the young fledge, and both parents neglect their young after seven to ten days post-fledging. At this time the young may separate or remain together on their natal territory for up to several weeks.

Red-breasted, Red-naped, and Yellow-bellied Sapsuckers take one to two weeks longer than Williamson's Sapsuckers to gain their independence, even though they too attempt to feed on their own almost immediately after leaving the nest. This difference may be a function of needing to learn how to secure a more varied diet, including which trees will be most fruitful for sapwells.

It was fascinating to watch Red-breasted Sapsuckers in the Cascade Range of Washington during this transitional period. Four juveniles, with varying degrees of red wash on their gray feathers, had left their nest a few days earlier. The fledglings

Woodpeckers that consume acorns, such as Lewis's, Red-headed, and Acorn Woodpeckers like this one, prefer to pluck acorns from trees rather than taking them from the ground. They typically grab the acorn with their bill and fall or fly to let their weight free it from the tree.

spent much of their time clinging at or near the wells. Periodically, a parent would arrive with an insect in its bill. The juveniles begged, and at times one opened its bill toward its siblings, to attain prime position adjacent to the parent as it rolled its insect gift in thick sap before shoving it into the open bill of the nearest pleading youngster. Gradually, the young began pecking and licking at the well that had just been primed by a parent, lapping the sap or snapping up any lured insect. Between parental visits, the young continually appeared to compete for access to the most fruitful wells.

Eventually, each adult led two juveniles on rounds between various wells and eventually to fir trees twenty to fifty yards away, where they watched as their parents carved fresh honeycomb-like cells and licked the sap as it came to the surface. At one tree, the male appeared impatient at the begging of the young and alternated between pecking at them and feeding them. After being discouraged from begging, the reddest juvenile began pecking at the tree and after several minutes had carved a recessed cell just above those of its parent. As this cell began filling with sap, the youngster quickly lapped it up and began carving another cell adjacent to the first. The next day, two juveniles followed each parent upslope, away from the nest area and into denser stands of fir, ending my observations.

The more social *Melanerpes* woodpeckers may continue to look after their post-fledging juveniles for a couple of weeks to a few months. With the Acorn, Black-cheeked, and some other *Melanerpes* woodpeckers, young are led back to the nest to roost at night, but with most species, even closely related ones like the Red-headed Woodpecker, newly fledged young do not roost in cavities, making them vulnerable to predation.

Fledged Lewis's Woodpeckers spend more time exposed than any other species I have witnessed, and they are awkward at first, snagging their tails on pieces of bark as they hitch up or down trees. This, combined with their ungainly flight and their habit of exploring cracks and crevices at the nest tree for a couple of days, makes it unsurprising that they suffer a high mortality to American Kestrels in their first few days out of the nest. Even with these losses, they remain in the vicinity of the nest tree for ten days. Fly-catching is a crucial skill for Lewis's Woodpeckers, and for the first few weeks they have a difficult time returning to the same perch after a sortie, but after a month or so they become indistinguishable from adults in flight and landing.

Red-headed Woodpecker fledglings appear quite competent quickly, and I was surprised to witness one fly-catching for insects within several weeks of leaving the nest. Even with these advanced skills, the young still follow their parents for up to twenty-five days and make little effort to find their own food if they are part of the only or second brood. At this point the parents respond to pleas with agonistic postures, calls, and even chases. Fledgling Red-headed and Acorn Woodpeckers from a first batch will opportunistically beg for food, even after the second brood has

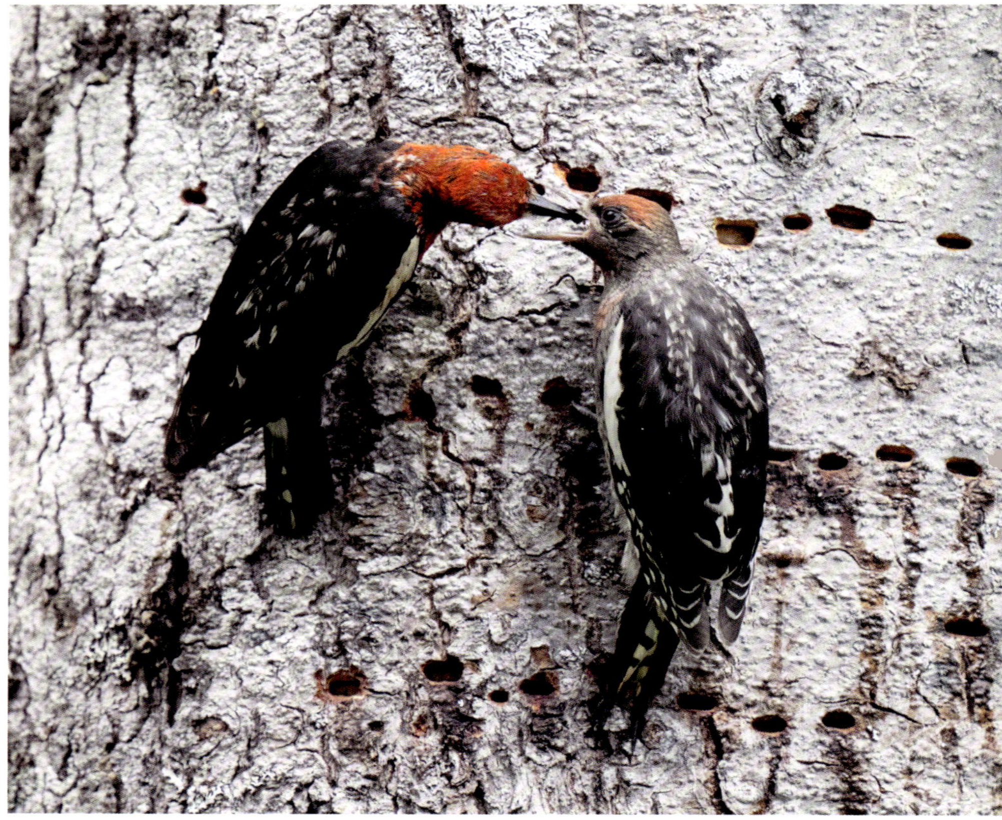

An adult Red-breasted Sapsucker feeds a juvenile an insect dipped in sap from a recently excavated well. This juvenile was creating its first wells alongside those of its parent.

Opposite: As fledging nears, adult woodpeckers, such as this Lewis's, often take more time between arriving at the nest and delivering food, which may encourage fledging or protect parents from the jabbing bills.

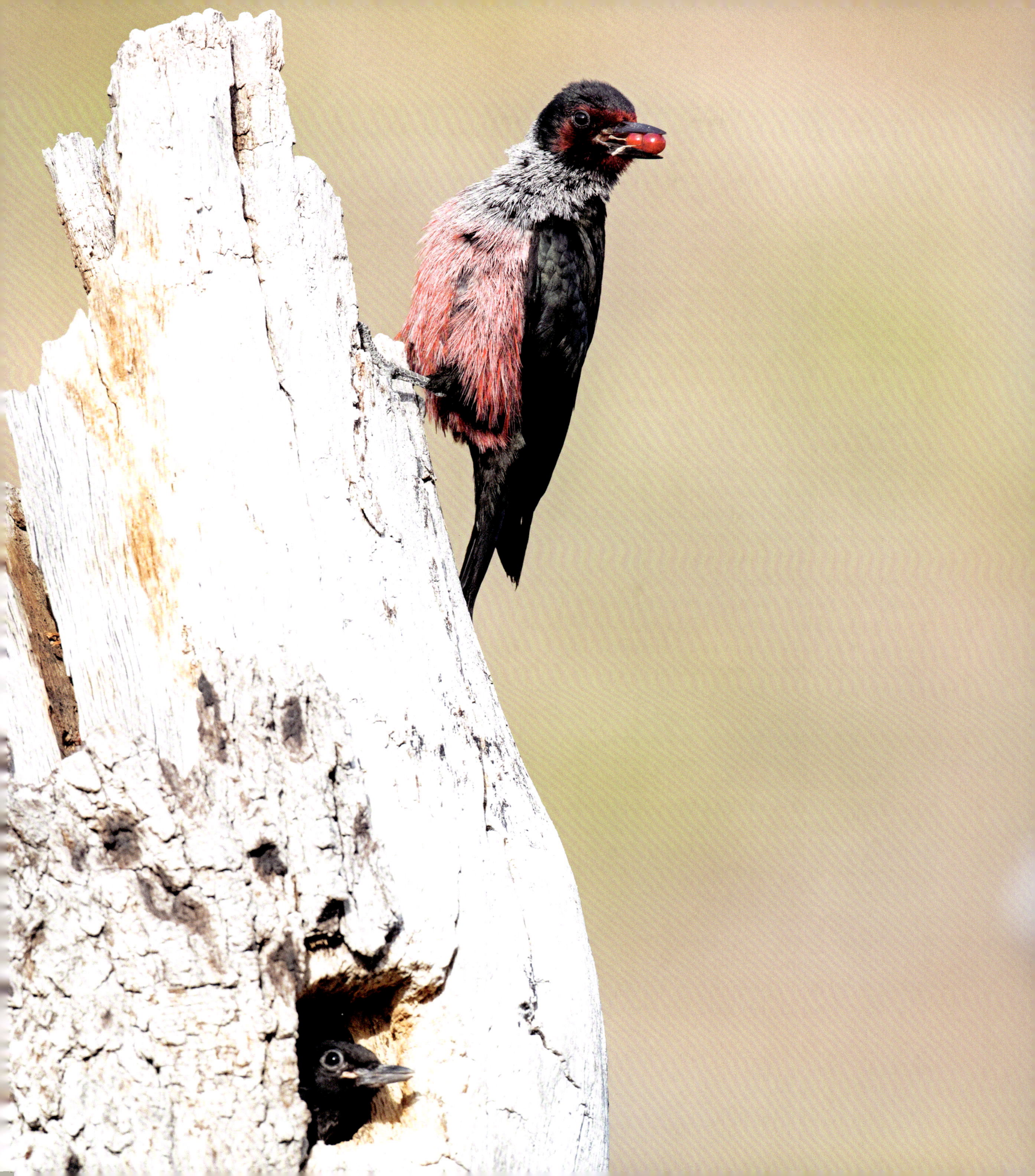

fledged more than a month later! It was amusing to watch first-brood young fly to recently fledged second-brood young and beg alongside them. The parents could tell the difference and chased them away, almost as if they were rival adults.

Puerto Rican Woodpeckers are a *Melanerpes* species that is closely related to the Red-headed Woodpecker, so I was curious to see how their post-fledging behaviors compared to their northern relatives. In Puerto Rico, I watched recently fledged young following their parents near a nest they left the week before. I was surprised to see two fledglings fly to the nest tree and enter the cavity, where they were fed by the parents as if nestlings. To my knowledge, returning to the nest tree and being fed there are two behaviors that have not been documented with this species. Puerto Rican Woodpeckers will feed their young for at least two weeks after they leave the nest, but the young may associate with the parents much longer. This is a similar pattern for others in the genus, especially with those in tropical habitats.

Northern Flickers feed their fledglings for two to three weeks. During the first week, the young gather at a snag or side of a tree to wait as parents collect food and return to feed them. During the second week, the juveniles follow the parents as they forage in trees. By the final week they join their parents on the ground, begging and foraging on their own when begging is not fruitful. Males appear to look after the young longer than females and attend them more closely. The survival rate for young is greater when they are moved to areas with more canopy cover as it keeps them safer from hawks.

Pileated Woodpeckers begin life outside the nest much as Northern Flickers, leaving recently fledged young together while adults forage. I watched young Pileateds at this stage for a couple of weeks in a mixed-conifer forest in southern Oregon. On the first day post-fledging, two juveniles perched on the same lichen-covered tree under the canopy of white firs, while the third fledgling perched near the foraging father on a third tree. The second day, one male youngster trailed its father on downed logs while the two females trailed their mother. During these forays, the young stayed close to the parents while they foraged but did not attempt to excavate on their own, although they would beg and lunge at the adults when hungry. Throughout the first week post-fledging, when the parents flew longer distances, they left the young together in a small group of trees. The fledglings occasionally postured toward or chased one another in an apparent attempt to retain the perch nearest to where the adults delivered prey. These juveniles would continue to be dependent upon their parents until the early fall.

Unfortunately, what we know about the likely-extinct Ivory-billed Woodpecker is based on limited information, with uncertain reliability. They likely had the longest period of dependency. The young are reported to have been fed exclusively by their parents for up to three months and may have foraged with them until the adults nested again the following spring. The juveniles are believed to have been strong fliers, but their specialized bills may have needed time to develop before they could feed as adults. Juveniles of the Pale-billed Woodpecker, in the same genus, are also believed to be dependent longer than other North American woodpeckers.

Most woodpeckers follow a similar pattern post-fledging, and during this period in prey-rich areas, a large number of woodpeckers can be found around concentrations of food or near water sources, as recently fledged young follow adults and learn to sustain themselves. Individuals of many species may disperse, sometimes with their young, from nesting areas for safety or to pursue seasonally abundant prey.

To see woodpeckers respond to a boom in prey, I visited a recently burned spruce forest in north-central Washington just south of the Canadian border. A wall of brittle, black tree skeletons

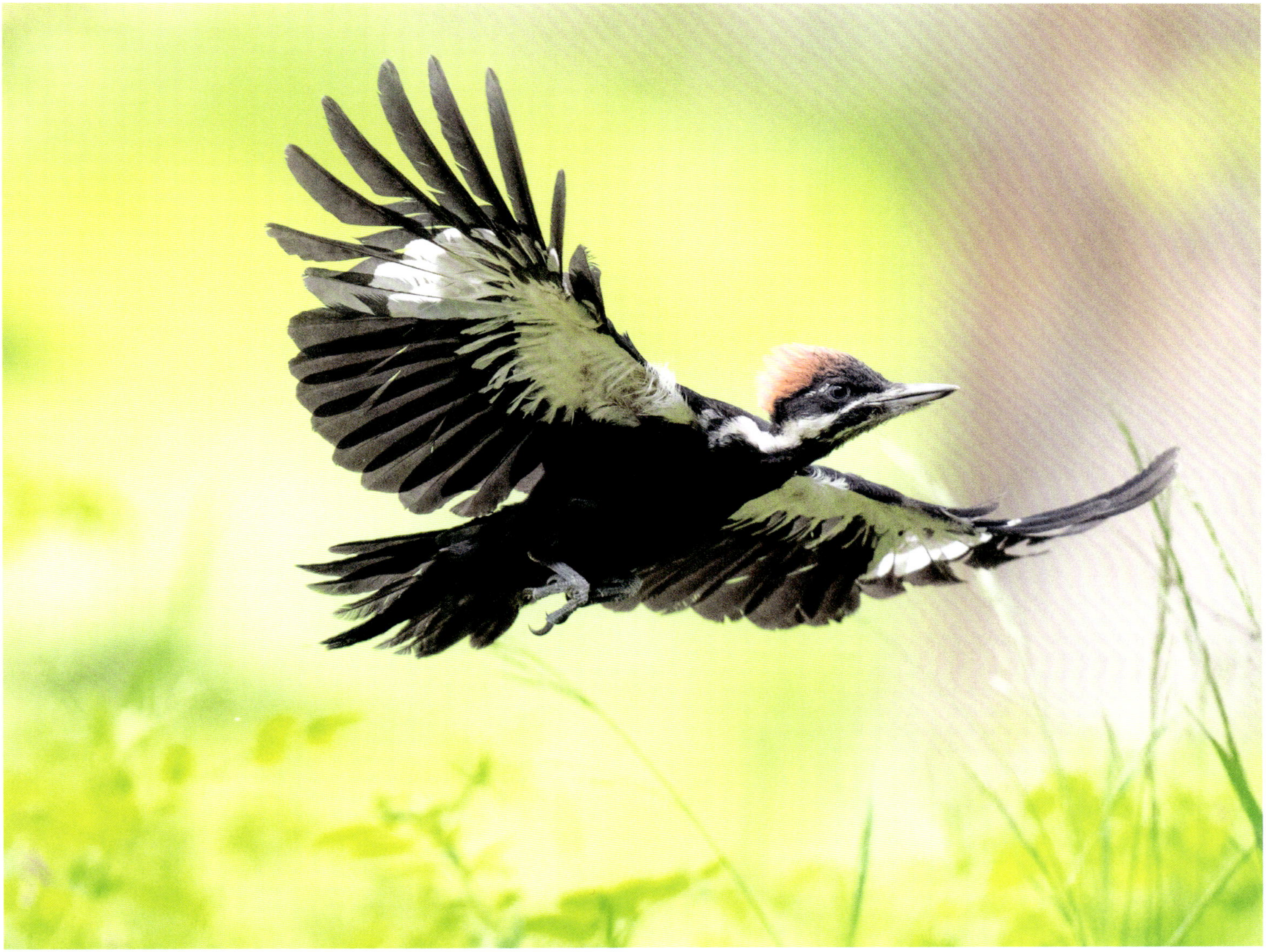

A day after leaving the nest, a juvenile female Pileated Woodpecker is already following a parent as it forages among downed logs.

loomed ahead, and rivers of charred remains flowed for miles across the valley, where spire-like Engelmann spruce and lodgepole pine once livened the landscape.

I stepped carefully, as the very foundation of this forest had disintegrated into ash. Without warning, the sooty vacuums would yank me down a foot into a void once occupied by sturdy root balls. Ubiquitous metallic crackling sounds had me feeling like I was walking under power lines rather than through a mountain forest. These were the sounds of beetles and their larvae chewing the last life out of the gasping stand and threatening to do the same to the adjacent scarred live trees. Even though I knew such fire was a necessary and natural part of maintaining healthy forests, the scene felt foreboding.

Hope arrived like a fast-flying arrow on dark wings. A woodpecker's black head, adorned with a golden cap, snapped out a rattling snarl. The Black-backed Woodpecker then confronted an American Three-toed Woodpecker by flying to the opposing side of a thin black tree. The two animated birds acknowledged each other, bills open and churring, showy yellow head feathers erect and wings spread as they spiraled up the tree. The Three-toed then lowered its wings and retreated.

These birds represented the dawn for this forest, arriving shortly after waves of flames, yet thriving amid the superficial ruin, fed by the life within the scorched trees. Each of these woodpeckers had a mate, and together they were raising young that would inherit the role of helping these formerly overgrown stands transition into richer habitat. Fueled by beetle larvae consuming dead and dying trees, these woodpeckers began creating nests and shelter for themselves and other animals. In order to continue assisting the transformation of this forest, these fledglings needed to survive the first four weeks out of the nest, the period when these two species suffer the greatest mortality.

Over the next few days, it was fascinating to watch Black-backed, American Three-toed, and Hairy Woodpecker fledglings as they followed their parents into denser stands of trees and toward an unburned area at a vulnerable time when none had quite mastered hunting on their own. The young of each species appeared to have been out of the nest for one to two weeks and would likely be on their own within the next two to three weeks.

The behavior of all three was similar in that the young were capable of controlled flight yet stayed close to foraging adults within the canopy of a partially burned segment of the forest. In each case, one adult appeared to be looking after one youngster. The Black-backed and American Three-toed Woodpecker juveniles often flew to the vertical face of the same tree as a parent, where they remained for dozens of minutes until they could no longer see the adult foraging from tree to tree.

The American Three-toed Woodpeckers foraged higher on the trees than the other two species, and in the densest part of the stand, at the edge between lightly burned and unburned spruce and lodgepole pines where they found cooler temperatures and avoided predators. The juveniles flaked bark from the tree with sideways strikes as their parents had. One juvenile showered itself in dirt after directing a blow at an old robin's nest, making it clear it still had a bit to learn.

The Black-backed adults made much more noise, both their distinctive "rattle and snarl" call and their deeper and more forceful pounding lower on the more severely burned trunks under the canopy. They also moved between trees more frequently than the American Three-toed.

The Hairy Woodpecker juvenile seemed to have been intentionally left perched on the limb of a burned snag at the edge of the stand while its parents hunted in the severely burned areas. The juvenile foraged for thirty to fifty minutes at a time without any interaction from a parent, and it probed cracks and crevices with its tongue and bill and came up with a winged ant. Upon hearing a parent approaching, the suddenly animated juvenile flew aggressively toward the incoming bird.

It was surprising to see these three families from three different species, each with two to three young, forage in a space of one to two acres below the nest of an American Goshawk, perhaps their most dangerous potential predator. While the three were relatively close, they mostly avoided one another through their subtle differences in foraging preferences and a seeming understanding of their relative dominance.

A male Black-backed Woodpecker delivers a beetle larva to its male nestling in a burned tree on its penultimate day in the nest. Abundant beetle larvae make burned conifer forests a boon for Black-backed Woodpeckers, who respond with greater nesting density over the first few years post-fire.

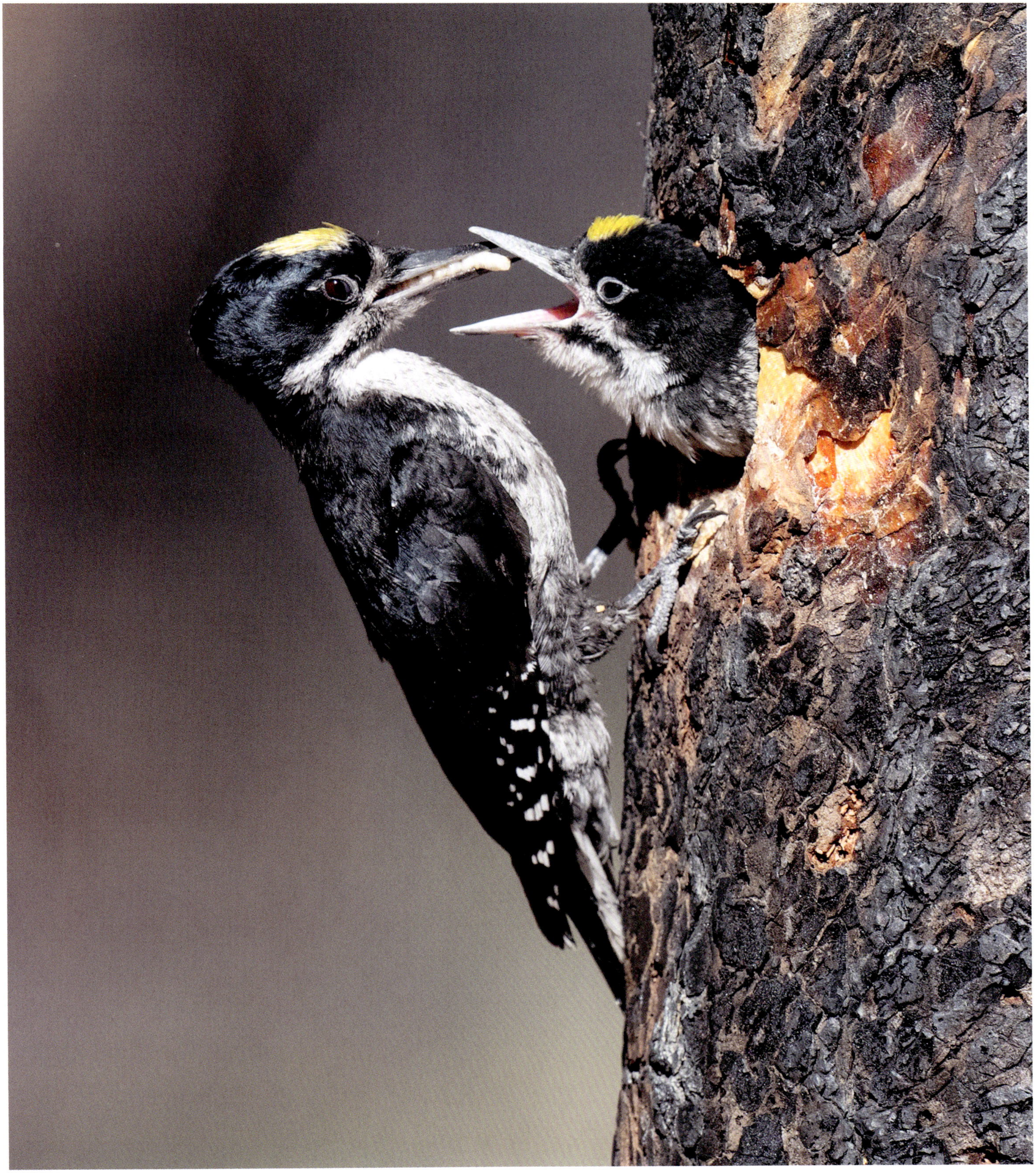

WOODPECKERS AND FIRE

Fire has always been part of restoring and creating habitat, especially in dry forest ecosystems. Woodpeckers benefit first and play a keystone role in preparing that habitat for use by other species.

Oak, pine-oak, and mixed-conifer habitats are richest when giant fire-resistant trees dominate over open understories. In a healthy system, flames periodically remove smaller trees—their competition for sun, water, and nutrients. Over the years, such burns leave a mosaic of habitats with varying degrees of grassy openings, canopy coverage, and tree species composition and age. This diversity helps meet the needs of a variety of species over a large and ever-changing landscape and allows individuals to move relatively short distances to find food and shelter.

Pyrodiversity is the landscape-level spatial and temporal variation in fire impacts and connectivity between burned and unburned forest within fire-maintained landscapes. It results in the greatest conservation benefit for dry-forest specialists, including many of North America's most at-risk species, such as Red-cockaded, Lewis's, and Red-headed Woodpeckers; Williamson's Sapsuckers; White-headed Woodpeckers; and irruptive species such as American Three-toed and Black-backed Woodpeckers. Unburned or lightly burned areas adjacent to high-severity burns may provide invaluable refuge from heat and predators as well as foraging habitat for some woodpeckers.

Since the early 1900s, forest managers harvested the largest, most fire-resistant trees and attempted to extinguish fire quickly, creating swaths of overgrown forests cluttered with thin, undersize, single-age stands and a plethora of dry, woody debris, degrading the habitat and increasing the risk of uncharacteristically severe fire. Fortunately, fire is now used as a tool to improve habitat and increase fire resilience, although we must prioritize creating new snags, critical elements that allow woodpeckers and other cavity nesters to breed.

Woodpeckers benefit from fire in different ways. Most species, including American Three-toed, Arizona, Black-backed, Downy, Hairy, Pileated, Red-bellied, White-headed, and especially weak excavators like Red-headed Woodpeckers; Northern Flickers; and particularly Lewis's Woodpeckers, benefit from the fact that wood-decay fungi soften trees that are killed by fire, making them easier for woodpeckers to excavate. Lewis's, Acorn, and Red-headed Woodpeckers are the most notable beneficiaries of more open habitats that make fly-catching insects more productive and increase crops of acorns, nuts, and berries, which allows the birds to produce more young and at times forestall migration.

Natural burns have historically arrived in late summer, an ideal time for wood-boring and bark beetles to swarm in and feast on the vulnerable trees, mate, and lay eggs. These swarms provide vital food for Black-backed, American Three-toed, and Hairy Woodpeckers and for dispersing woodpeckers of many species. Post-fire prey populations typically peak for the first two to three years, but sometimes more, depending upon the region, forest type, timing of the fire, and beetle species.

While woodpeckers are opportunistic and adaptive, some show subtle differences in where they forage relative to one another, especially

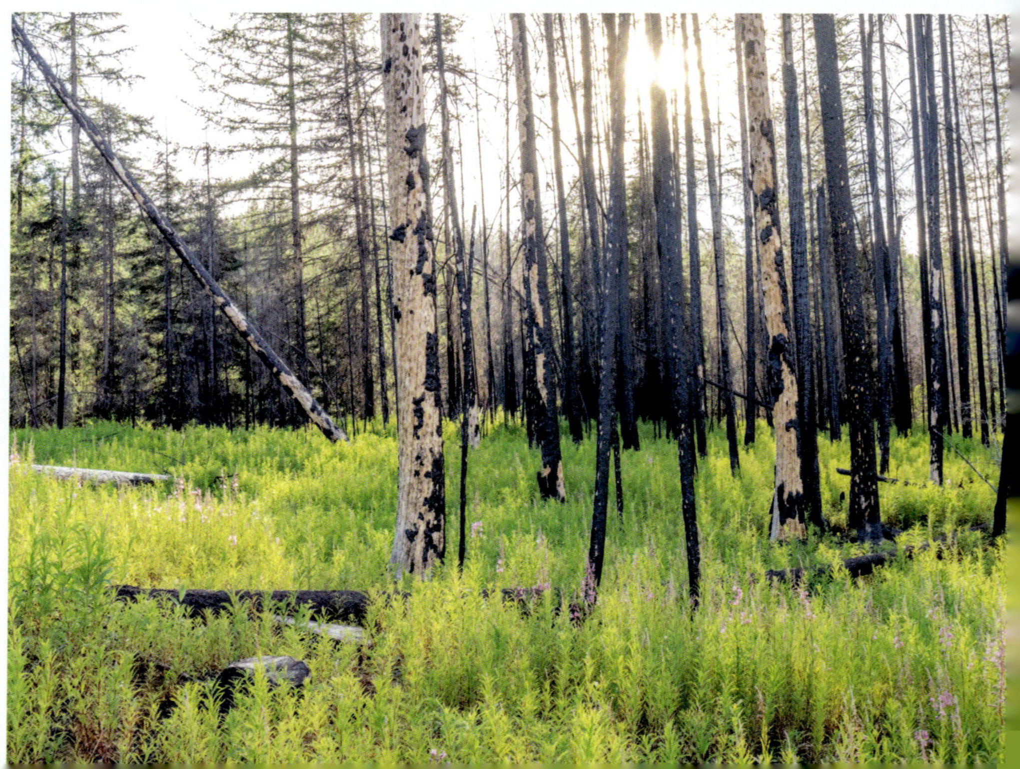

Fire is a necessary and beneficial ingredient in maintaining the health of our forests. In addition to creating more habitat diversity in the long term, it also creates new snags for nest sites for all species and provides an exponentially greater food supply for woodpeckers that eat beetles and their larvae. The pale, barkless patches on the trees result from Black-backed and American Three-toed Woodpeckers removing the bark to access beetle larvae.

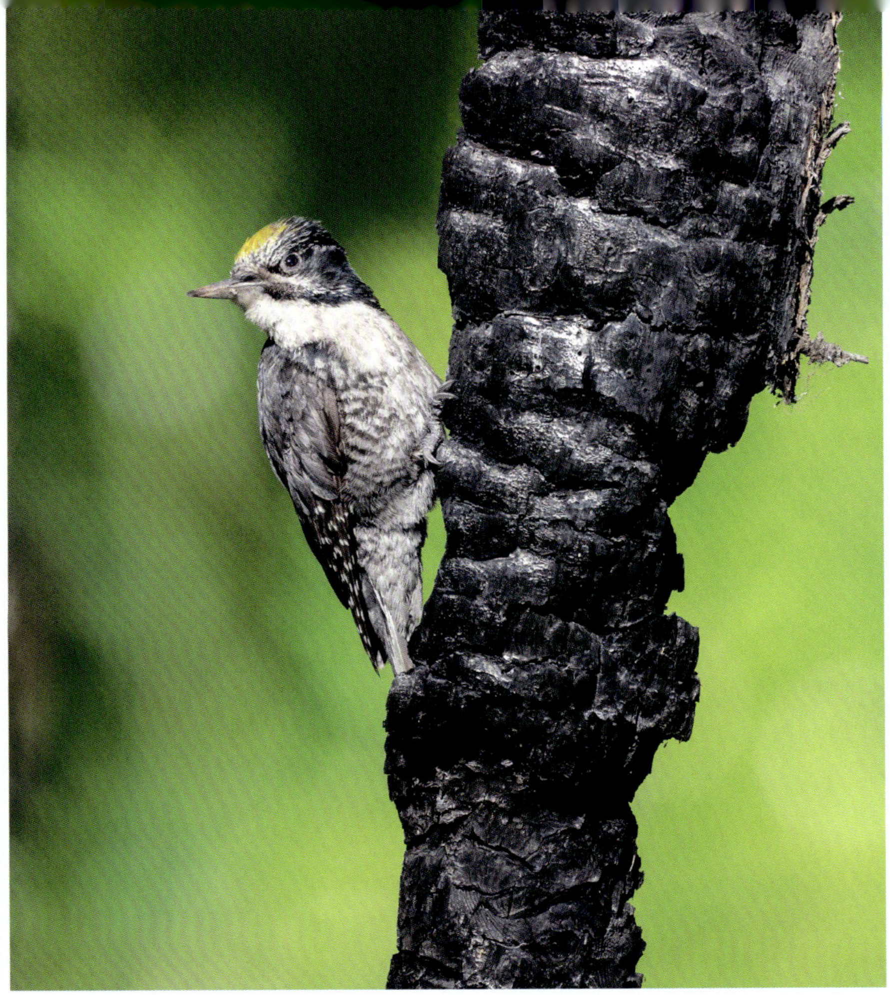

Wood-boring beetles converge upon recently burned forests where their larvae attract large numbers of woodpeckers and are of special importance to Black-backed and American Three-toed Woodpeckers (shown here).

when they co-occur. Black-backed Woodpeckers select more wood-boring beetle larvae on more severely burned trees compared to American Three-toed Woodpeckers, which consume comparatively more bark beetles in less burned trees, higher up and in denser stands. Lewis's, Red-headed, and Acorn Woodpeckers and Northern Flickers quickly benefit from the changed habitat of grassy, shrubby, or herbaceous openings punctuated by tall, isolated hunting trees and snags, allowing these fly-catching woodpeckers to hunt the winged insects that prefer the more open habitat. Flickers find their favored ant prey more easily on the open ground.

Our management of the pre- and post-fire landscape, exacerbated by a warming climate, has made fires more severe. The hotter, drier weather adds to the fuel already built up from decades of fire prevention. Shorter winters allow beetle pests to complete more breeding cycles in a single year than before, killing more trees. Our prevention of all fire, combined with the removal of the largest, most fire-resistant trees, has made forests vulnerable to the slightest spark from lightning or, more commonly, human activities. Under such conditions, conflagrations burn hotter, spread farther, and consume exponentially more trees.

Larger, higher-severity burns result in less pyrodiversity, reducing habitat complexity and the refuge from heat and predators provided by less severely burned or unburned trees, potentially reducing the number of woodpeckers the burn can sustain. Wood-boring beetles respond readily to fires occurring in the summer, but human-ignited fires are occurring more frequently outside of that period and may be less likely to attract beetles and provide food at the right time for dispersing woodpeckers. Pyrodiversity creates more opportunities for post-burn tree mortality and thus more cycles of beetles and rot to benefit more woodpeckers for a longer period of time.

Other management efforts can also do harm. Removing fuel pre-fire or salvaging logging within three to five years of a fire can take too many large trees and snags needed by woodpeckers for nests and feeding substrate. Machinery and roads used in salvage logging can disturb soil and watersheds and forestall the recovery of healthy habitat. To make matters worse, rather than creating snags from what timber remains, burns are often salvage logged, the timing and severity of which can prevent woodpeckers from nesting and negatively impact other elements of habitat recovery. Such operations frequently only retain green trees and rarely enough snags.

Woodpeckers, as a keystone species, can help bring life into a burned forest as long as a large number of softened snags and an ample food supply allow woodpeckers to thrive. A large number of woodpeckers feast in burns, drawing unusually large numbers of uncommon species such as American Three-toed and Black-backed Woodpeckers. Black-backed Woodpeckers can breed in exponentially higher concentrations, producing significantly more young that can disperse into new territories and await the next forest disturbance, and may even help control some smaller insect outbreaks.

DISPERSAL

Most adult woodpeckers remain on their breeding territories year-round. As young leave the nest, the movements of adults are limited, but as the young mature and become more agile, adults forage over a larger part of their home range. Provided there is ample food, most remain relatively close to the previous year's breeding location year-round and retain that breeding territory the next year. There are examples of adults dispersing when food supplies are low or after previous nesting efforts have failed, but we do not know how often this happens as this stage is so little studied.

Young are typically at least partially dependent upon their parents for three to five weeks, although this period can reportedly be as short as one to two weeks for some sapsuckers and as long as several months for Pileated Woodpeckers. Most juveniles continue to forage very close to natal sites through late summer and even into fall, provided food is plentiful.

Very little is known about how far juveniles disperse for most species. The supply of food and available potential nest sites are important factors. There are examples of individual woodpeckers dispersing dozens of miles or more, but with few studies on most species, we cannot say for certain what is typical for each. Sometimes woodpeckers may travel much farther in what is akin to a multiyear dispersal before they establish a territory or find a mate in a subsequent year.

In the most prey- and snag-rich habitat, such as a recent burn, first-year American Three-toed or Black-backed Woodpeckers may nest extremely close (<1 mile) from the nests from which they fledged. In other situations, a Black-backed Woodpecker may disperse more than thirty miles. Juvenile Northern Flickers from migratory populations may disperse more than one hundred miles from breeding sites, which is likely much farther than a resident Northern Flicker would be expected to disperse.

Acorn and Red-cockaded Woodpecker life histories have been studied more than any other woodpeckers, and their cooperative breeding groups include juveniles that will need to disperse in order to have a chance at contributing to the gene pool. Red-cockaded Woodpecker breeding groups sometime retain juvenile males as nonbreeding helpers, but most females disperse up to 3.6 miles by their first fall, depending upon the quality of habitat and availability of suitable nest cavities. Some males disperse later.

Acorn Woodpeckers also include juveniles who stay on as nonbreeding helpers, with males more likely to remain as part of the group. Individuals usually disperse in same-sex groups of one to three birds, with females usually leaving by the end of the third year and those males that disperse, by the fourth or fifth year. Females disperse farther and more aggressively compete for vacancies in other groups that may be nine miles or more from their natal site.

Berries are an important supplemental food to the Red-naped Sapsucker's diet of sap and ants, particularly during fall and winter. This individual can be identified as a juvenile due to its lack of a solid black bib and a male because of its red chin.

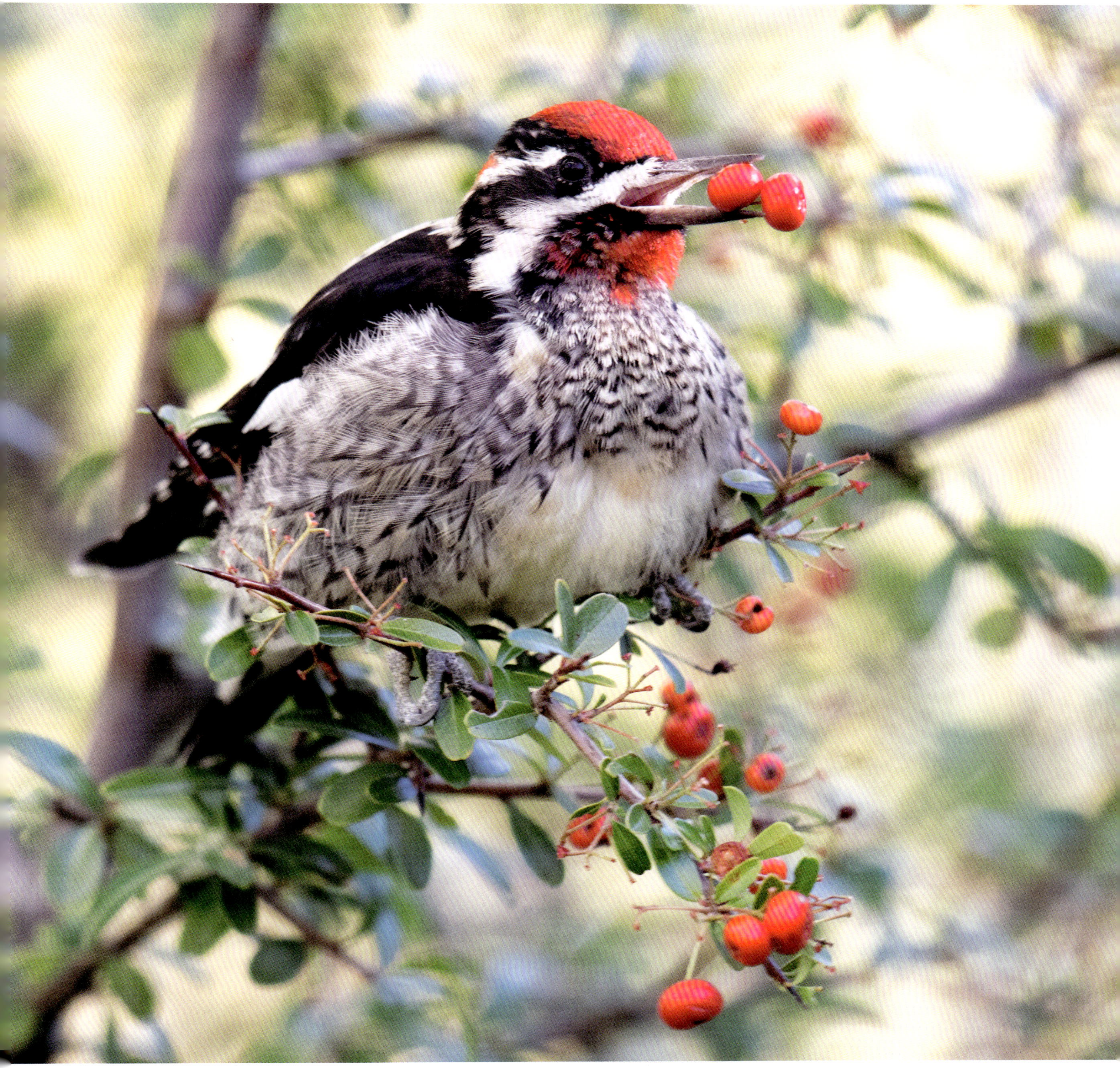

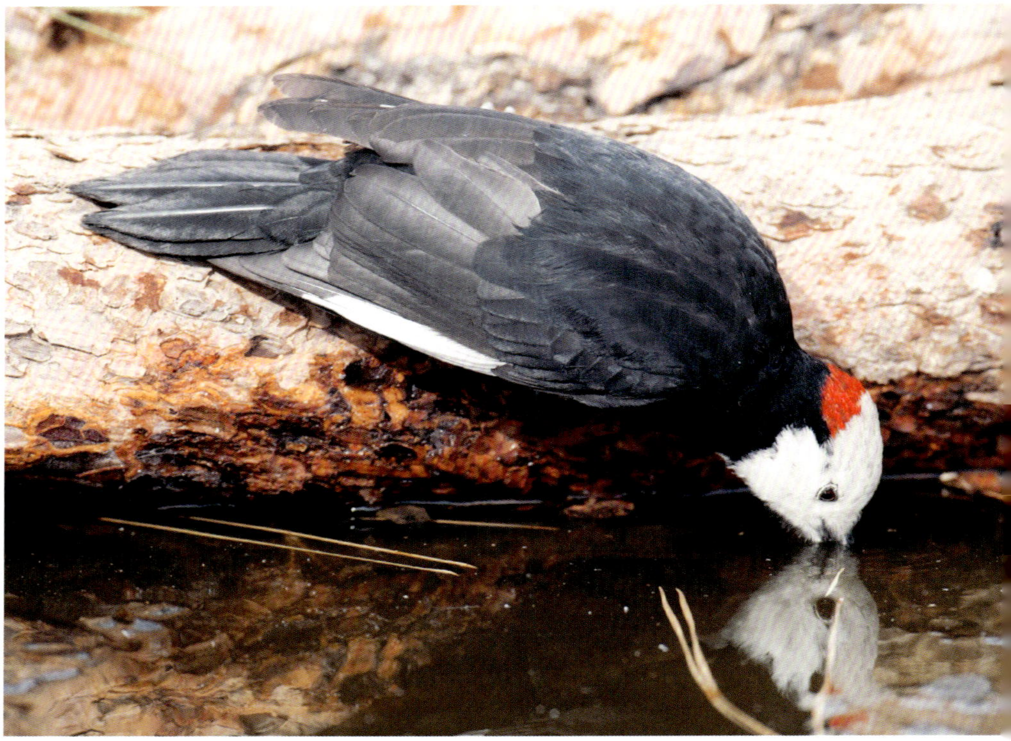

SHARED FALL RESOURCES

During the late summer or early fall, when juvenile woodpeckers are no longer dependent upon adults but have not dispersed, the combination of both adults and one or more broods of juveniles from one or more species can result in an unusually large number of woodpeckers in a small area.

Water

In all but the subtropical and tropical parts of North America, woodpecker dispersal occurs at a hot and dry time of the year, early summer into fall, which also coincides with their consumption of drier foods such as acorns, seeds, and nuts. While water is utilized year-round, these factors, plus the greater number of woodpeckers at this time of year, make this a more commonly witnessed event at this time.

Most woodpeckers are not often seen drinking or bathing. They acquire most of their water through the foods they eat or from rainwater or dew collected from cavities and depressions in trees, on leaves, or even snow on limbs, which presents much less risk than venturing to the ground and exposing themselves to predators.

Most of my drinking and bathing observations have come in the hot days of late summer and fall, when I have had the chance to witness Northern Flickers; Red-breasted and Williamson's Sapsuckers; and Hairy, Pileated, Lewis's, and Acorn Woodpeckers bathe and drink from ponds, puddles, streams, and lakes. I also know that Golden-fronted, Red-headed, and Gila Woodpeckers do the same.

Making a visit to the ground is dangerous, so it was not surprising to see Lewis's and Acorn Woodpeckers go on this adventure exclusively in groups. On several occasions I observed Lewis's Woodpeckers assemble on branches above a puddle. With a few watching, one would tentatively work its way, foot by foot, lower and closer to the water source. Each would do this in turn, but normally at least one waited until a satiated bird flew into a watch position before dropping down for a drink.

Acorn Woodpeckers are similarly careful. I watched a group fly to a small island in a pond, where most waited in the willows as sentries, and others inched their way down partially submerged sticks, ready to flee at any alarm. I have also seen White-headed and Acorn Woodpeckers drink from much safer raised public drinking fountains and Arizona and Ladder-backed Woodpeckers sip from birdbaths.

Some woodpeckers derive much of their water from the food they eat, but many such as the White-headed Woodpecker (male pictured here) frequently drink from streams, ponds, knot-holes, birdbaths, and even sprinklers.

Food

Three North American woodpecker food bounties—fruit, insect outbreaks, and large acorn (or mast) crops—often reach their peaks just as the young of several beneficiary species are beginning to disperse from their natal territories. Depending upon the quantity of food, this timing may reduce the distance they need to travel to establish winter or even spring territories.

RIPENED FRUIT

In the United States and Canada, apples, pears, dogwood, and other fruit begin to express their climax color just as juvenile woodpeckers are transitioning to adult plumage and provide easy food for the inexperienced young.

The bountiful apple harvests of Washington state's eastern Cascades attract both tourists and local pickers, but I was drawn to one particular group of trees near an oak grove where several species of woodpeckers nested. I was hoping to see Northern Flickers, but I was not expecting to see the Lewis's Woodpeckers that joined them in working whichever tree was farthest from humans. The Lewis's would perch on a branch within lunge reach of the apples and poke at the softest parts of the fruit. The flickers scavenged on the ground for what the Lewis's dislodged. The flocks of Lewis's included both juveniles and adults beginning their fall dispersal. While there were a lot of skirmishes when woodpeckers were too close, sometimes several shared a single tree.

Juvenile Gilded Flickers follow their parents away from nest sites shortly after fledging and remain dependent for a few weeks. While they sometimes eat saguaro pollen, which this adult male is collecting, they primarily consume ants and termites.

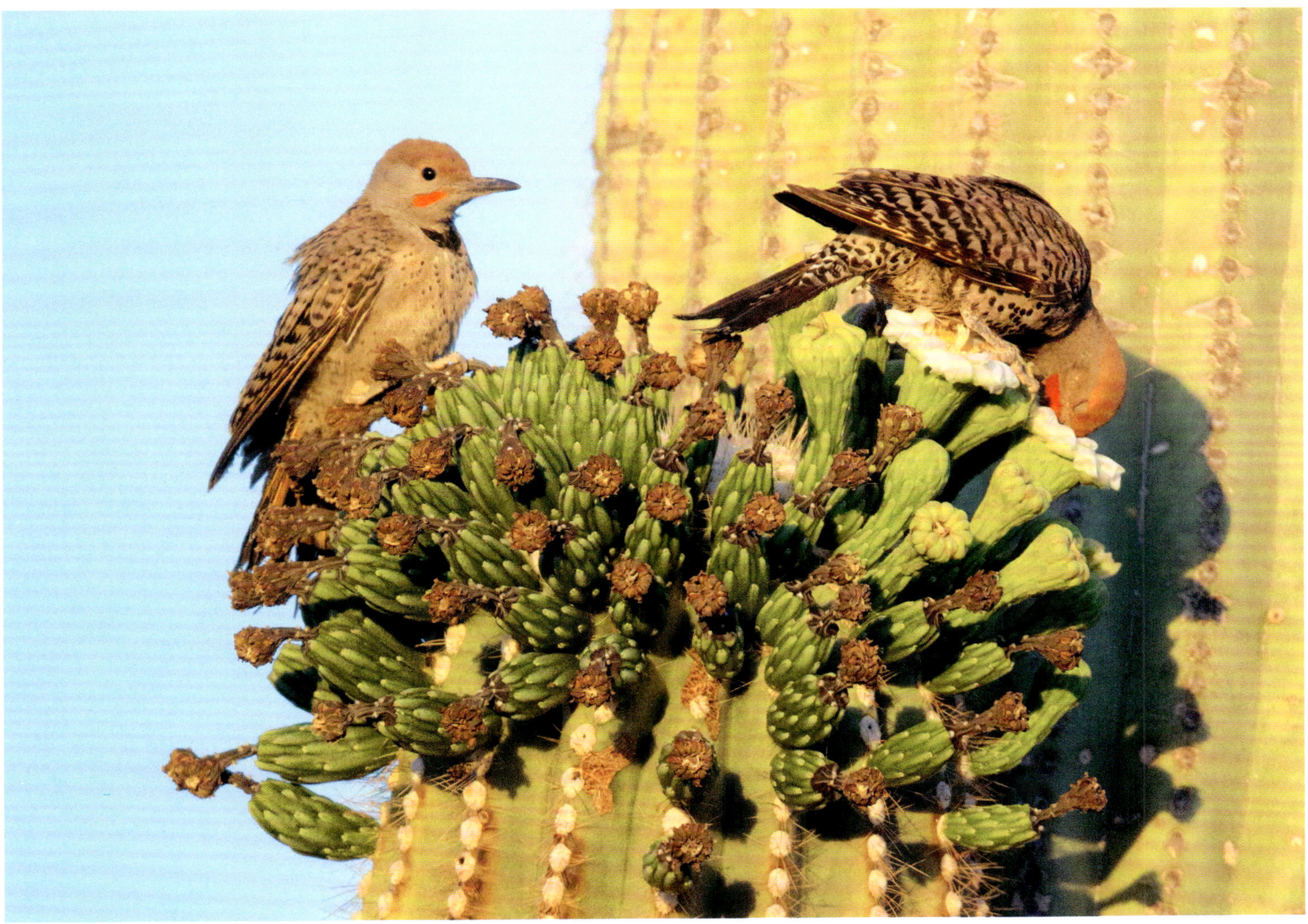

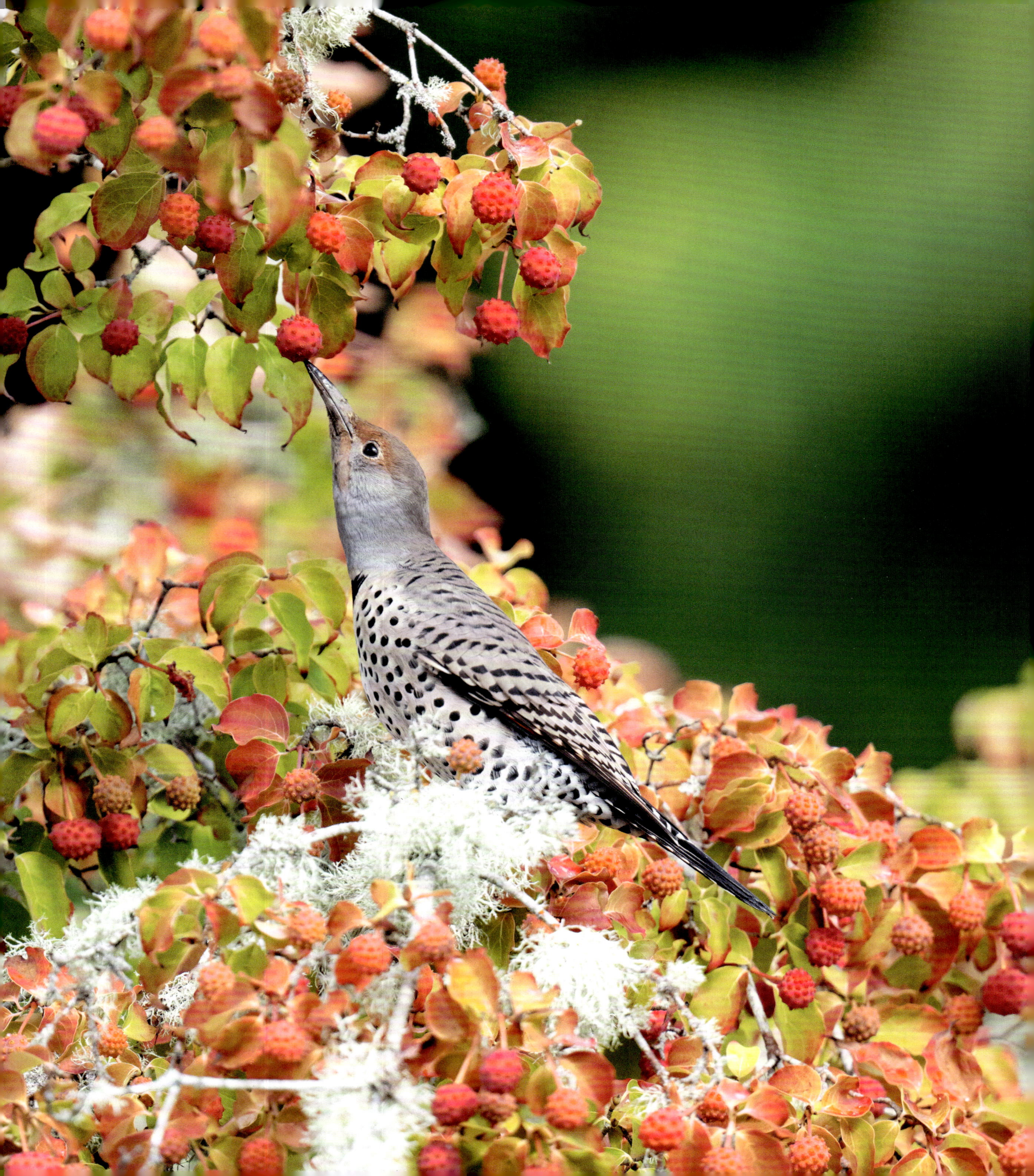

In a backyard on the west side of the Cascades one October, I observed hundreds of red puckered balls of dogwood fruit hanging among leaves transitioning from green to pink, creating a compelling canvas for juvenile woodpeckers emerging into their own best reds to join adults of the brightest crimsons. Red-breasted Sapsuckers, Northern Flickers, and Pileated Woodpeckers joined in the feast. Each would climb on a branch with a cluster of ripe fruit and project its bill through the leaves of its shady hideout to pluck the choicest fruit.

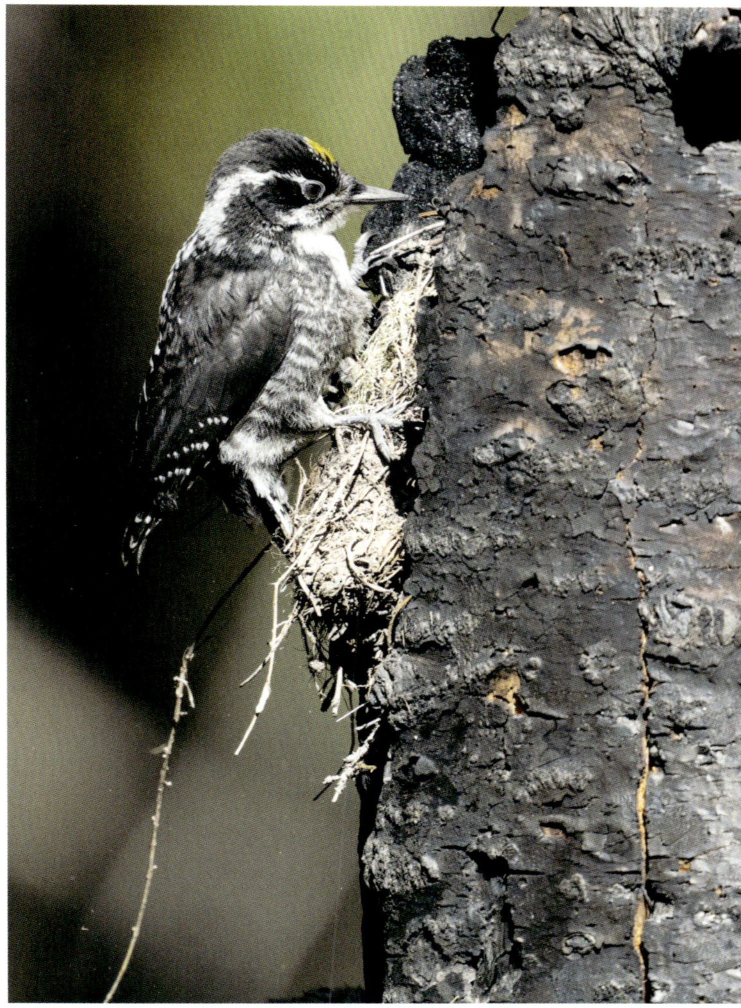

BEETLE OUTBREAKS

Beetle outbreaks often occur after trees are killed or damaged by drought, flood, or fire. Forests infested with wood-boring beetles provide an abundance of food to woodpeckers in the first three to five years after a fire, resulting in exponentially more fledglings per acre than in unburned areas (ten to twenty times more for Black-backed Woodpeckers in some instances). This allows newly independent juveniles of some species to eventually nest very close to where they fledged.

It is not known exactly how American Three-toed or Black-backed Woodpeckers find areas with such concentrations of bark and wood-boring beetles. It is known that some beetles, including the wood-boring beetles, somehow sense smoke and heat, allowing them to find fires, but how do the woodpeckers find the beetles? Research suggests that when young Black-backed Woodpeckers disperse, they may travel up to thirty miles seeking foraging habitat, and some may come close enough to outbreaks to find signs of their favorite prey. Some have speculated they hear the prey, while others hypothesize that they can detect concentrations of beetle pheromones under the right conditions.

American Three-toeds reproduce in unburned spruce and lodgepole habitat and Black-backed in mixed-conifer forests. If they find the concentration of their target prey in dying, dead, or diseased forests, they will find a plethora of food and breeding sites and have the luxury of wandering smaller home ranges, defending more concentrated territories, and producing more fledglings. American Three-toeds are reported to be able to nest within three hundred yards of the nests from which they fledged in such circumstances, and I observed three Black-backed nests within six hundred yards of one another in a recent burn. These two species will nest in areas without insect outbreaks, but in such places they require larger territories and home ranges to meet their food needs.

Juvenile woodpeckers instinctively react to insect prey, although it takes a while for them to master where to look and how to best access their targets. This juvenile male American Three-toed Woodpecker found himself covered in a cloud of dirt after pecking at an old bird nest.

Opposite: Northern Flickers like this adult female feed primarily on ants on the ground, but they opportunistically take advantage of ripe fruit and berries, including those from native and ornamental dogwood.

MAST YEARS

Every few years, trees or shrubs produce a banner crop of fruit, nuts, or acorns—a *mast year*. Oak woodland and savanna habitats can produce a rich bounty of food for several species of woodpeckers at the very time that their young are dispersing. An Acorn Woodpecker nesting group will work together to collect and store as many acorns as they can, but Red-headed, Lewis's, and Red-bellied Woodpeckers show little tolerance for their young once they begin collecting and storing their acorns or other mast and will drive them away from such places with nearly as much vigor as they would an unrelated bird.

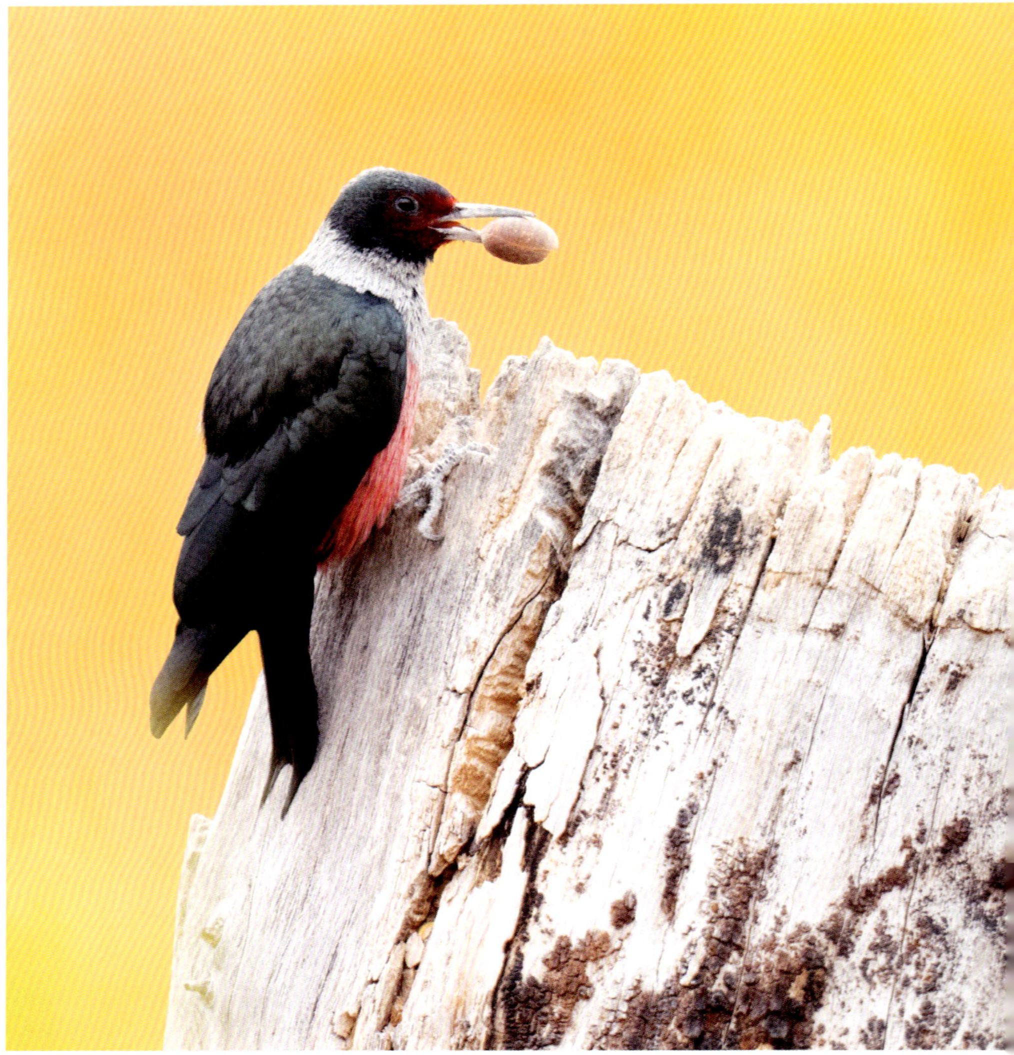

A Lewis's Woodpecker arrives at an anvil to break up an acorn and store smaller pieces in cracks and crevices in the anvil snag and elsewhere.

On this cold Minnesota morning, a gust of wind swept through the pin-oak stand, and the response sounded like a hailstorm, as dozens of hard acorns ripped through the layers of orange and yellow leaves, then thumped and bounced on the ground. I eyed dozens of birds either disappearing into the canopy or dropping from the trees in awkward, weighty flight with a prize in their bills. In a frenzy, Blue Jays, Red-headed Woodpeckers, and Red-bellied Woodpeckers worked diligently to collect as many acorns as they could before wind knocked their prizes to the ground, where they might be lost to squirrels or turkeys or be devoured by the woodland floor.

Within the canopy, Red-headed Woodpeckers, bills open like clamps, yanked acorns from twigs with a quick twist, or grabbed them with their claws to apply their full weight as they flapped, falling backward into flight. Acorn-laden birds dropped precipitously from treetops into labored flight toward snags, where they attempted to process or store acorns.

If these birds were fortunate, their home ranges may have had more than one species of oak with harvestable acorns, with the fruit of each tree falling at a different time and creating multiple windows for collecting food, improving their rate of survival. More oak species also increase the chance that at least one will have a productive crop.

OAK HABITATS

Oak habitats provide plenty of year-round food, with a plethora of insects in the spring and summer and acorns in the fall that, if stored, last into the winter. Periodic fire helps maintain large openings and creates snags that can be infected by rot and serve as easily excavated nest sites for several species of woodpeckers.

Mature oaks appear like sculptures scattered across a canvas of golden grass. Oaks are slow-growing, fire-resistant trees that can live more than a thousand years, often reaching over one hundred feet tall. In centuries-old live oaks, the gnarled limbs can stretch one hundred feet laterally from eighteen-foot-wide trunks. This keystone tree provides critical food, shelter, and nutrients to hundreds of other plants, mammals, insects, and birds. Oak species can be found from southern Canada through most of the United States, throughout Mexico, and even into western Cuba. Canopy cover can vary from 10 to 30 percent for a savanna to as much as 50 percent in woodlands before transitioning to forests, where the largest oaks succumb to the shade of faster-growing conifers.

Open oak habitats are of particular importance for four North American woodpeckers: two of our most threatened, the Lewis's and Red-headed, and the Acorn and Nuttall's. All four are indicator and keystone species because they provide nest cavities for birds such as Western and Eastern Bluebirds, American Kestrels, Northern Pygmy-Owls, Western and Eastern Screech-Owls, Wood Ducks, Oak Titmice, and even flickers.

Oaks attain the zenith of their form, size, age, and ecological value in oak savanna habitats. Their long taproots allow them to rise up out of the surrounding warm, dry grasslands as islands of rich arboreal habitat surrounded by equally valuable grasslands or sagelands, providing winged inhabitants access to the bounty of each. Here, grass and oaks come together to form our most insect-rich habitat and provide food for fly-catching woodpeckers such as Lewis's, Red-headed, and even the Acorn Woodpecker, all of which may utilize this technique to capture 30 to 70 percent of their food during some parts of spring and summer.

Open areas are also rich in berry foods such as currants. During winter, some woodpeckers that do not excavate for insects or their larvae rely on stored mast of acorns for the vast majority of their food, making these areas among the preferred fall and wintering habitat for Red-headed Woodpeckers in the East and Lewis's Woodpeckers in the West. Acorn and Nuttall's Woodpeckers rely on mature oak habitats year-round. Acorn Woodpeckers thrive where there are multiple oak species as a hedge against devastating acorn crop failures. Nuttall's glean and probe for insects year-round.

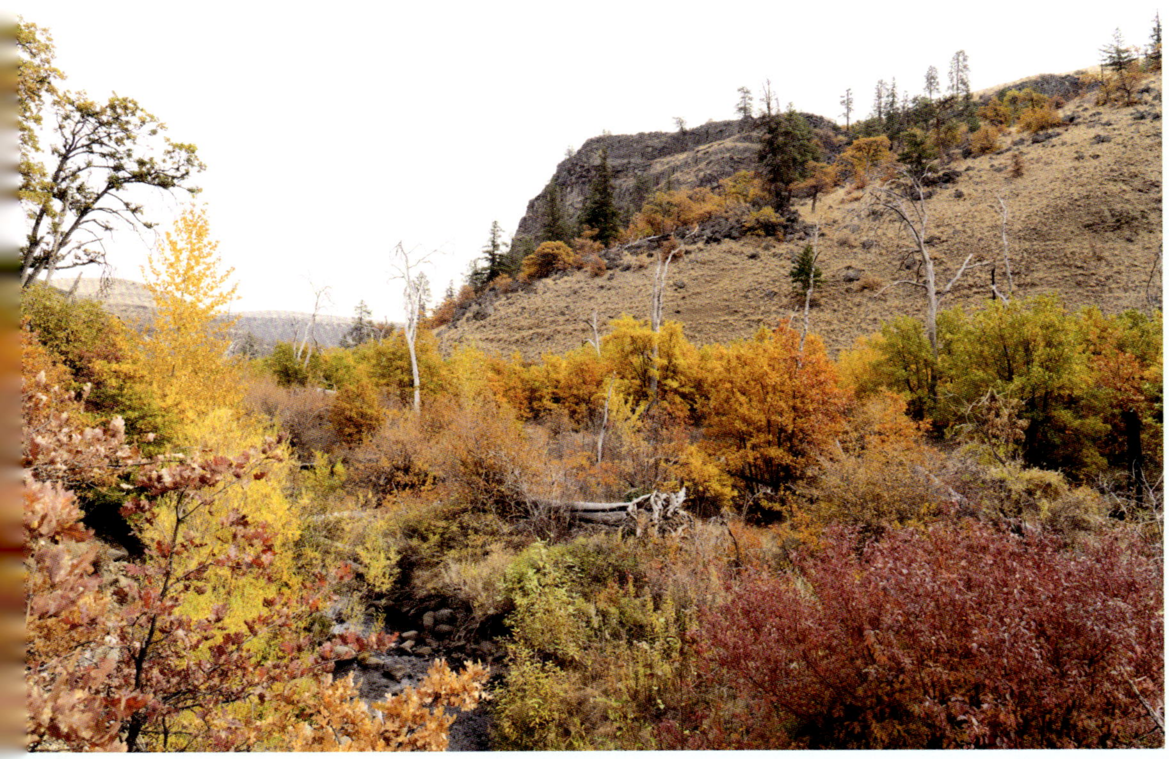

Healthy oak habitats are threatened by fire suppression, which increases the risk of uncharacteristic fire, grazing that prevents regeneration, and agriculture and climate change and the resulting lowering of the water table beyond the reach of oak taproots.

STORAGE BEHAVIORS

Several woodpecker species store hard mast (acorns, nuts, seeds, and grain) temporarily when they find a rich food source. For Acorn, Lewis's, and Red-headed Woodpeckers, and, to a much lesser degree, Red-bellied, their ability to store enough food may impact whether they remain on a territory as residents or leave for the winter, how early they breed the next year, and how many clutches and young they will produce. This draws them to oaks and the acorns they provide during fall to sustain them until spring, when insects are plentiful again.

Many woodpeckers bring the hard mast to an *anvil*, a sturdy surface where they can break up larger items without the resulting smaller bits being lost in the process. The anvils I have seen are relatively flat surfaces on the top of broken-off trees, especially those recessed slightly so that the broken pieces form a small pile. Cracks in trunks, branches, or bark furrows are also used to hold the food being processed. White-headed and Gila Woodpeckers use similar anvils to break up large insect prey or seeds and nuts to feed their nestlings.

Red-bellied Woodpeckers are for the most part nonmigratory generalists that opportunistically store mast, so it is not surprising that they utilize the most rudimentary storage strategy of these four *Melanerpes* woodpeckers. As *scatter hoarders,* they store their collected berries, seeds, corn kernels, nuts, acorns, and other food over a broad area that may be adjacent to the collection site or several dozen yards away. They store these foods in cracks, holes, and crevices in wood and use their long tongue, which extends more than an inch and a half beyond the bill tip, to push the pieces two to three inches into the openings and pound them in with their bills, making them more difficult for other birds to access. They do not guard or otherwise protect these locations.

Lewis's Woodpeckers are able to forestall risky migrations by being more sophisticated *larder hoarders* that concentrate their food storage in a small area that they defend against other woodpeckers, jays, nuthatches, chickadees, squirrels, and chipmunks. Several individuals may share a storage tree when such trees are limited, but they will defend the part of the tree or branch they use from others of their species.

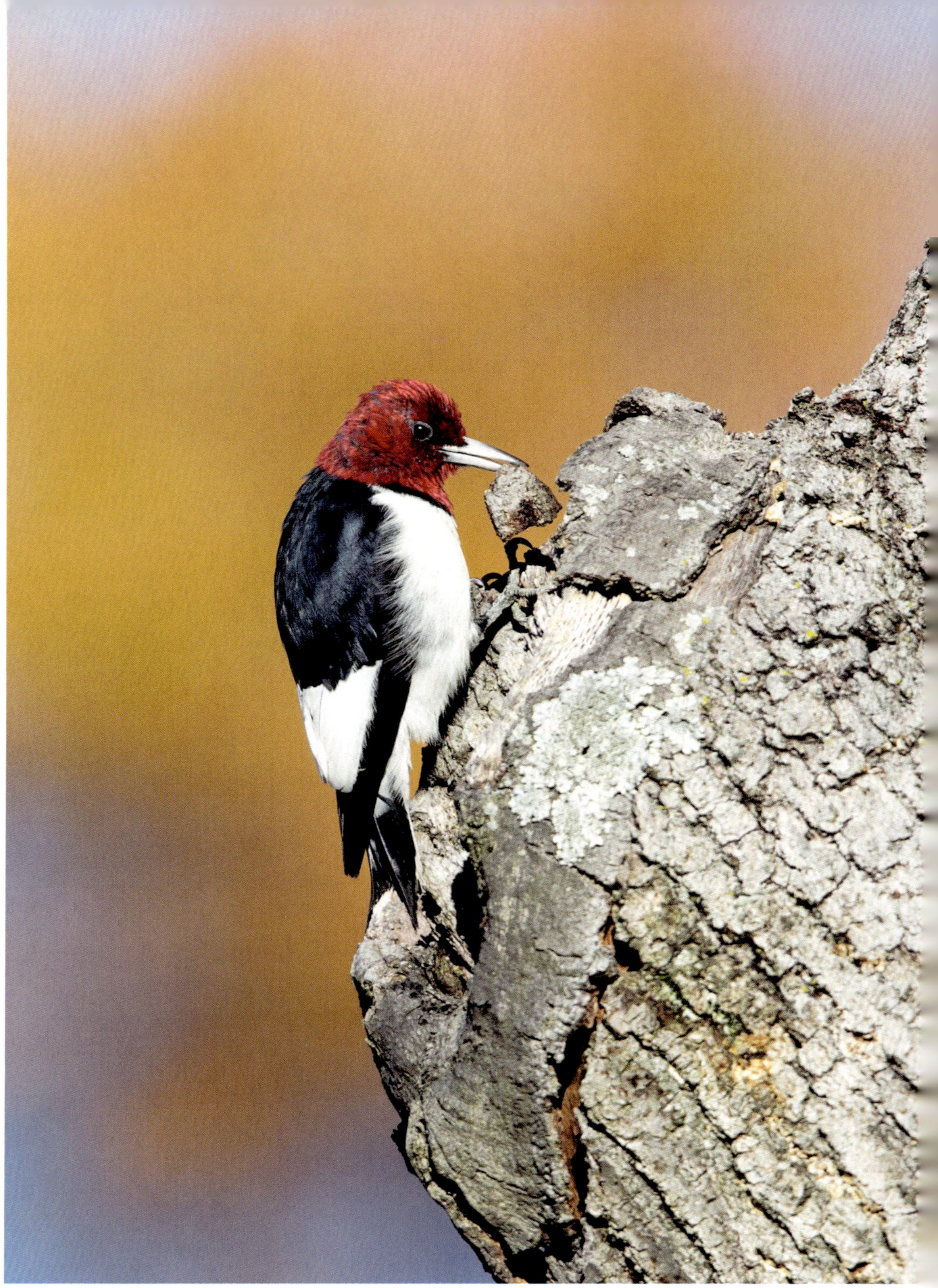

Like Lewis's Woodpeckers, Red-headed Woodpeckers break acorns into pieces before storing them for long periods of time. Red-headed Woodpeckers are the only species that covers its cache, as this one is doing with a piece of bark.

When mast such as acorns, nuts, and grains are suddenly abundant, Red-headed Woodpeckers collect as many as they can and store them in cracks and holes in a concentrated area. Later they break their prizes into pieces and move them to more secure locations over a wider area.

Lewis's utilize existing cracks, crevices, and holes in several adjacent or nearby snags, and pieces of nuts and acorns fill every opening during good mast years. The same storage trees are used in subsequent years. Mated Lewis's pairs sometimes contribute to and manage a larder together, but more commonly each takes care of its own.

The availability of storable food, much more than weather, determines where individuals and often large groups of Lewis's Woodpeckers will overwinter, wander, or migrate. I have seen some migrate from areas with moderate winters but low acorn crops, and others stay all winter in areas with freezing nights and occasional snow after a strong acorn crop. When they do overwinter, they spend a great deal of time managing their storage, turning, flipping, or even moving food to better-fitting locations. It is likely these manipulations reduce decay and spoilage. They may prefer to eat other foods, but the hard mast from these storage trees comprises most of the food they will consume until spring, when non-wood-boring insects are available, at which time the nut stores are largely neglected. These same storage areas are sometimes used during the nesting season to hold insects, fruit, and other food for shorter periods during times of plenty.

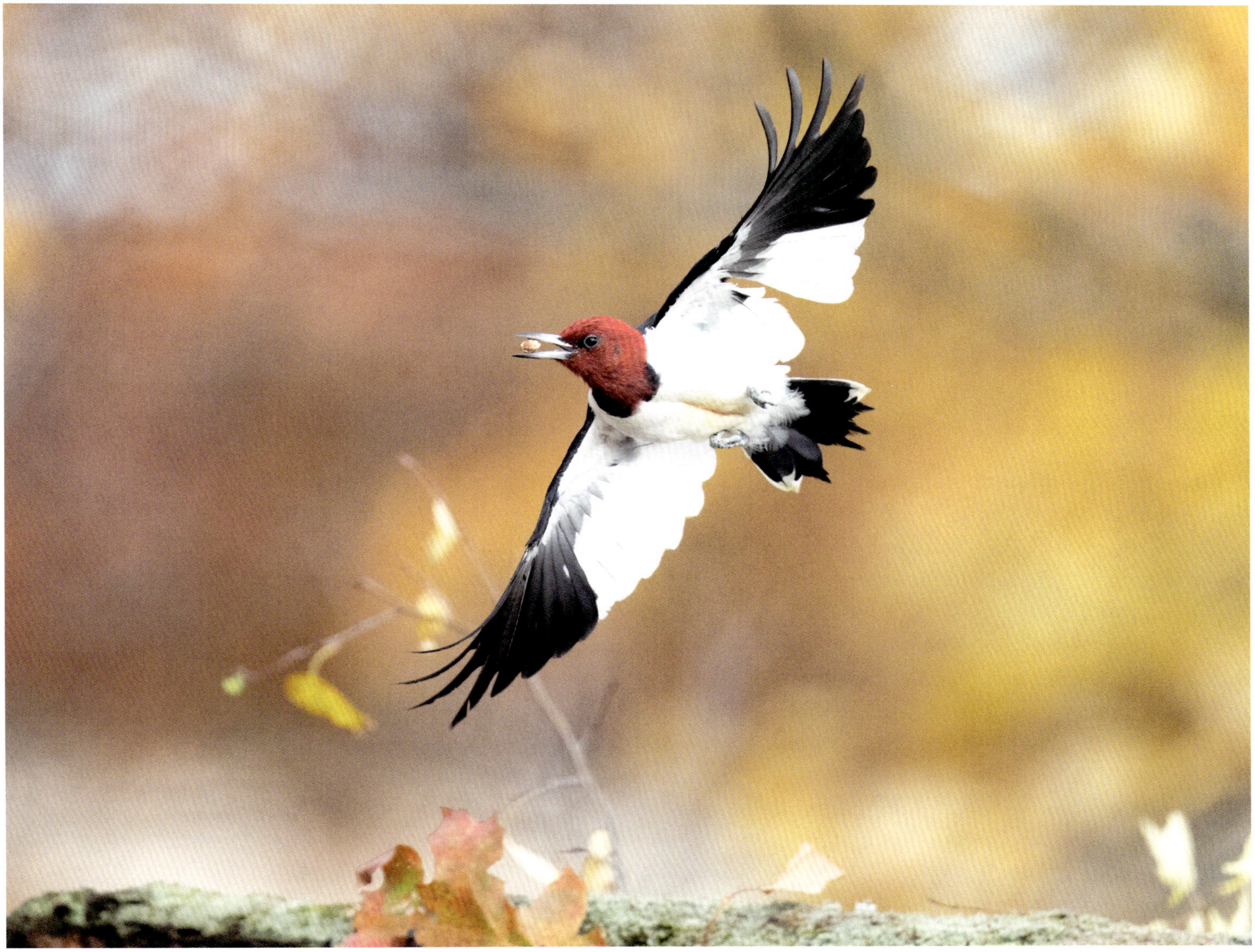

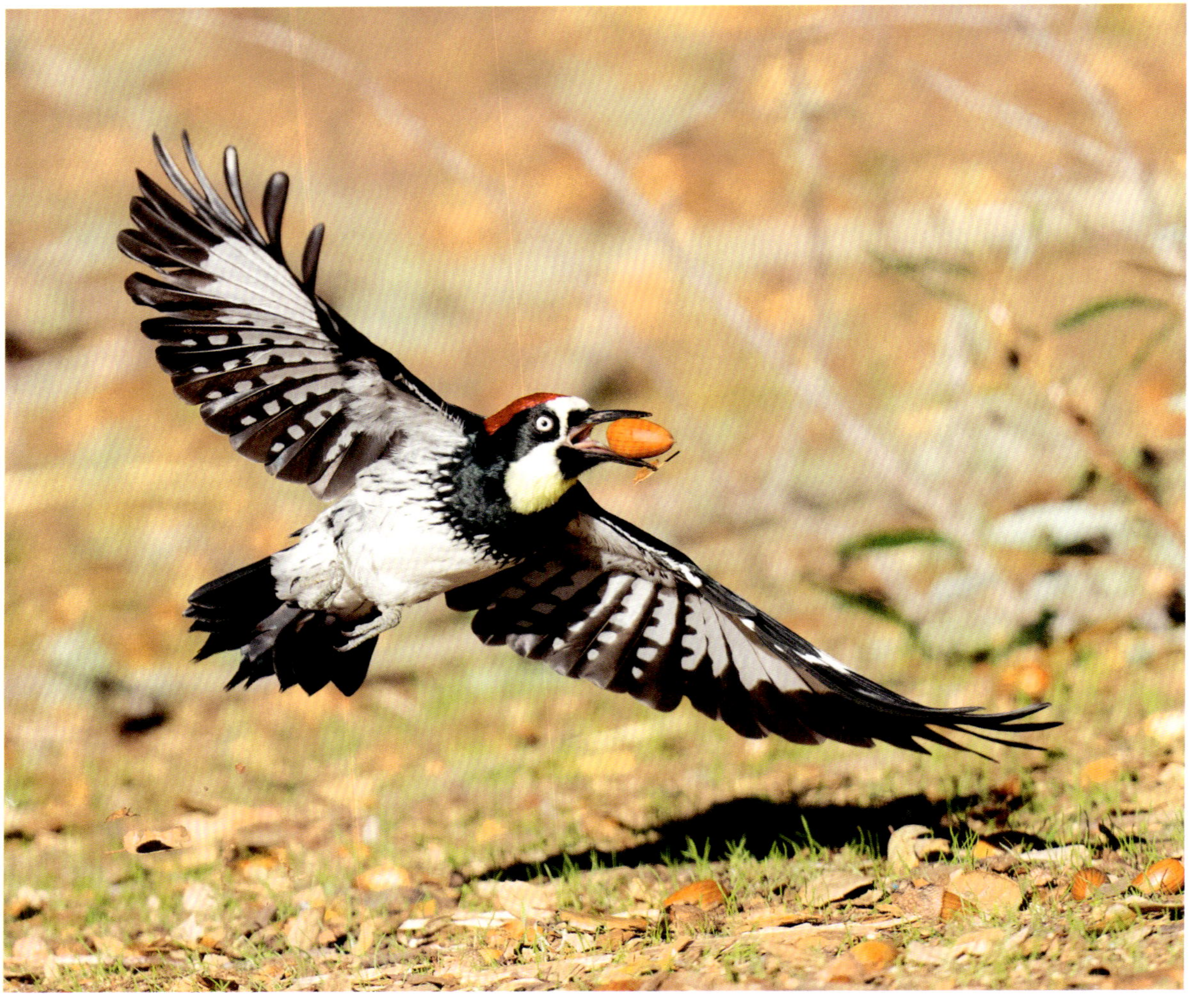

After Acorn Woodpeckers have exhausted the acorns in the trees, they may pluck freshly fallen ones from the ground to eat or to store in their granaries.

Opposite: Hundreds or thousands of acorn-size holes in the thick bark of an oak or pine tree form Acorn Woodpecker granaries, like the one this male is peeking out from. The breeding success of the family group is correlated with how many acorns they can collect, so it serves them to continue carving holes; some granaries may include up to fifty thousand!

Red-headed Woodpeckers are unique in that they use both the scatter hoarding and larder hoarding behaviors. When a bounty of nuts, seeds, or acorns appears, they are strictly larder hoarders, storing their prizes in existing cracks, crevices, furrowed bark, and knotholes in almost the exact same fashion as the Lewis's Woodpeckers, in a concentrated group of snags or trees, and defend these sites against all others, including their recent young.

Once most of the readily available mast has been stored, and the competing mast collectors have dissipated or dispersed, Red-headed Woodpeckers become scatter hoarders. At this time, whole nuts or acorns are taken to anvils to be broken up, and those and the other stored segments are moved to new sites over a wider area where they are pounded in so tight that non-woodpeckers would have a hard time removing them.

Larder hoarding allows Red-headeds to focus on collecting as much as they can when competitors are most active, and the scatter hoarding makes storage sites less vulnerable to a massive loss in a single raid and reduces the spread of fungal decay.

The Red-headed Woodpecker's storing behaviors differ from those of the Red-bellied and Lewis's in two other ways. Red-headed Woodpeckers frequently conceal important larders by covering

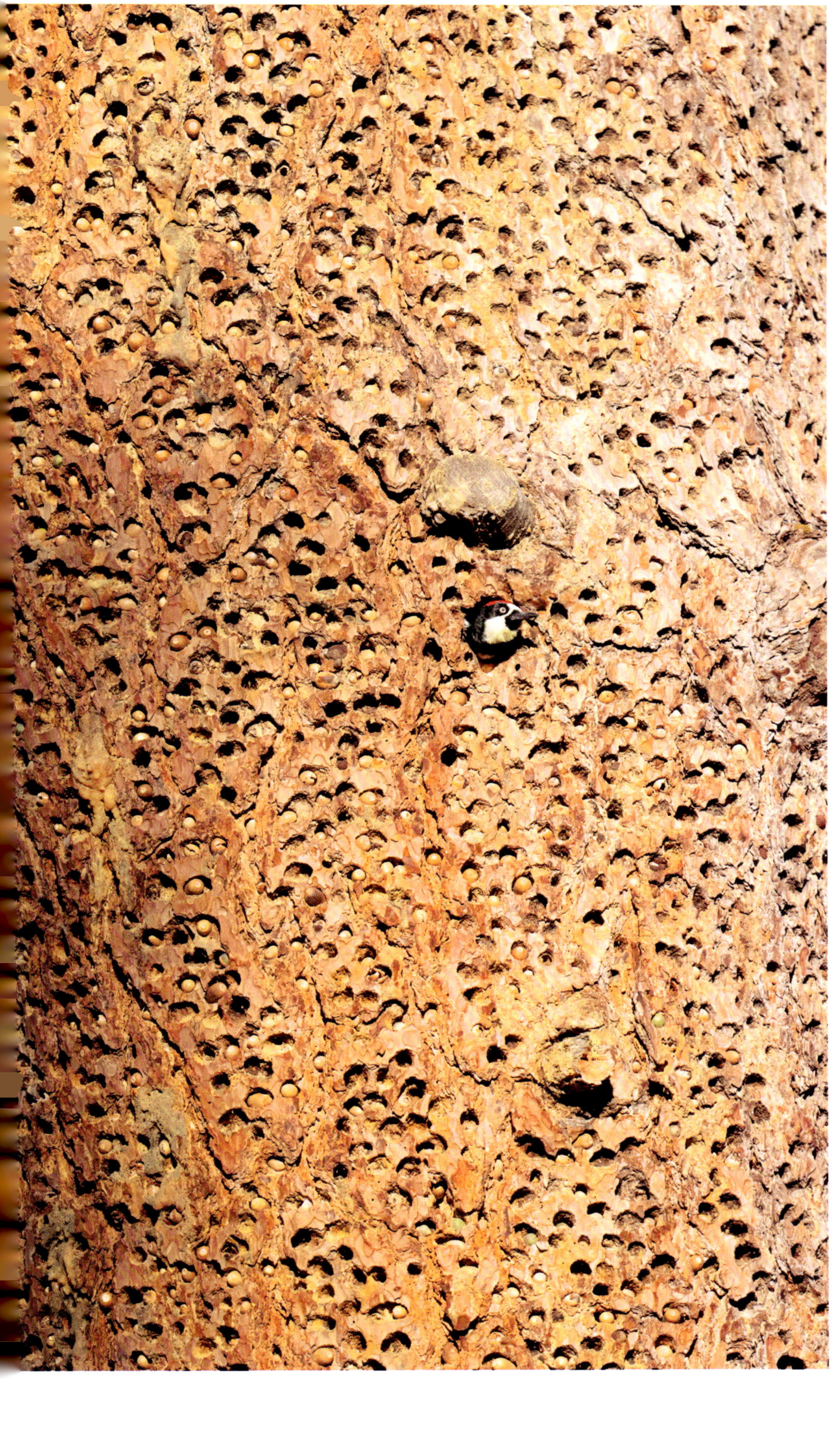

them with bark, moss, or wood chips, and sometimes they utilize wet wood chips, which harden over their treasures. When insects are available, they sometimes store them live by wedging their bodies into the cracks of trees in such a way that the legs wave in the air but cannot be used to gain freedom.

Acorn Woodpeckers are far more reliant upon oaks than any other woodpecker species, with population density and stability positively correlated with number of oak species, which decreases the probability of a total acorn crop failure. Given this dependence, it is not surprising that Acorn Woodpeckers deploy the most highly developed larder hoarding approach, creating custom-size holes for each individual intact acorn rather than relying upon existing cracks, holes, or bark furrows. These larders, called granaries, could be made up of fifty thousand holes! This strategy allows them to stay on their territory year-round rather than risk their lives in migrations or other peregrinations.

Each family of Acorn Woodpeckers has one primary granary tree and several secondary trees to which they add storage holes during winter. This multiyear process may allow a family group to carve several hundred acorn holes per year. When granaries are filled, the birds will store acorns, nuts, and seeds in cracks in trees, in previous nest and roost holes, knotholes, crevices, birdhouses, and even in human-made buildings. One Acorn Woodpecker stored tens of thousands of acorns weighing over seven hundred pounds in the wall of a California house before the hoard was discovered by a pest inspector. Studies have shown that the larger the accessible granaries, the more successful the subsequent breeding season. These impressive collection efforts are undertaken even though the woodpeckers will never be able to consume all of them.

As fall winds to a close, woodpecker family groups have separated or weakened their bonds as they move or change behaviors in preparation for winter.

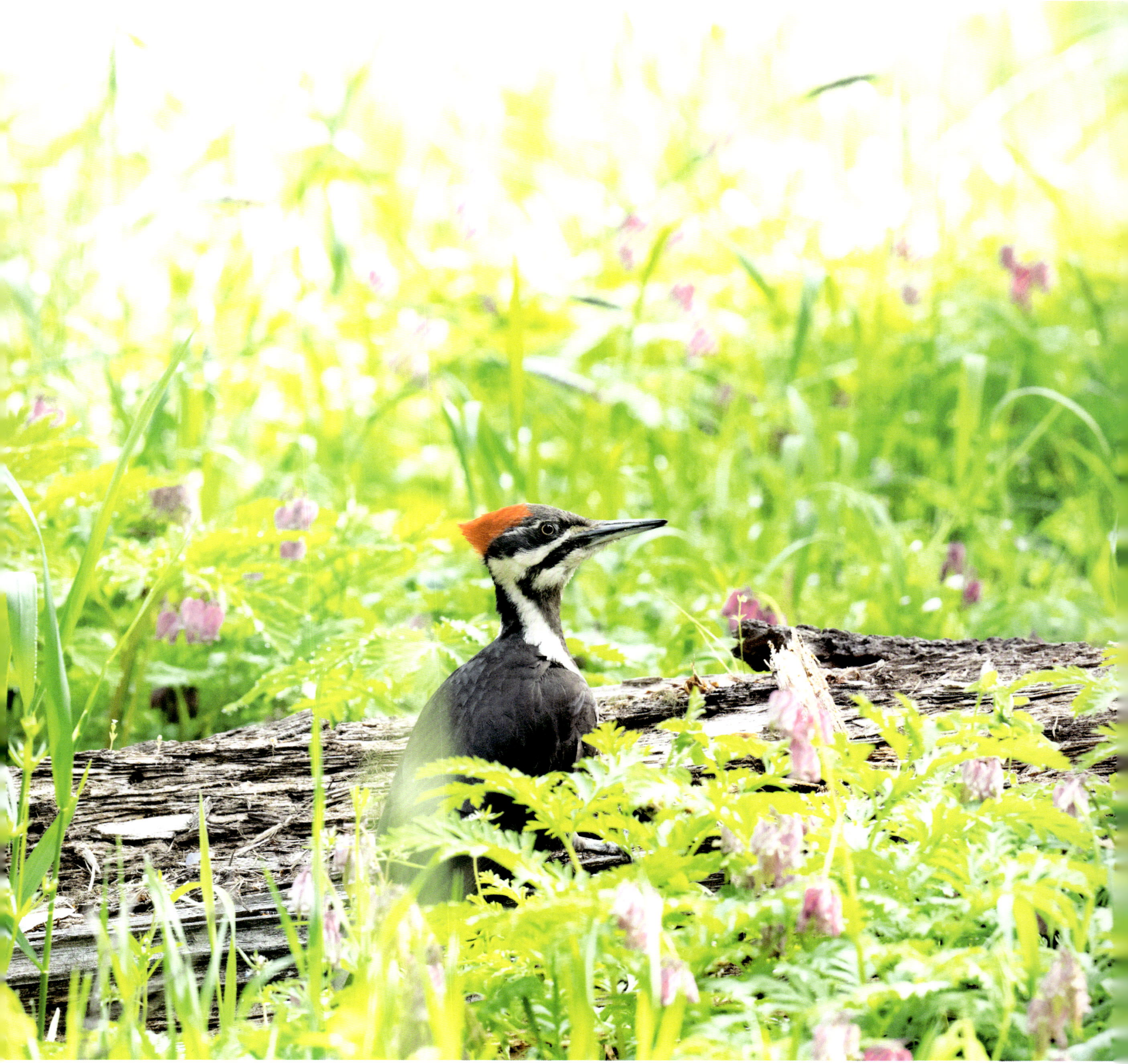

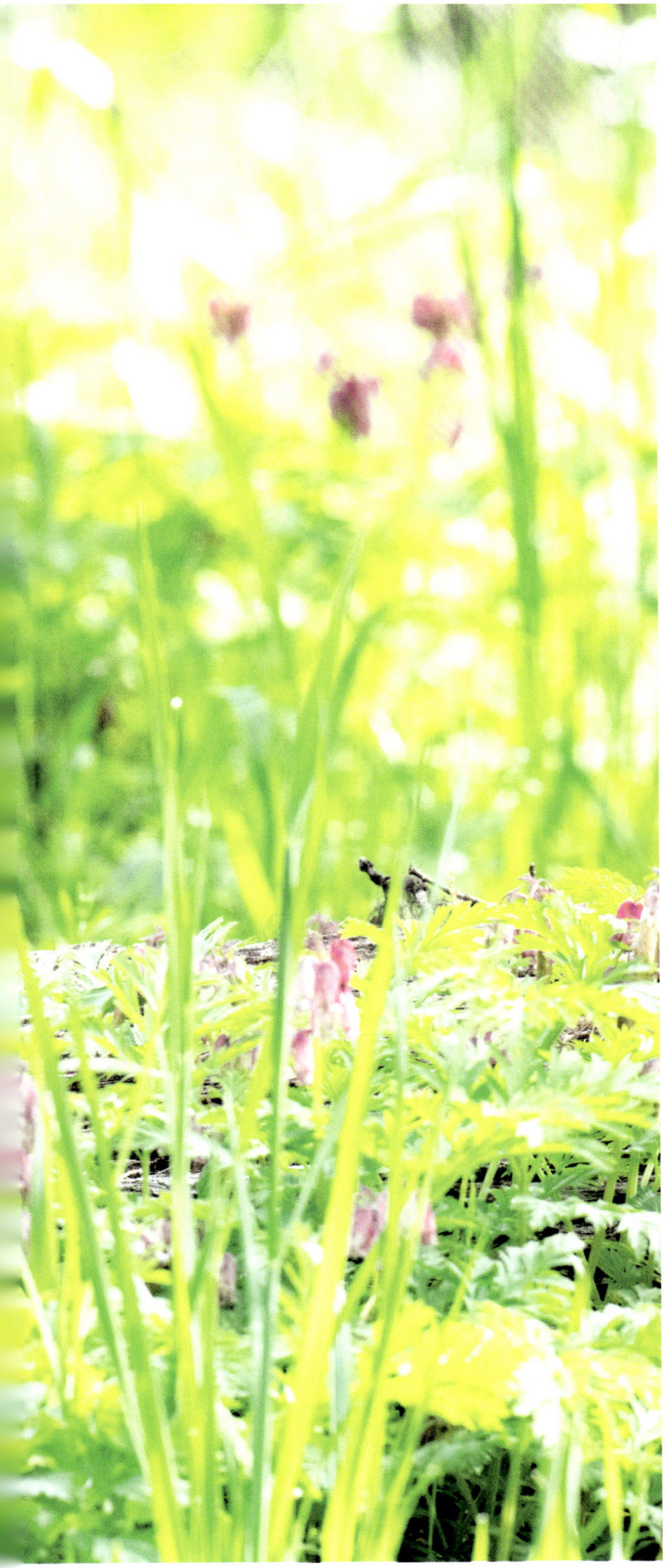

Left: Most woodpeckers do not forage on the ground, but Pileated Woodpeckers, like this female, frequently feed upon carpenter ants hiding within downed logs. Northern Flickers, Acorn Woodpeckers, Lewis's Woodpeckers, and Red-headed Woodpeckers also make frequent trips below the trees.

Below: A female Pileated Woodpecker regurgitates insects from her crop for her male fledgling. During the first few days after leaving the nest, some fledglings are left alone or in small groups in safe places while the parents forage.

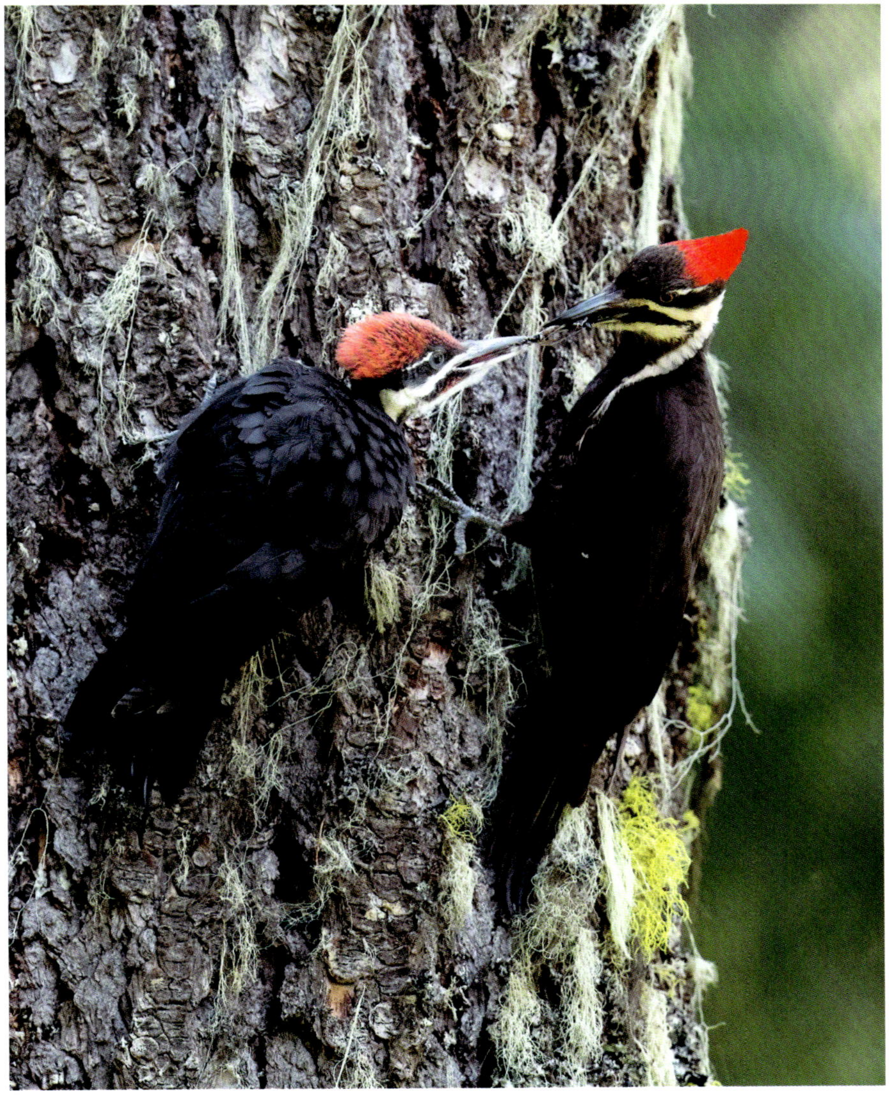

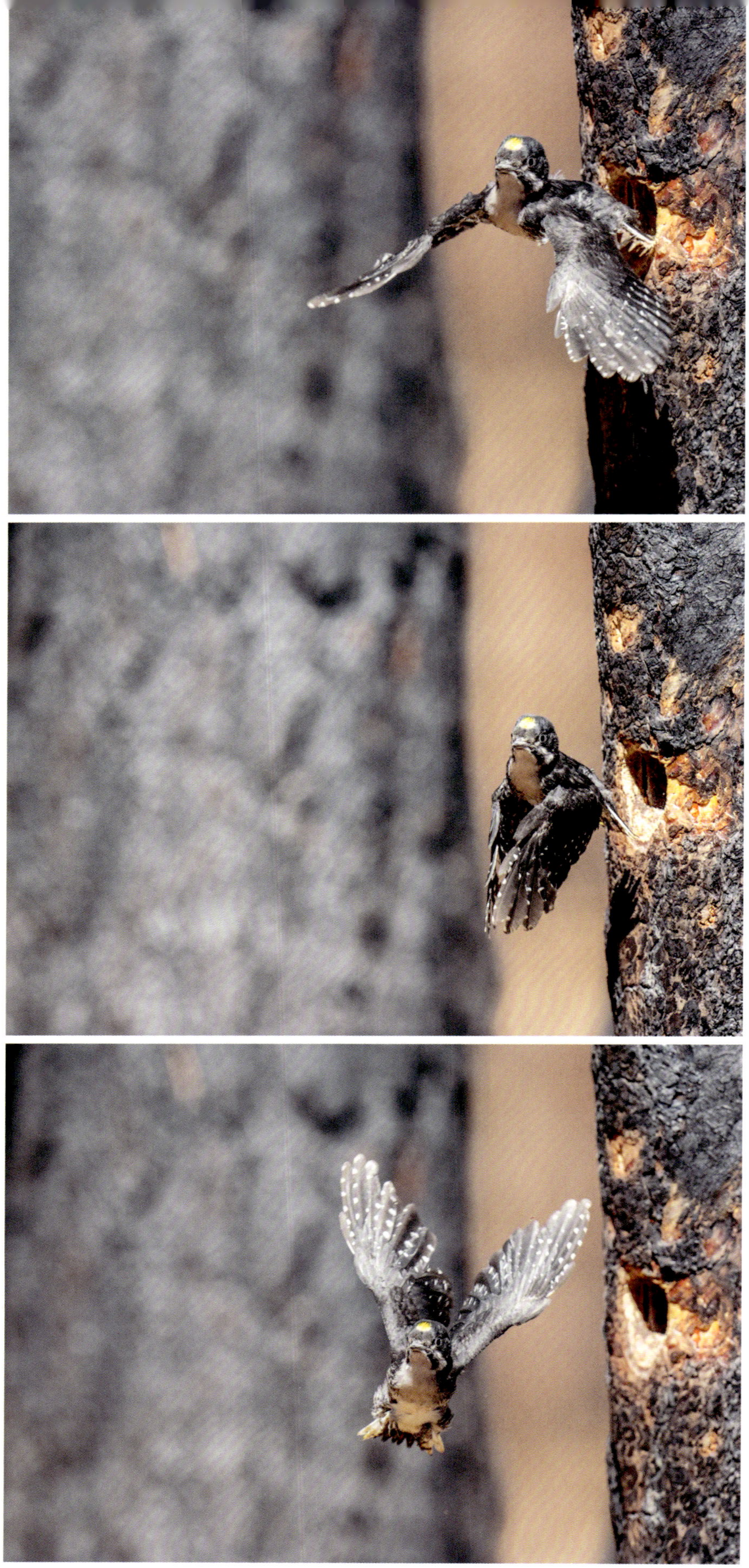

Five minutes after its sibling was fed, a Black-backed Woodpecker unexpectedly pushes its body beyond the lip of the cavity and immediately begins a downward stroke to break its fall before gaining control and turning toward the trees into which it would disappear.

Opposite: Gray-breasted Woodpeckers are communal breeders, often making it difficult to discern familial relations, especially when they are feeding together upon cactus fruit. The adult male woodpecker on the right is feeding alongside what appears to be a young male, which may be his offspring.

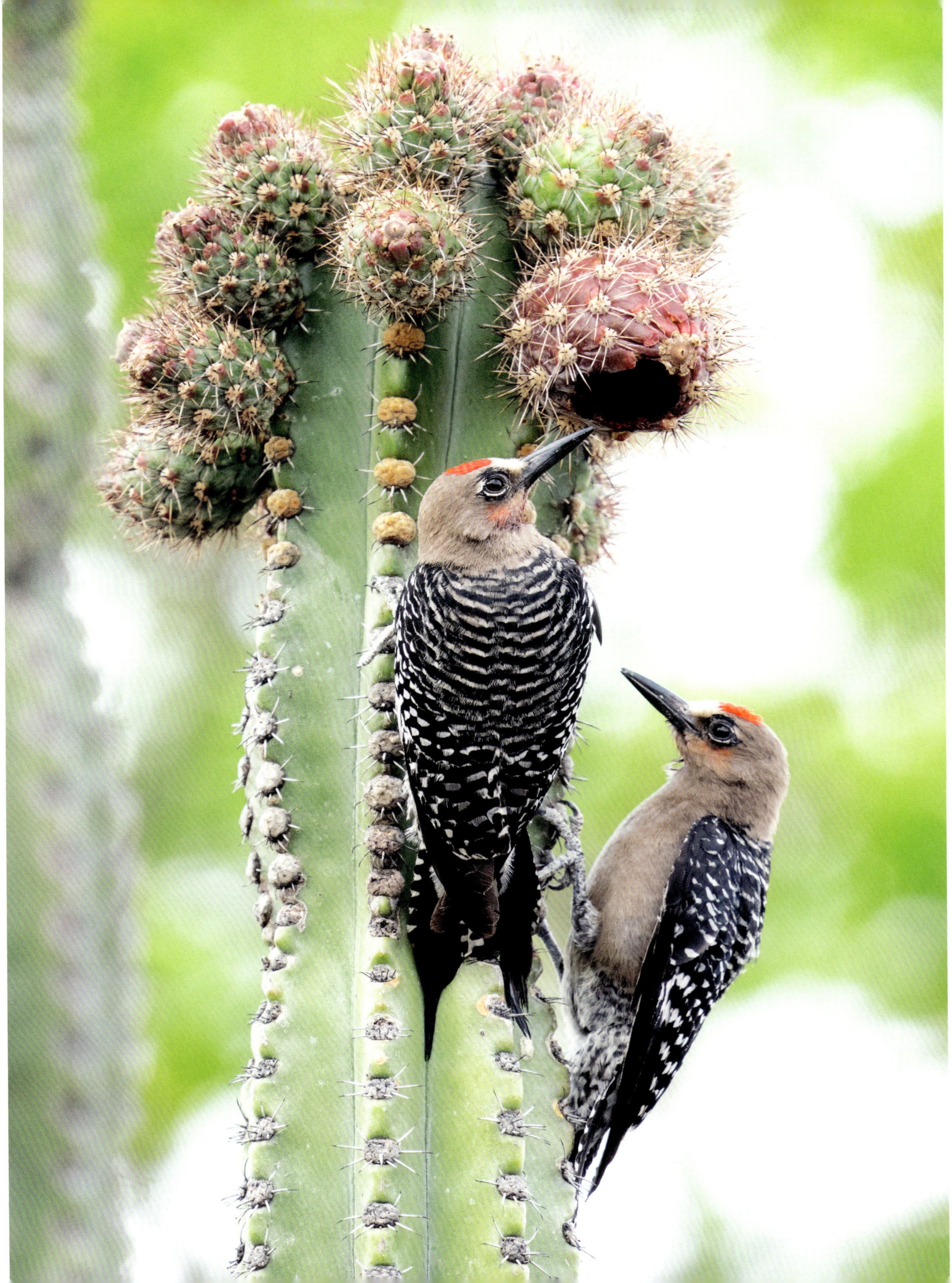

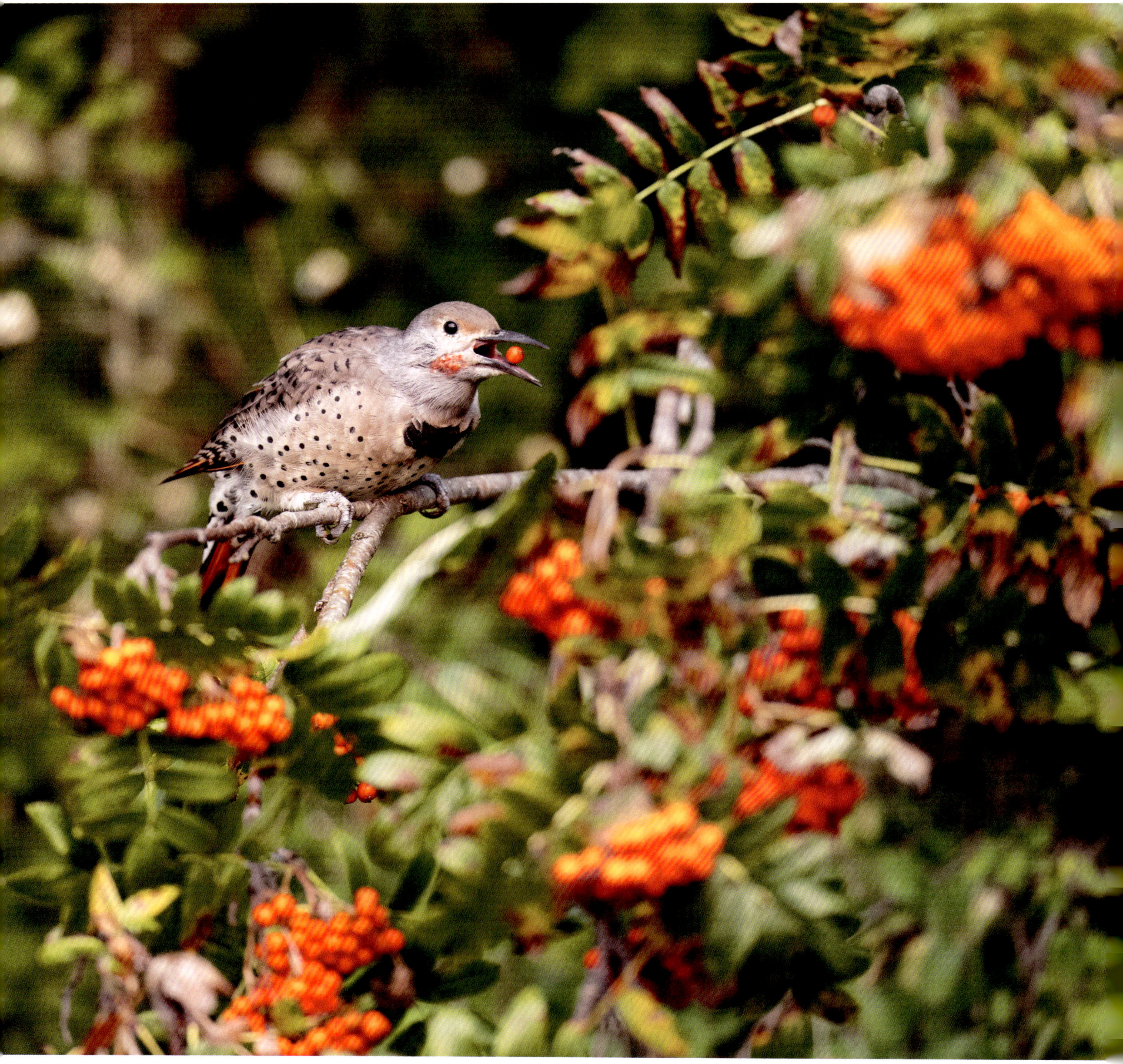

Juvenile woodpeckers of some *Melanerpes* species will return to their nest to be fed by their parents, even after they have fledged. These two juvenile Puerto Rican Woodpeckers had left the nest days earlier, then returned, begged, and were fed before leaving the area with their parents.

Opposite: The berries of Cascade mountain ash and other fruiting plants ripen just when juvenile flickers, like this Northern Red-shafted, are becoming independent from their parents, giving them some easily accessed food.

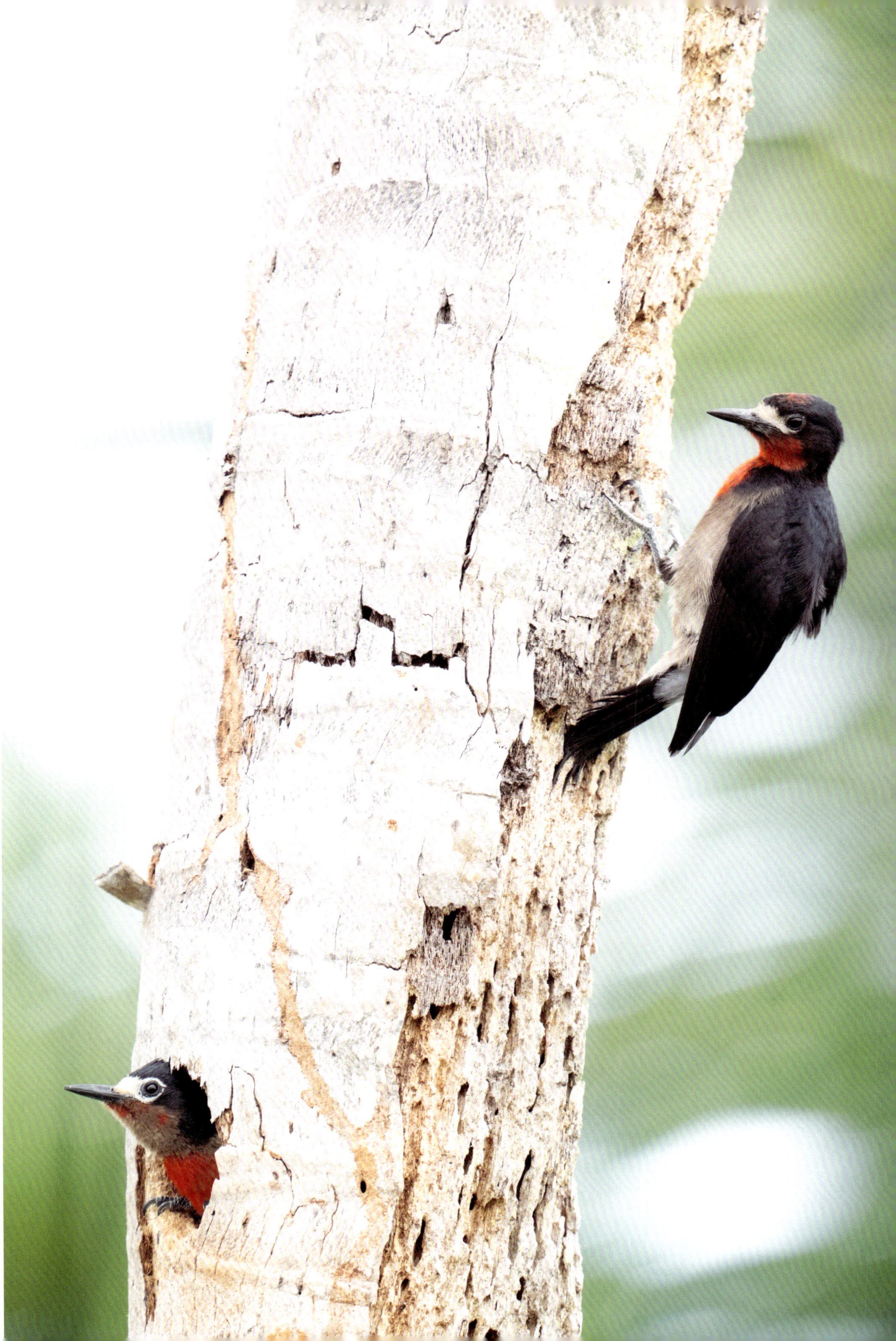

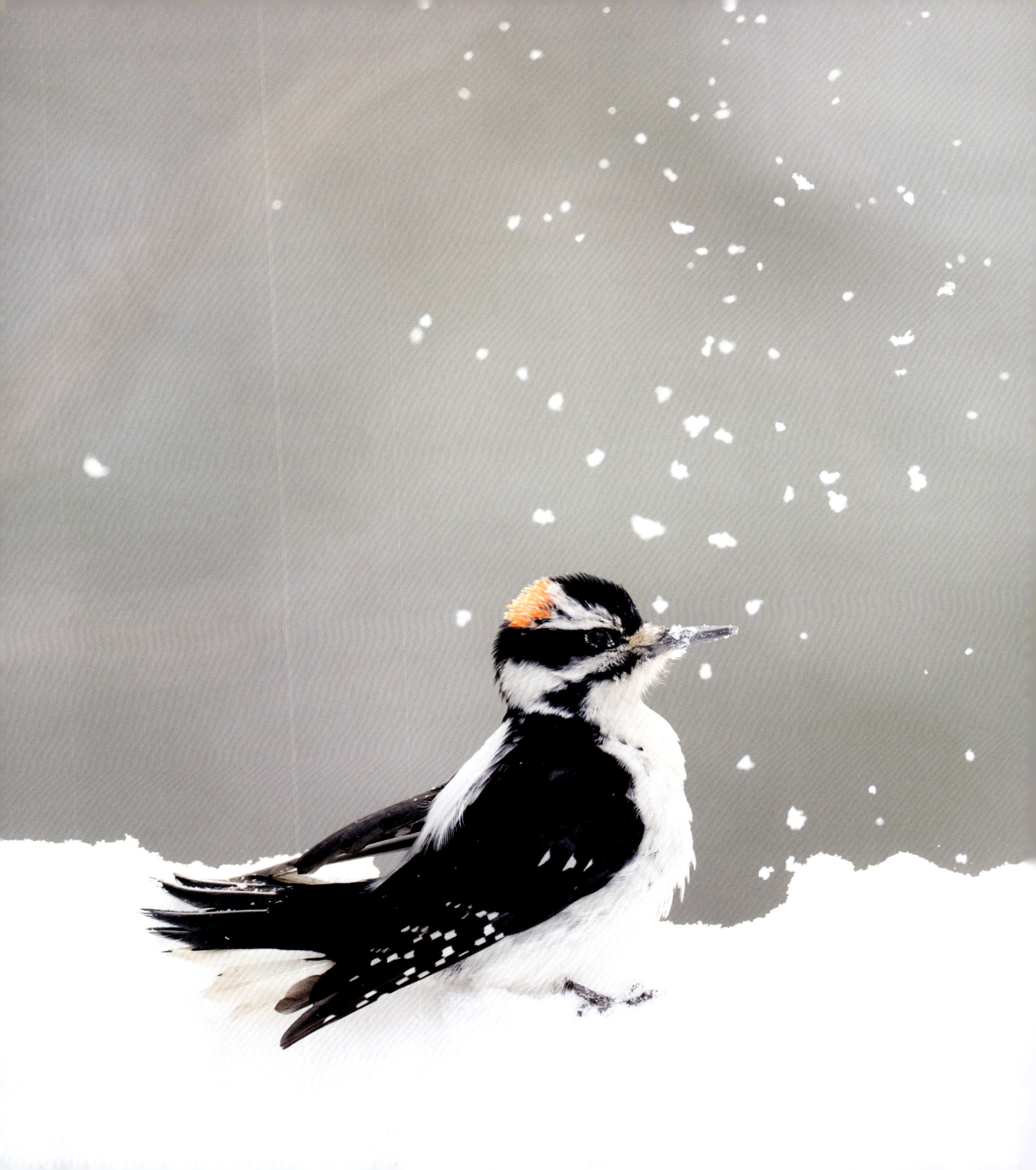

5

Surviving Winter

On a trip across the sagelands of eastern Washington, I was surprised to see twelve mustachioed Northern Flickers in silky plumage perched together on the blue-green sage, an unusual host. They reminded me of the Christmas decorations of my youth, with their seemingly unrealistic and fanciful colors. One had the red mustache behind the bill expected of the Red-shafted race but sported the red crescent of the Yellow-shafted on its nape. Another featured the black malar of the Yellow-shafted mismatched with the red underwings of the Red-shafted. Still another had a mustache that was a slurry of black and red with orange underwings.

It was the holidays, so I should not have been surprised. This tableau was not the result of artistic license taken by a designer, nor the mistake of artificial intelligence; these were simply flickers sporting variations of the colors found in the western and eastern races, a relatively common winter sight in the Northwest when migrants from the intergrade zone in the colder and snowier north, where the two races mix, flood south to find more accessible food.

The dozen wandering woodpeckers balanced on the sagebrush as I watched them, and their snow-dotted malars, underwings, and undertail feathers flashed as they raised and lowered wings, tails, necks, and breasts to maintain their tenuous perch in the snowy wind.

Snow flies as a male Hairy Woodpecker forages. One of the hardiest species, the Hairy lives year-round in some of the coldest places in North America.

Finding sustenance and shelter in winter is the final and most difficult test for woodpeckers in their annual struggle to survive. Shorter periods of daylight and longer stretches of cold leave less time to safely accomplish all that keeps them alive. For many species, favored foods are dormant or inaccessible, forcing them to adapt, move, or perish. Woodpeckers must find or create a cavity in which to shield themselves from the weather, or find shelter underneath limbs or perhaps a building's eaves. Their ability to accomplish these things will help determine how long they will live.

Adaptation may involve braving some of the earth's harshest conditions, breaking through the shields of their quarry, utilizing more of their home range, or even moving to new areas to secure alternate food. Moving could be as simple as wandering in search of unpredictable seasonal abundance, or it could mean risking migration hundreds of miles to a warmer place with longer days but unfamiliar habitats, only to endure a perilous return journey in spring.

Each species, depending upon its biology and geography, varies in its ability to withstand the immensity of the challenge, and populations of each species make choices dictated by instinct or intelligence that may be surprisingly divergent from year to year. Normally antagonistic woodpeckers may weaken pair bonds while becoming more tolerant of those of their own and even other species in places of plenty or in the process of securing nourishment.

MIGRATIONS AND PEREGRINATIONS

Winter's energies center around finding enough food. Woodpecker staples, including arthropods such as insects, are tunneled within trees or underground. From Alaska through Canada and the northern states, sap is less sugar-rich; vegetative fare such as buds and fruit, unstored nuts, seeds, and grain are unavailable; and the ground itself is covered by snow. In southeastern Mexico and on the Caribbean islands, much of winter falls within the dry season, when arthropods and other fruit are more difficult to find. Most young of the year must quickly adapt to these realities mere months after becoming independent and usually before they have established their own territory.

Most woodpeckers have food preferences and varying degrees of flexibility based on their morphology and feeding strategies that allow them to take advantage of fluctuations in the availability of different foods. Some species pursue core foods throughout the year, with changes in supplemental foods, while others dramatically alter their diets, especially during winter.

Most northern woodpeckers must adjust to winter by changing their diet, utilizing more of their typical home range, wandering outside it, or migrating. Only those that can rely on excavation of insects over gleaning, fly-catching, or eating fruit will remain in the harshest places.

The majority of woodpeckers remain on their home range during winter, some expand the range up or downslope, others execute opportunistic peregrinations in search of sustenance, and a minority migrate. Those that migrate must head to wintering areas between late August and early October and return to breeding areas between March and May. As with much of woodpecker life history outside of the breeding season, we still have much to learn about woodpecker migration, including whether they migrate alone or in groups, although both situations have been observed. Northern Flickers and Lewis's Woodpecker are among the species that have been observed migrating in large but often loose single-species groups, and Red-breasted and Williamson's Sapsuckers have been reported migrating together in informal flocks. More surprisingly, small groups of Red-headed Woodpeckers have been recorded migrating together, including on occasion with Blue Jays.

Those that do migrate must expend energy for long journeys across ever-changing landscapes, where the presence of food, water, shelter from weather and predators, and even habitat along the way or at their destination is uncertain. They must face the same uncertainties during their return. Much of their success will depend on factors they cannot control.

Steam rose from the trail in front of me, and an impossibly complex tangle of branches, vines,

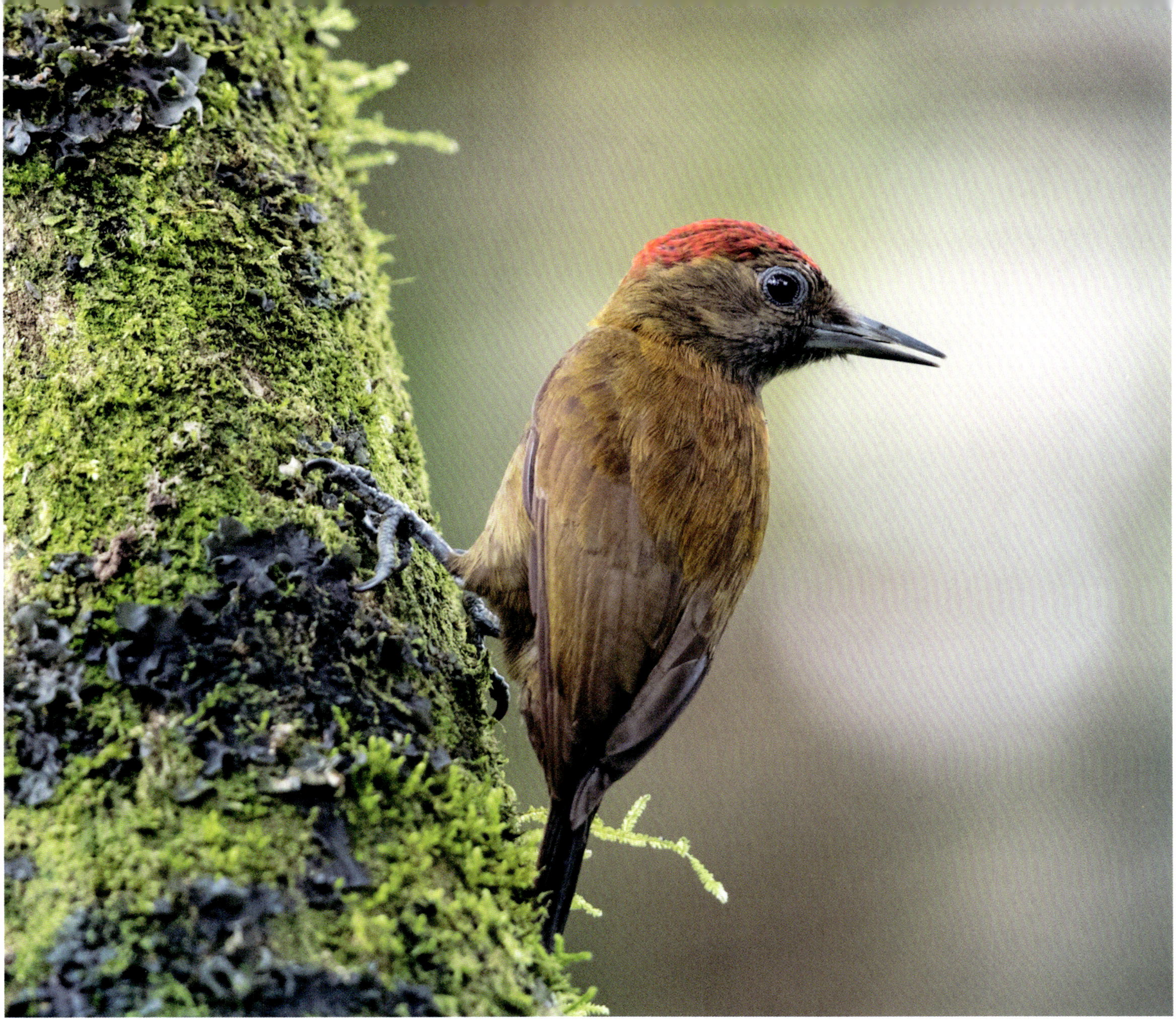

The Smoky-brown Woodpecker resides year-round from sea level to 4,900 feet in some Mexican oak forests, where they excavate for small wood-boring beetles and their larvae.

and leaves formed a virtual green wall with sunlit windows through which I anticipated spotting woodpeckers ahead. Along this trail in the tropical broad-leafed forest of Veracruz, Mexico, I would eventually spot Smoky-brown, Lineated, and Pale-billed Woodpeckers, and the bronze-winged race of the Golden-olive Woodpecker, all reinforcing my feeling of being far from home on this winter's day—until I heard the catlike mew of a Yellow-bellied Sapsucker. It was as if I suddenly heard English, and I immediately felt closer to home. Soon I spotted this winter visitor, tending its sapwells lit by a ray of sun.

This reminded me of an earlier trip to Cuba. I was feeling a world away watching Cuban Green and West Indian Woodpeckers and Fernandina's Flickers until my sense of place was disrupted by the contact call of a Northern Flicker and a surprise encounter with a Yellow-bellied Sapsucker on a palm tree!

Winter is a time of surprises, when the world feels smaller and more connected, and birds show up in places that surprise novices and experts alike. While I should have been anticipating these encounters, winter movements of birds are complex and not fully understood, and their timing is changing with the changing climate.

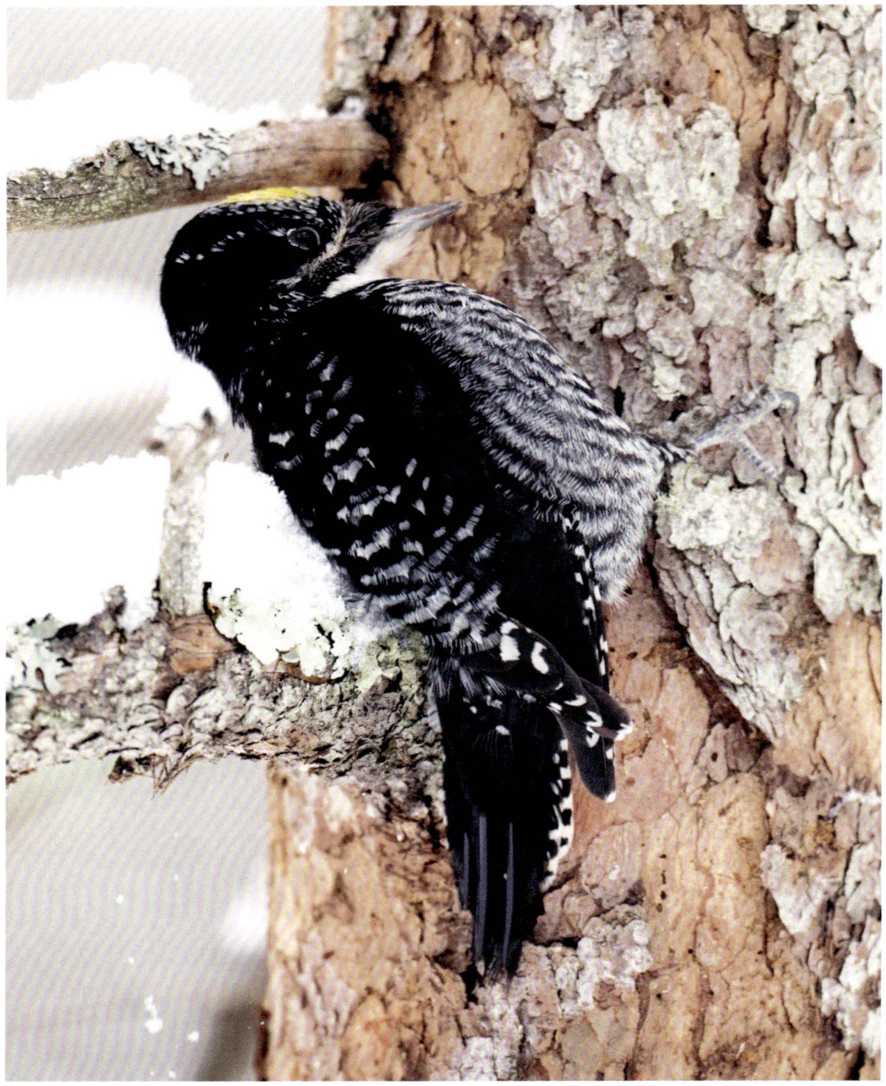

WINTER SURVIVAL STRATEGIES

Most woodpeckers reside on home ranges year-round, but several species that live and breed in Canada, the northern United States, and at higher elevations migrate. Southern woodpeckers benefit from a more stable food supply year-round and generally remain on home ranges.

Since woodpeckers within a particular genus share like morphology and food preferences, they tend to respond to winter's challenges in similar ways. By examining select woodpeckers, we can better understand the range of responses that allow species in each genera to survive winter.

Picoides and Dryobates

All members of these closely related genera are year-round residents within their home ranges. Some have been observed foraging on more of their home range during winter in response to food outbreaks but do not migrate, even though they range farther north than any other woodpeckers.

Subsurface excavation is the best strategy for a woodpecker to use in harsh environments, as these foods, especially beetles, ants, and their larvae, can be found year-round. The best example of this is illustrated by the species that breed at the highest latitudes and elevations. The diets of all members of *Picoides* and *Dryobates* are heavily weighted toward arthropods. The American Three-toed, Black-backed, Hairy, and Downy Woodpeckers live year-round in the highest mountain forests and northern boreal forests of Alaska and the Yukon by focusing their winter foraging on the subsurface excavation of wood-boring insects and their larvae.

Black-backed Woodpeckers excavate deep into trees, where they find more than 75 percent and as much as 90 percent of their favored foods year-round, a valuable capability during cold and snowy winters. The American Three-toed has similar insurance against severe weather.

Black-backed, American Three-toed, Hairy, and, to a lesser extent, Downy Woodpeckers take advantage of insect outbreaks after storms, floods, and fire that provide rich forage during winter. Historically, fire season lasts from July through October, and beetles arrive shortly after, exactly the right time for dispersing woodpeckers that might winter and spend the next three years building their numbers before insect populations

The absence of a fourth toe allows the American Three-toed Woodpecker to lean farther back and exert more force upon the edges of bark to scale them off snags or trees and reveal beetle larvae or their tunnels. This particular male is from the eastern boreal race (*bacatus*), distinguished by less white on the back and face.

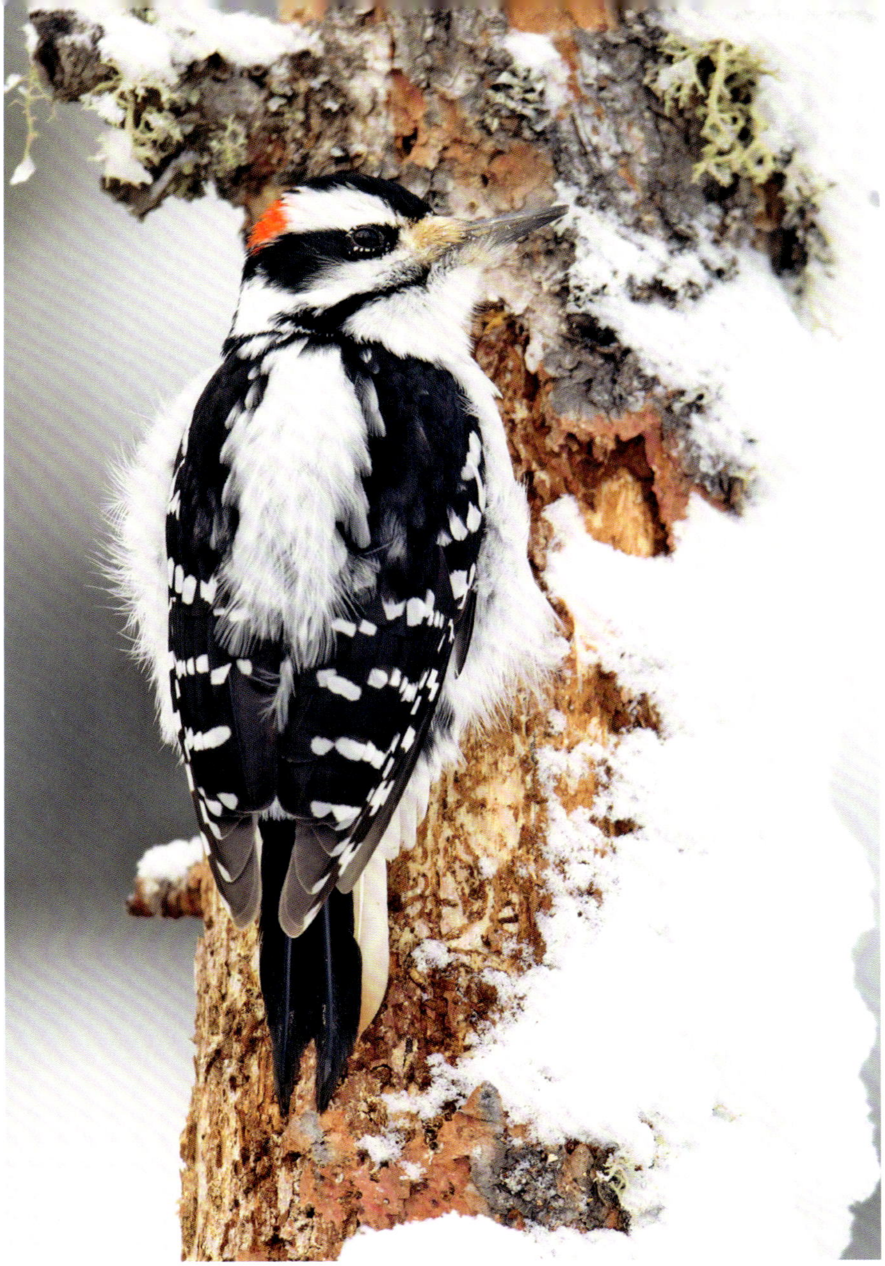

Their strong ability to excavate for insect prey beneath the bark allows Hairy Woodpeckers like this male to stay on their territories through the winter. In contrast to the American Three-toed and Black-backed Woodpeckers, the Hairy tends to chisel holes into the tree to reveal insect tunnels.

plummet. Studies have shown that American Three-toed Woodpecker populations can be as much as thirty times as dense in burned stands as in comparable unburned areas. Together these practices allow them to stay on home ranges year-round, with some long-distance movements in response to food outbreaks, severe weather, food shortages, or the dispersal of young that can bring individuals from the most extreme places downslope or south from time to time.

The remaining *Dryobates* woodpeckers can remain year-round on their home ranges. White-headed and Strickland's Woodpeckers do so in mountain conifer forests through their versatility in accessing food, whether by excavating, prying, flaking off bark, probing, feeding upon exposed arthropods, or pursuing vegetable matter such as seeds or sap.

In some areas, White-headed Woodpeckers feed almost equally on pine seeds (ponderosa, Jeffrey, sugar, and Coulter) and insects over a year, with the ratio fluctuating based on availability of each. Ants, spruce budworms, beetles, scale insects, and sap make up most of their diet during spring and summer, but they forage more on pine seeds from late summer through winter (depending upon species availability), which in Oregon may be as much as 90 percent of their diet during this time. During winter they may feed more on incense cedars and on trunks of trees (especially the females) and less on branches, and they also take suet from feeders.

Dryocopus

Dryocopus woodpeckers, the Lineated and Pileated, are powerful excavators that carve large holes in trees to uncover ants, wood-boring beetles, and their larvae, then snare them with their long, barbed tongues. This allows them to reside on their home ranges year-round.

Pileated Woodpeckers are even able to winter in forests as far north as the Yukon thanks to their ability to secure ants, especially carpenter ants, throughout the year. These insects compose more than half of their food during winter. In some places carpenter ants may be nearly 75 percent of their winter diet. They are known to locate carpenter colonies during summer and fall while they are also feeding upon beetle larvae, termites, other ants, and insects as well as fruit, but focus on harvesting from these known colonies during winter.

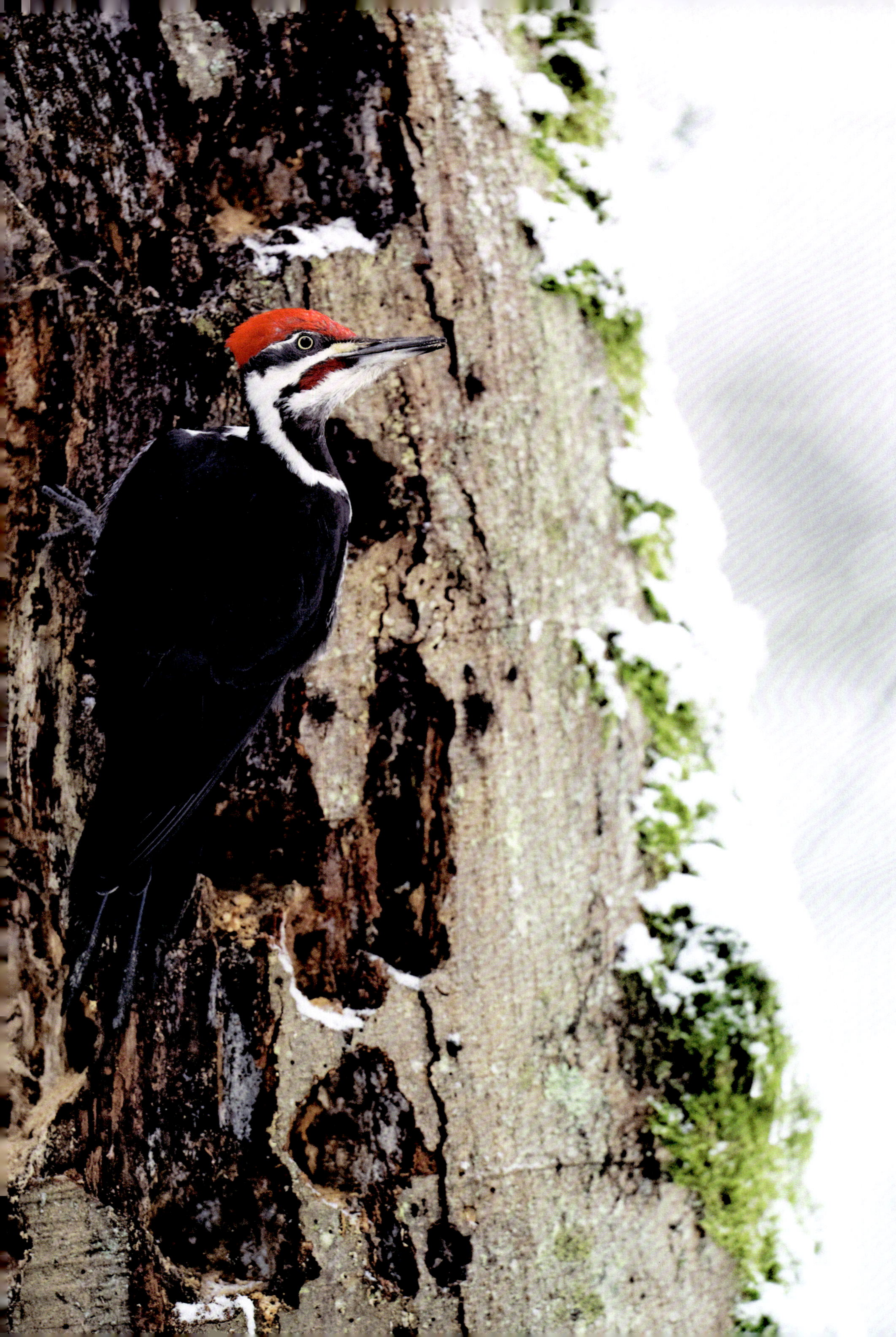

Pileated Woodpeckers specialize in carpenter ants and appear to keep track of colonies during the winter, when other foods are more difficult to find. They excavate large, rectangular openings in the trees until they locate tunnels and then lap the insects and larvae up with their tongues.

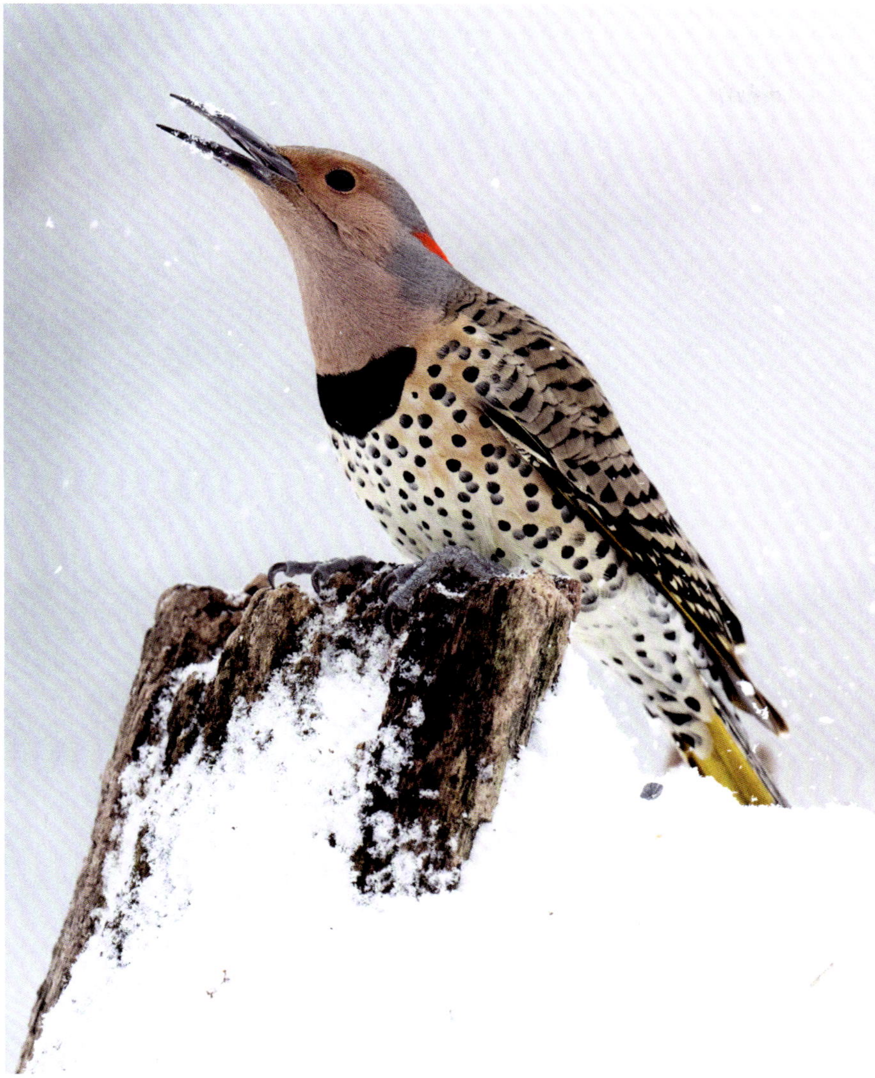

Persistent snow cover limits the areas in which Northern Flickers can winter, as they specialize in ground feeding on ants and other invertebrates and must migrate to areas where this prey can be accessed.

Colaptes

Within this genus, which includes flickers and Gray-crowned and Golden-olive Woodpeckers, only some populations of the northernmost species, the Northern Flicker, are partially migratory. The individuals that breed from the Sonoran Desert south are resident year-round.

Northern Flickers are ant specialists, but they are poor excavators, preferring instead to hammer the ground and utilize their long, sticky tongue instead of their tree-breaking abilities in pursuit of ants and their larvae. Deep and persistent snow in Alaska, most of Canada, and in some mountain areas makes it nearly impossible to forage for ants on the ground, necessitating the northernmost populations to migrate and the highest elevation ones to move downslope, where they are able to pursue subterranean prey. In these areas, they also pursue insects, fruit, and suet when available. During winter in some northern areas, fruit may make up the bulk of their diet.

When moving south or downslope, Northern Flickers do not usually cross the Continental Divide, keeping Yellow-shafteds to the east of the Rocky Mountains and most Red-shafteds in the Rockies and to the west. The intergrade zone where the two mix—in the rain shadow of the Rocky Mountains from Texas north to southeastern Alaska—has remained relatively stable in size for over sixty years. Birds probably from the intergrade zone in British Columbia are commonly seen in Pacific states and provinces during winter. Female Northern Flickers tend to winter farther north than males, although there is a lot of overlap.

Most of Canada and many northern states from eastern Montana to Maine have few, if any, flickers during winter, whereas many locations from southeast Alaska through Mexico have migratory flickers joining residents, and some places in southeast California, Arizona, New Mexico, Texas, and northern Mexico will host their only flickers of the year.

Colaptes species in the tropics don't have to migrate but sometimes change their foraging priorities. The Fernandina's Flicker thrives in the dry season (November–April) by investing more time foraging for insects, worms, grubs, and seeds on the ground, which is usually flooded and inaccessible during the wet season (May–October), when they are forced to focus on arboreal ants in decaying palms.

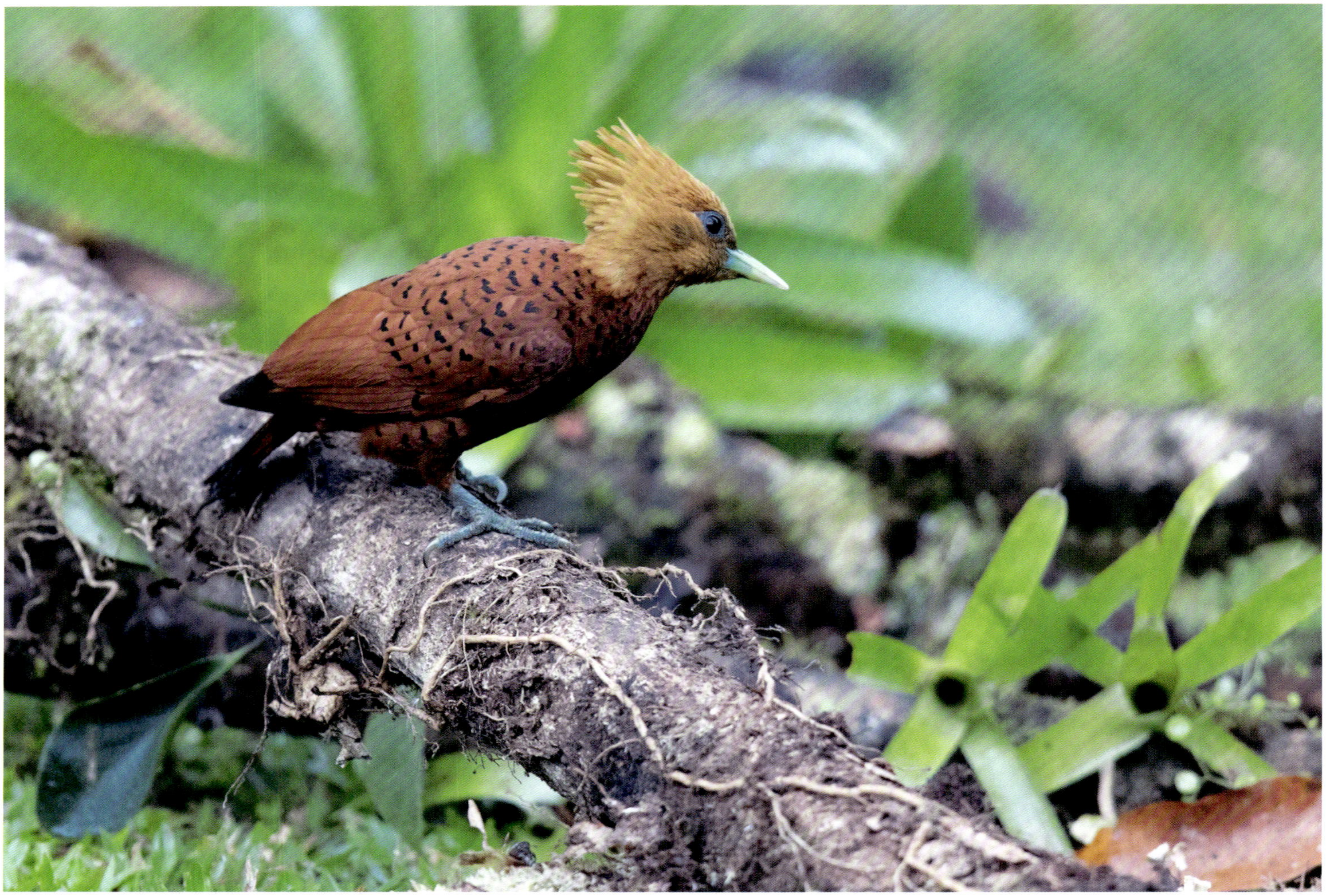

Celeus

The lone North American member of genus *Celeus*, the Chestnut-colored Woodpecker, is an ant and termite specialist that breeds from southeastern Mexico through Central America in lowland humid evergreen and semideciduous forests. During winter, when food is more challenging to find, they are known to forage in mangroves and coastal scrub.

Sphyrapicus

This genus of four sapsuckers includes North America's most migratory woodpeckers. All are short- to medium-distance migratory birds that breed in parts of Canada and winter as far south as Mexico. Those that migrate leave the coldest parts of their ranges during late summer or early fall before the flow of xylem sap (less than 5 percent sugar) slows or freezes and the birds are challenged to find enough remaining insects and fruit to survive. The sapsuckers that live farthest north or at the highest elevations travel the farthest, though not all species make a true migration, rather moving south or downslope with the weather or even remaining resident year-round. Once they can begin tapping the more sugar-rich (as much as 25 percent) phloem tissue of the trees in the early spring, they return to breeding areas. During winter these woodpeckers supplement their diets with insects, fruits, plant matter, and even suet that may not be accessible in breeding areas. With sapsucker species, the females usually migrate farther than males, and many of both genders winter in habitats distinctly different from those used for breeding.

Ants and termites form the core of the diet for the Chestnut-colored Woodpecker (female pictured here). Sometimes, when food is in shorter supply in their typical tropical moist forest habitat, they wander into mangroves and coastal scrub.

The Yellow-bellied Sapsucker, the world's most migratory woodpecker, breeds farther north (east-central Alaska) and winters farther south (Caribbean islands and Panama) than any other sapsucker. Most migrate at night, often in large numbers, and leave wintering areas between early September and late October, returning between early March and mid-May.

These sapsuckers rely on sap for up to 100 percent of their winter diet to as little as 20 percent between May and August. In the northernmost parts of their winter range, they feed on arthropods they pry from bark, like *Dryobates* woodpeckers do, as well as frozen fruit, frozen sap, and even suet from feeders. During colder weather they require more sap and will rip bark from trees to start wells. Farther south, most of their foraging time is spent drilling and tending sapwells. They supplement this with soft inner bark and associated sap referred to as *bast*, as well as any fruit, arthropods, and seeds they can find. In winter they have been observed storing fruit, nuts, seeds, and acorns.

Yellow-bellied Sapsuckers often choose winter habitats very different from the young stands of quaking aspen, birch, and maple or riparian mixed-conifer forests from their breeding areas and end up in places such as pine-oak woodlands of the Sierra Madre and among the tropical plants and gardens in the Caribbean.

Most Red-naped Sapsuckers are short-distance migrants that fly south from their breeding range into Southern California, Arizona, New Mexico, and Mexico, while some remain within their breeding range in Arizona and New Mexico. They return between late March and early May.

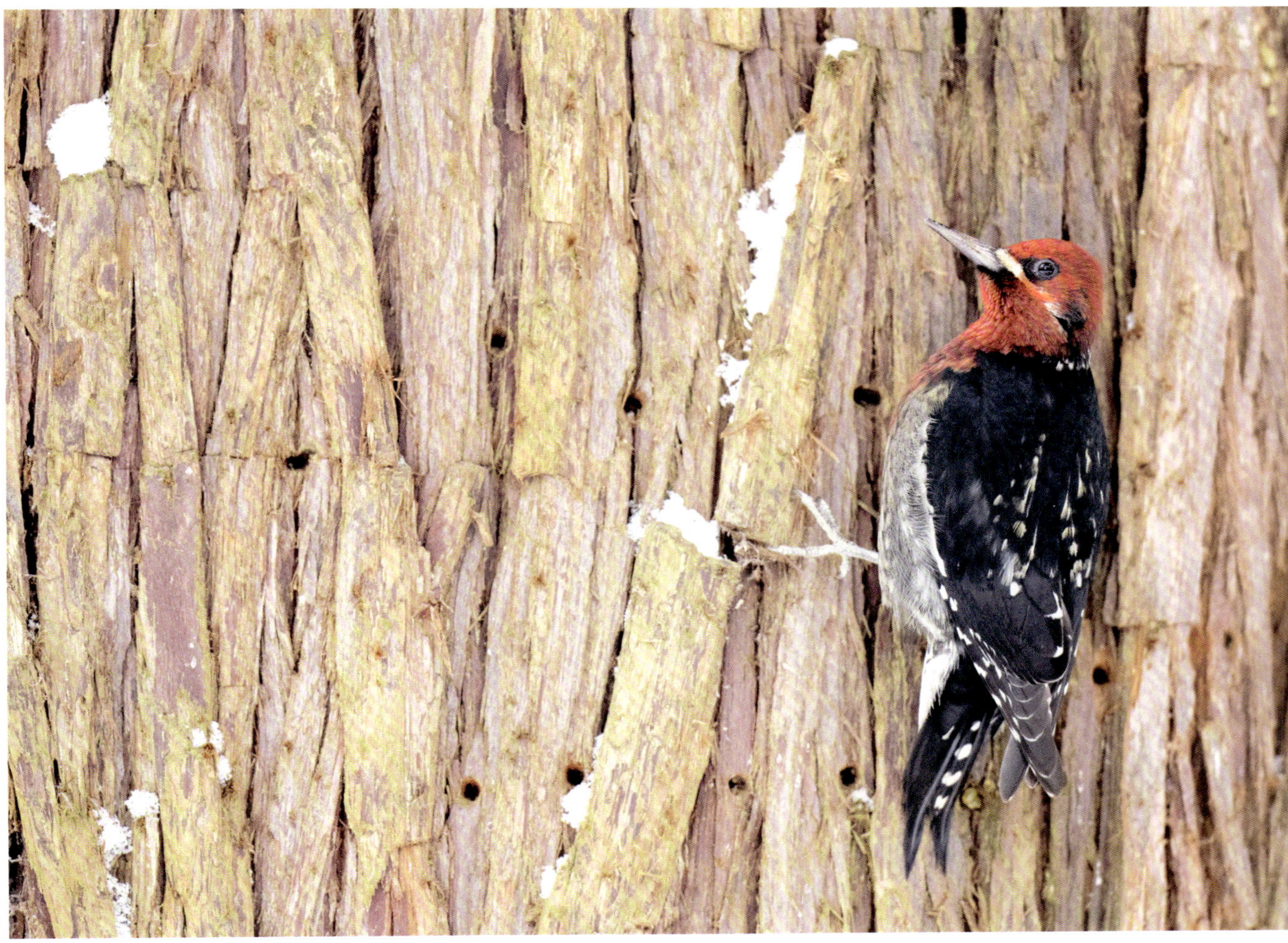

Red-breasted Sapsuckers sometimes migrate or wander to lower elevations during the winter when sap freezes. This particular bird likely moved from the Cascade foothills into the Seattle area during a snowstorm when the zylem wells in a western redcedar still provided accessible sap.

This species of mountain aspens and mixed conifers occupies a wider range of habitats outside the breeding season, including lower-elevation habitats such as pinyon pine and juniper during migration and pine-oak and riparian woodlands, as well as orchards. Their winter diet is centered around sap and bast, which they take from xylem wells in conifers instead of the phloem in deciduous trees they utilize during the breeding season.

The Red-breasted Sapsuckers breeding in wet temperate forests from southeast Alaska west of the Cascade crest through Oregon are of the *ruber* subspecies and are mostly resident along the coast, with mountain birds moving to coastal valleys to avoid colder temperatures, which stops sap flow, and deeper snows of the mountains and interior. The *daggetti* subpecies breeds from central Oregon through southern California and western Nevada. While they are resident in some coastal areas of Oregon and northern California, as well as interior parts of southern California, most California birds migrate to the coast and northern Baja in Mexico.

Red-breasted Sapsuckers are short-distance migrants that fly from inland and high-elevation breeding areas to coastal and lower-elevation winter areas between late August and early October, perhaps traveling in groups of three to four birds. They move back to breeding areas from mid-March into May, two to three weeks before Red-naped Sapsuckers. Wintering areas may not have the same old-growth characteristics necessary for breeding, as unfrozen sap is prioritized over other foods.

Williamson's Sapsuckers have a complex, mysterious migratory pattern that appears to be changing with a warming climate. Birds of the Rocky Mountain subspecies (*nataliae*) migrate more consistently and farther than the Pacific Slope subspecies (*thyroideus*) and winter as far south as Jalisco and Michoacán, Mexico. Some Williamson's Sapsucker individuals remain in breeding areas year-round. Migrating birds often flock with a post-breeding dispersal from mature conifer forests into atypical habitat such as oak scrub, pinyon-juniper, deciduous riparian, and nonnative conifers from September through October in northern areas. During winter they can be found outside their typical mature conifer habitat in low- to mid-elevation mixed oak and deciduous riparian forests. Females winter at lower elevations than males, and in Arizona I have seen females in urban desert parks tending xylem wells in palm trees and males in nearby mountains feeding on berries. All birds appear to arrive back in breeding areas between late March and May, with males arriving up to two weeks before females.

The Williamson's Sapsucker's diet centers around sap and phloem fibers with added arthropods before nesting, ants during nesting, and a return to sap and phloem fibers supplemented by fruit in winter. In the Arizona winter, Williamson's focus heavily on madrone and juniper berries as well as others from both exotic and native plants. On their way back to breeding areas in Colorado, they are known to set up territories at lower elevations until sap flow allows them to move upslope.

Melanerpes

Melanerpes woodpeckers are weak excavators, but they are our most omnivorous genus and take advantage of the seasonal abundance of surface insects, fruit, mast, and sap. Most live in warmer climates with smaller fluctuations in potential food than in colder regions, allowing them to simply utilize more of their home range, as the Gila does when it moves upslope or north, or by altering their diets like the tropical Jamaican and West Indian Woodpeckers. Surface insects, fruit, and sap are difficult to find during winter in Canada, the northern United States, and the interior, forcing northern individuals to move or migrate.

Some Red-headed, Lewis's, Acorn, and Red-bellied Woodpeckers winter farther north or at higher elevations where arthropods are more difficult for weak excavators to access and where alter-

The Red-naped (pictured here), Red-breasted, and Yellow-bellied Sapsuckers were once considered a single species and can be difficult to distinguish, complicated by the fact that they sometimes hybridize where their ranges overlap. The Red-naped is most easily distinguished from the Yellow-bellied by the red nape patch on the back of the neck and from the Red-breasted by the black band that separates the red throat (partially white on female) from the breast.

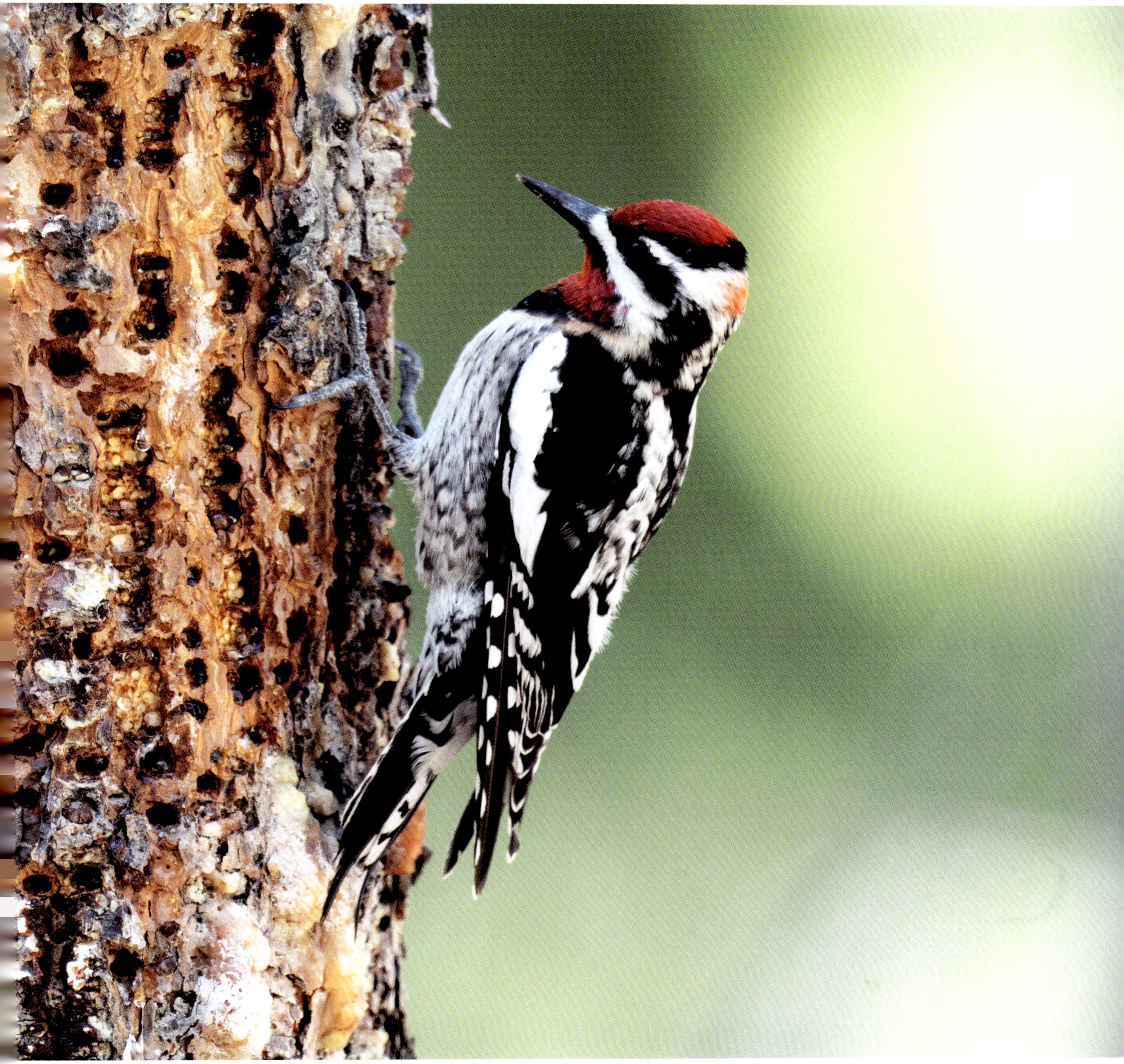

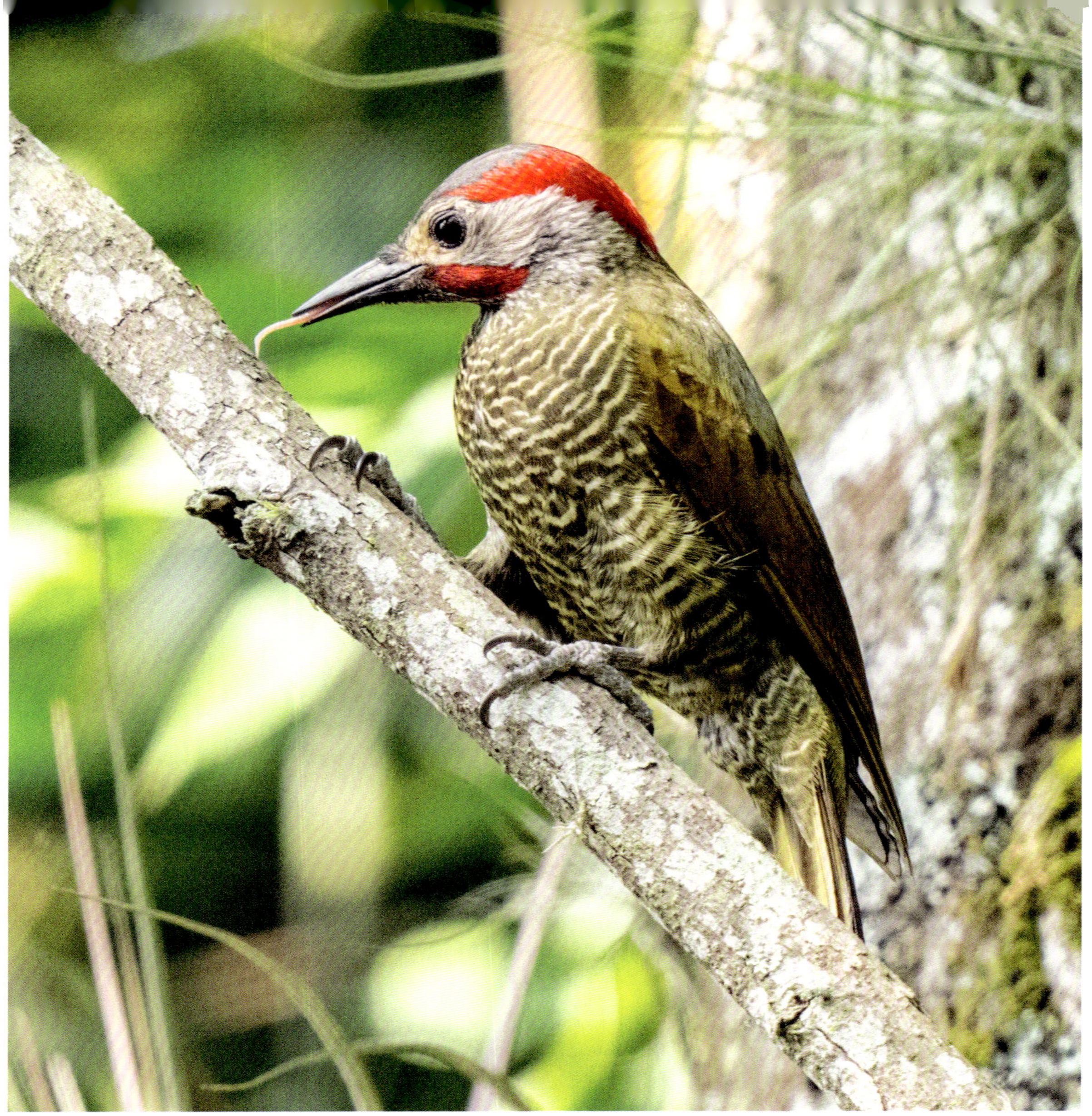

native foods are not readily available. Red-bellied Woodpeckers are master generalists, and all but the northernmost populations reside on breeding territories year-round. Some become more vegetarian, and each has adapted to feeding upon stored mast or migrating in order to survive the winter.

The Lewis's and the Red-headed execute very similar survival strategies. These two breed farther north than any other *Melanerpes* and are the North American woodpeckers most reliant on fly-catching. As a result, they are the most migratory.

For both species, migrations vary by year, region, and even population. Both are mostly migratory at the northernmost and interior parts of their distribution and mostly year-round residents in southern areas. In many parts of their ranges, both demonstrate unpredictable facultative winter movements with populations that migrate in some but not all years based upon climatic conditions and the amount of hard mast available. Because of fluctuating annual food supplies, wintering locations may vary from year to year, with some places hosting many birds one winter and none for the next several.

These two species move from breeding areas in open pine, oak, or burned habitats with plenty of flying insects and well-decayed snags to winter in areas with abundant food such as acorns, nuts, grain, seeds, or fruit. In late August, members of each species begin low-altitude slow migrations,

The Bronze-winged Woodpecker is recognized by the American Ornithological Society (AOS), and thus by this book, as a northern Mexican subspecies of the Golden-olive Woodpecker. However, the International Ornithological Congress (IOC) recognizes it as a separate species based upon its significantly different call, as well as some marking variations.

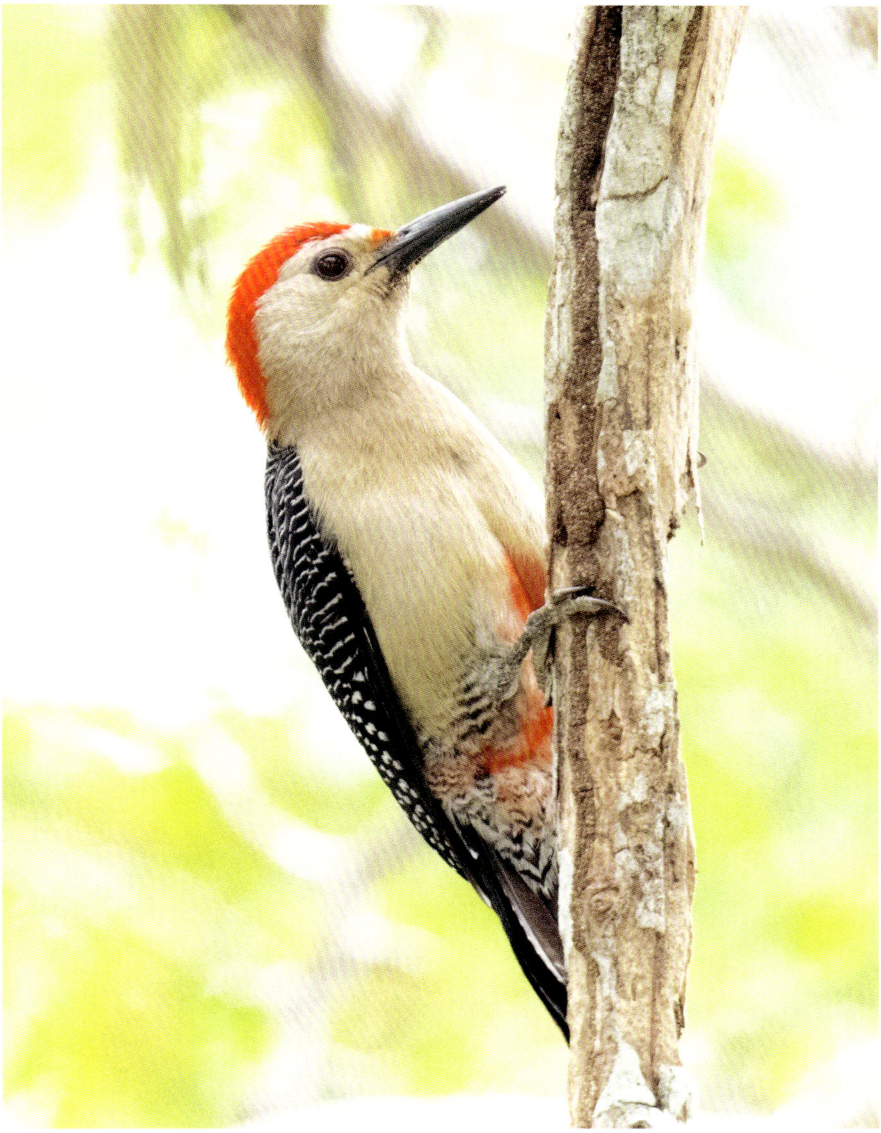

The AOS considers the Velasquez Woodpecker a subspecies of the Golden-fronted Woodpecker. Their calls are the same but the Velasquez's plumage lacks yellow or orange, making it look more like a Red-bellied Woodpecker. On the other hand, the IOC recognizes this as a separate species at the time of this writing, although this designation is under review.

allowing them to spot and take advantage of food along the way, which may encourage them to stay. Through the winter they feed on mast, especially acorns, as a last resort and utilize this diet when other foods are not available. For all the commonalities in their winter behaviors, there are some differences.

Lewis's Woodpeckers are predictable migrants from the northern and interior parts of their range. Migrating birds move to the southern part of their breeding range, from southern Oregon to Baja California, as well as New Mexico and Arizona, but many winter in parts of California, Nevada, Arizona, and northern Mexico, where they do not breed. Multiple family groups of Lewis's often congregate in winter, feeding on acorns, nuts, or fruit for as long as it lasts, which may or may not encompass the entire winter.

Lewis's Woodpeckers more frequently breed in mountain pine forests and in some areas begin their migration by moving to higher elevations with family groups before returning to lowlands and flying toward winter areas from August through October. Nomadic flocks of up to 150 birds, comprising smaller subgroups of a dozen or so, travel between 60 and 620 miles during these times. They depart wintering areas in mid- to late April and arrive on breeding grounds in early May.

Red-headed Woodpeckers are predictable migrants only in the western Great Plains, the northern Great Lakes, and the far northeast of their range. Most migrants move south, but some move east, and even some southern birds may move north. They are known to follow watercourses on their return in fall. Adults and juveniles tend to follow the same paths year to year and may possess sun compass orientation. Adults usually precede juveniles in migration.

Wintering Red-headeds are generally found within the breeding range, with the exception of some birds in a narrow band in Texas. When mast is plentiful, winter territories can be small and populations can be dense, but more frequently birds are found singly. Their fall migration extends into early November, a few weeks later than Lewis's. They migrate silently during the day, in loose flocks, sometimes along with Blue Jays. During their spring migration, they are believed to migrate at night and arrive on breeding grounds between March and May.

Other Genera

The genera *Nesoctites* (Antillean Piculet), *Xiphidiopicus* (Cuban Green Woodpecker), and *Campephilus* (Pale-billed Woodpecker) each have one living member in North America that resides on stable home ranges year-round in tropical habitats.

SOCIAL TOLERANCE

Woodpeckers appear to adjust their social dynamics in complex ways in their efforts to secure food in winter. Some may weaken pair bonds and become more territorial to the exclusion of their mates and forage farther and in different strata while tolerating others of their own and other species.

While the energies of the spring and summer relate to nest sites, winter efforts are focused toward securing and, in some cases, competing for food against any animal showing interest, whether mammals, birds, or even their own mates. Contrasting encounters between Lewis's and Acorn Woodpeckers demonstrated for me the differences in the way woodpeckers respond to competitors when food is abundant during fall compared to when it is scarce during winter.

One late fall day in an oak woodland in eastern Washington, I watched several Lewis's Woodpeckers plunge headlong into the canopy of two adjacent oak trees to pluck acorns before breaking them on anvils and storing them in cracks in trees. A smaller Acorn Woodpecker landed at the top of one oak to the vocal protest of the Lewis's but remained unmolested to seek its own acorns and even fly-catch between the trees.

In contrast, on an early winter day in southern Oregon, I observed a wintering Lewis's Woodpecker moving acorn pieces between cracks in an oak snag. Several other Lewis's were tending their own caches in nearby trees, occasionally squawking when a neighbor came too close. An Acorn Woodpecker moved to within fifty yards of the trees, triggering two Lewis's to give chase, driving the intruder back into the denser woods a hundred yards away. I witnessed them driving away squirrels and scrub-jays with the same intensity. Although the larger Lewis's Woodpecker appeared dominant in this encounter, studies have shown that dominance is usually contextual and birds that are defending nests or food sites are often dominant over larger species when they are defending their own food stores or nest sites.

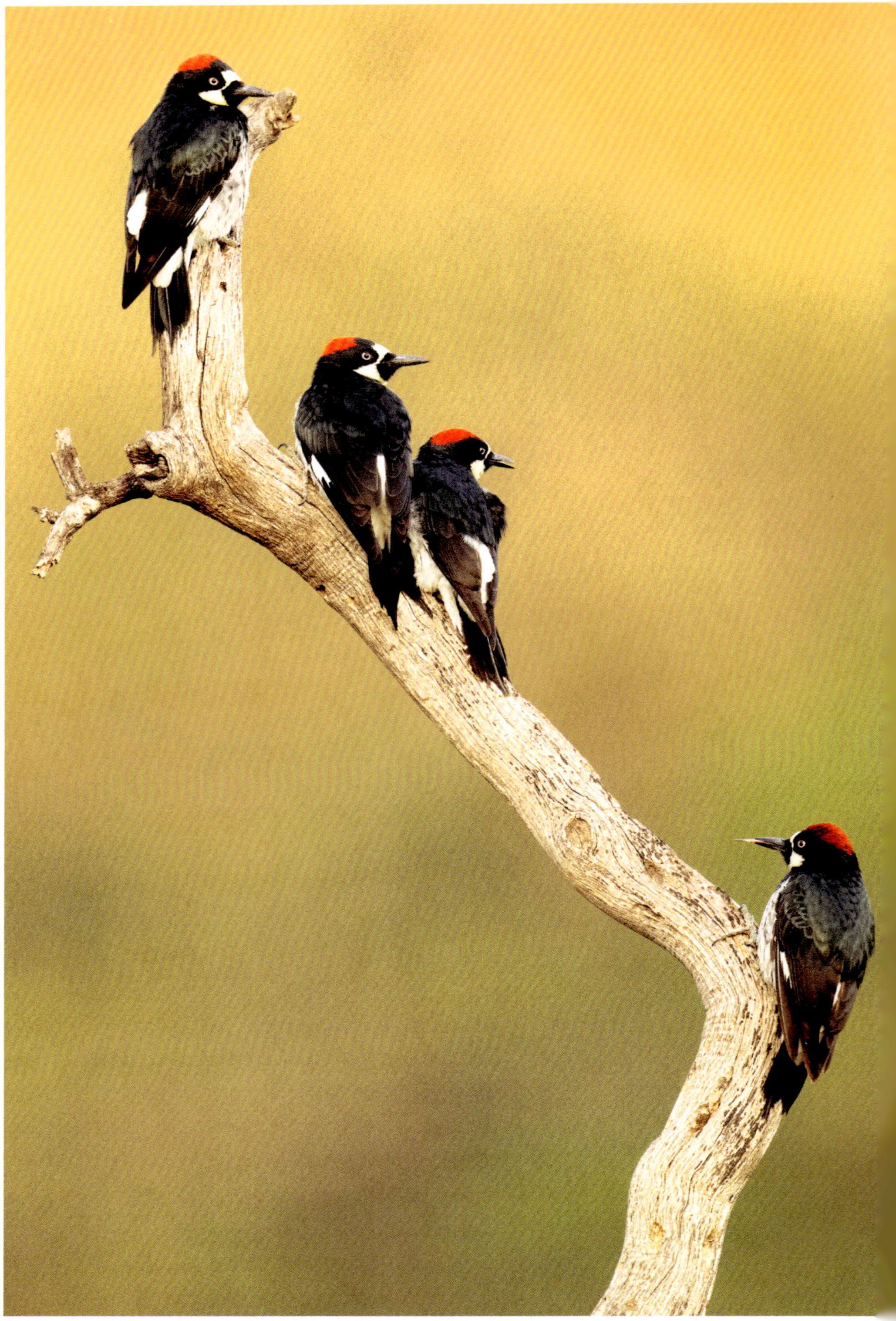

Cooperative breeders such as these Acorn Woodpeckers are the most tolerant of having others in close proximity.

Given the winter hostility I witnessed, it is surprising when woodpeckers tolerate or even cooperate with other birds, including other woodpeckers, during winter feeding.

Feeding in a single-species flock is least surprising, particularly with cooperative or colonial nesting species. Northern Flickers and Lewis's Woodpeckers often forage together in groups of up to a dozen or more on fields, lawns, or fruit trees. Fruit trees attract single-species groups of Jamaican and Hispaniolan Woodpeckers, and fellow *Melanerpes* Puerto Rican Woodpeckers may form unstructured groups of up to a dozen as they forage in winter.

Different species feeding together is far less common, especially when one species does most of the work. Some woodpeckers are very protective of their winter food sources. Anecdotes of Red-headed Woodpeckers driving Pileated Woodpeckers or other woodpeckers from their feeding areas and Red-cockaded chasing Downy Woodpeckers from theirs are to be expected, but it is more surprising when they behave commensally.

Pileated Woodpeckers often attract uninvited beneficiaries to their trench-like feeding cavities and the many arthropods they scatter during excavation. Hairy, Downy, and even Red-bellied Woodpeckers are known react to the calls of Pileated Woodpeckers in winter and seize prey dislodged by their work. Northern Flickers and Williamson's Sapsuckers are also known to forage in Pileated Woodpecker excavations.

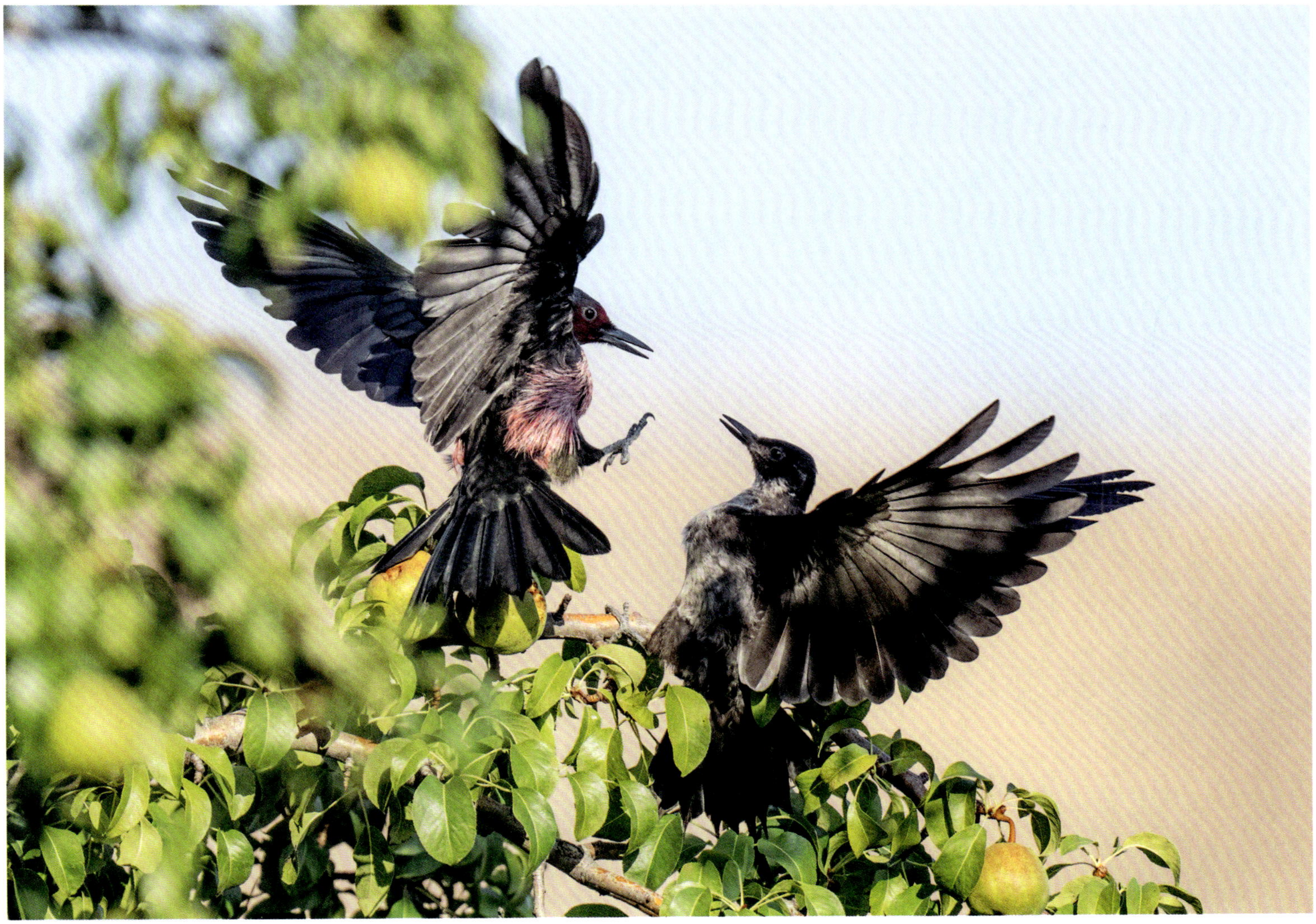

Lewis's Woodpeckers often nest and winter in loose colonies with little conflict, although these two confronted each other over a limited number of ripe pears.

CARIBBEAN ISLANDS

The Caribbean islands, emerald jewels sprinkled across the cerulean sea, host more than five hundred bird species, including twelve woodpeckers. The roughly 149,000 square miles of land comprise diverse tropical habitats, including savanna and pine and broadleaf forests.

Woodpeckers are mostly nonmigratory species that execute only short flights between trees, making them relatively poor dispersers, with seven endemics (Fernandina's Flicker, Antillean Piculet, and Cuban Green, Hispaniolan, Puerto Rican, Jamaican, and Guadeloupe Woodpeckers) found on only one major island and the West Indian Woodpecker found on a few. This is despite the fact that less than 125 miles of sea separates Cuba from Mexico and most of the islands from one another. Even the three continental species found here—Hairy Woodpeckers, Northern Flickers, and wintering Yellow-bellied Sapsuckers—have limited distributions.

The Fernandina's Flicker and Antillean Piculet require particular habitats, whereas other endemic woodpeckers are generalists adapted to exploit most forest types from sea level to mountaintops. All are keystone species that create cavities for many species, including some threatened or endangered species, such as the Red-crowned Parrot, Black-billed Parrot, Yellow-billed Parrot, Hispaniolan Parrot, as well as the Cuban Pygmy-Owl, Puerto Rican Owl, Bare-legged Owl, Fernandina's Flicker (a primary and secondary cavity nester), and the Antillean Piculet (also a primary and secondary cavity nester).

The adaptability of most of these woodpeckers makes them resilient, especially on the larger islands with forest habitat, as long as plenty of snags remain. The Puerto Rican and the "near-threatened" Guadeloupe Woodpecker, which live on only one small island, are more vulnerable. Fortunately, these have adapted by creating cavities in utility poles. Cuba's size gives it a better hedge against storms, but the "vulnerable" Fernandina's Flicker is a weak cavity excavator threatened by parrot hunters felling snags and native West Indian Woodpeckers aggressively evicting them from cavities in any remaining snags. The extinction of the Ivory-billed Woodpecker in Cuba due to logging and clearing forest for agriculture serves as a warning.

Roughly 13 percent of Caribbean birds are considered threatened, and more than 8 percent are "near threatened." The growing human population means forests are converted to urban and agricultural lands, which are only good for generalists, putting endemic island species at risk of extinction.

Utilizing more shade coffee and cocoa plantations, as well as silvicultural systems, can provide good habitat for native and migratory forest species, especially when contiguous to original plant communities. More education about challenges posed by tourist infrastructure, the cage-bird trade, and hunting is needed. We must slow climate change, which is increasing the severity and frequency of storms that threaten snags, inundate habitat with seawater, and alter the remaining degraded and fragmented habitat.

The Caribbean islands are generally tropical moist forests but contain a variety of habitats from wet savanna to dry forest. The most species-rich habitats combine mature palm trees and snags, openings, and tropical forest.

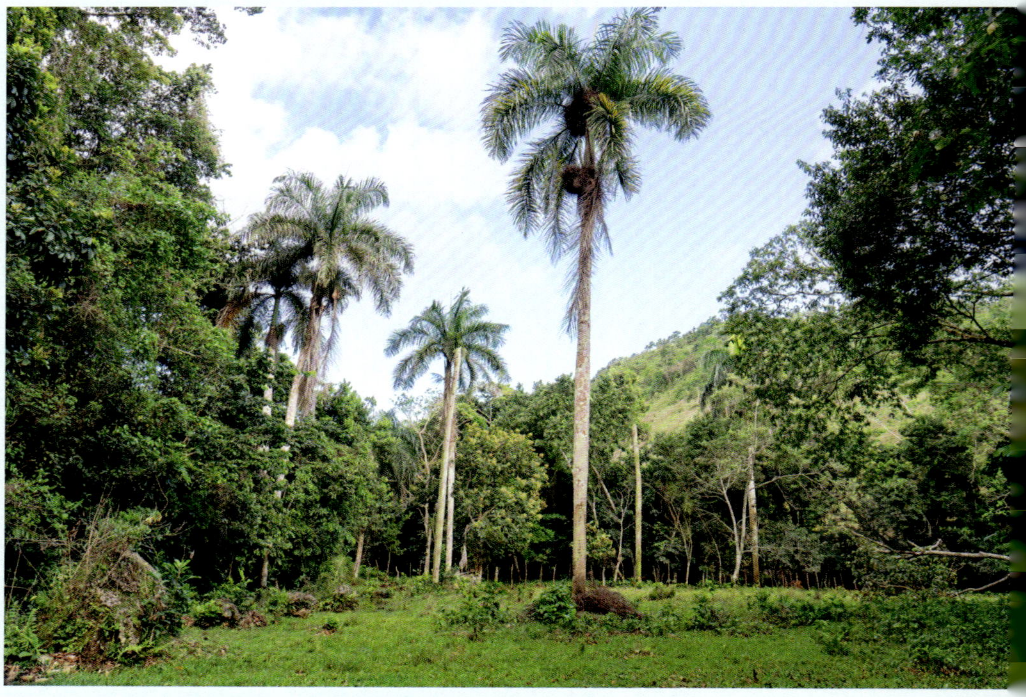

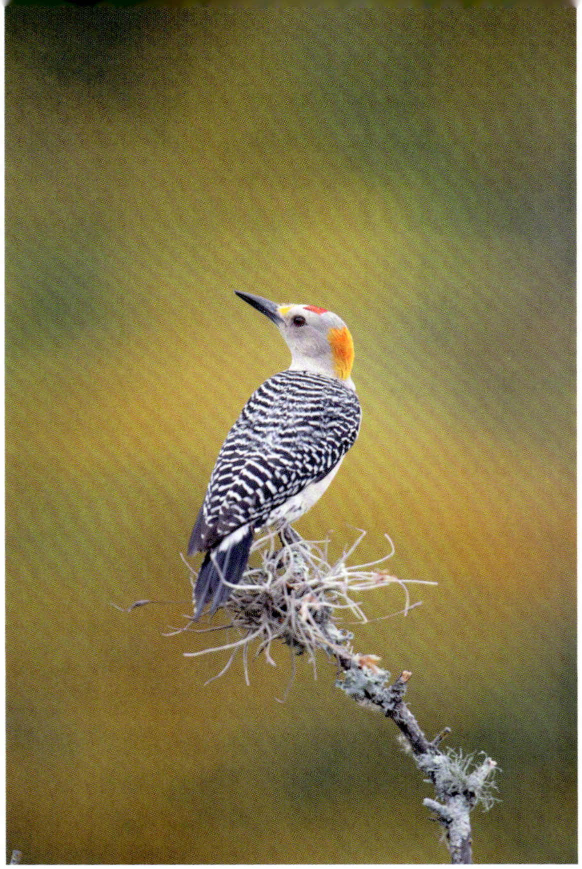

Gleaning and probing for insects, as well as consuming fruit, notably that of prickly-pear cactus, occupies most of the foraging efforts of the Golden-fronted Woodpecker. Although it excavates its own nest cavities and roost sites and will pound on wood to make prey reveal its location, it rarely excavates for subsurface prey.

Mixed-Species Foraging Flocks

Like the murmur of pebbles rolling under an incoming wave, soft calls and small shapes were racing as if in a current on the limbs beside and above me. Each ball of feathers alighted and pecked for mere moments on the smallest surfaces: Black-capped and Chestnut-backed Chickadees flitting high from bud to bud, a Red-breasted Nuthatch hanging from a moss-covered twig at eye level, and a Downy Woodpecker digging into a slender lichen-encrusted branch with his forceps-like bill, trailed by a Ruby-crowned Kinglet seemingly cleaning up what was left after gray and green slivers flew. And as quickly as it had rolled in, the flock I was observing in northwestern California was gone, giving me no chance to raise my lens.

Mixed-species foraging flocks, also called mixed-species feeding flocks, include woodpeckers as either a core member or an intentional beneficiary. These flocks form when birds of many different species, especially smaller insectivorous ones, forage as they move from place to place in a more or less cohesive unit. During winter, when food can be more difficult to find, these foraging groups can help woodpeckers locate rich pockets of prey. Woodpeckers' participation seems to increase as the temperature decreases.

The core flock members are the generalist gleaners that tolerate feeding alongside other species, but the components vary by habitat and region. Groups benefit from including other species with unique foraging strengths and feeding preferences, meaning they add to the group's skill set without competing as directly for food. The most effective groups include one species with better vision at different ranges, another with more sensitive hearing, others that are alert or more easily heard when they warn of predators, among many other possible unique attributes. Some members prefer to hunt high in the forest column, and others feed lower and target different prey. The group may include sentinels, loud and less social species that tend to perch higher, where they are better able to see and warn flock members of danger. Larger species, sometimes called attendants, join or follow a flock only after it has formed and passed through their territory.

Such flocks create a lot of noise, which probably attracts a lot of predators, but it could also distract a predator and certainly allows a group to more quickly spot or warn of danger. Collectively, a group is more likely to know what areas are depleted of prey and may be able to overwhelm the defenses of territorial birds. Each member also benefits from more eyes to spot prey and from easier access to the prey that is uncovered, dislodged, scattered, or driven out of hiding during the commotion.

Woodpeckers serve as both core members and attendants, benefitting by being able to focus on foraging while others alert them to danger and adding value to the group by uncovering or moving subsurface prey the others can't access. Most of the other birds do not compete directly with the woodpeckers for feeding sites or food, except when multiple woodpeckers are part of a mixed flock. Such groups can contain several species of woodpeckers, particularly in more tropical habitats. Woodpeckers that benefit from such groups tend to recognize and respond to the calls of the core participants.

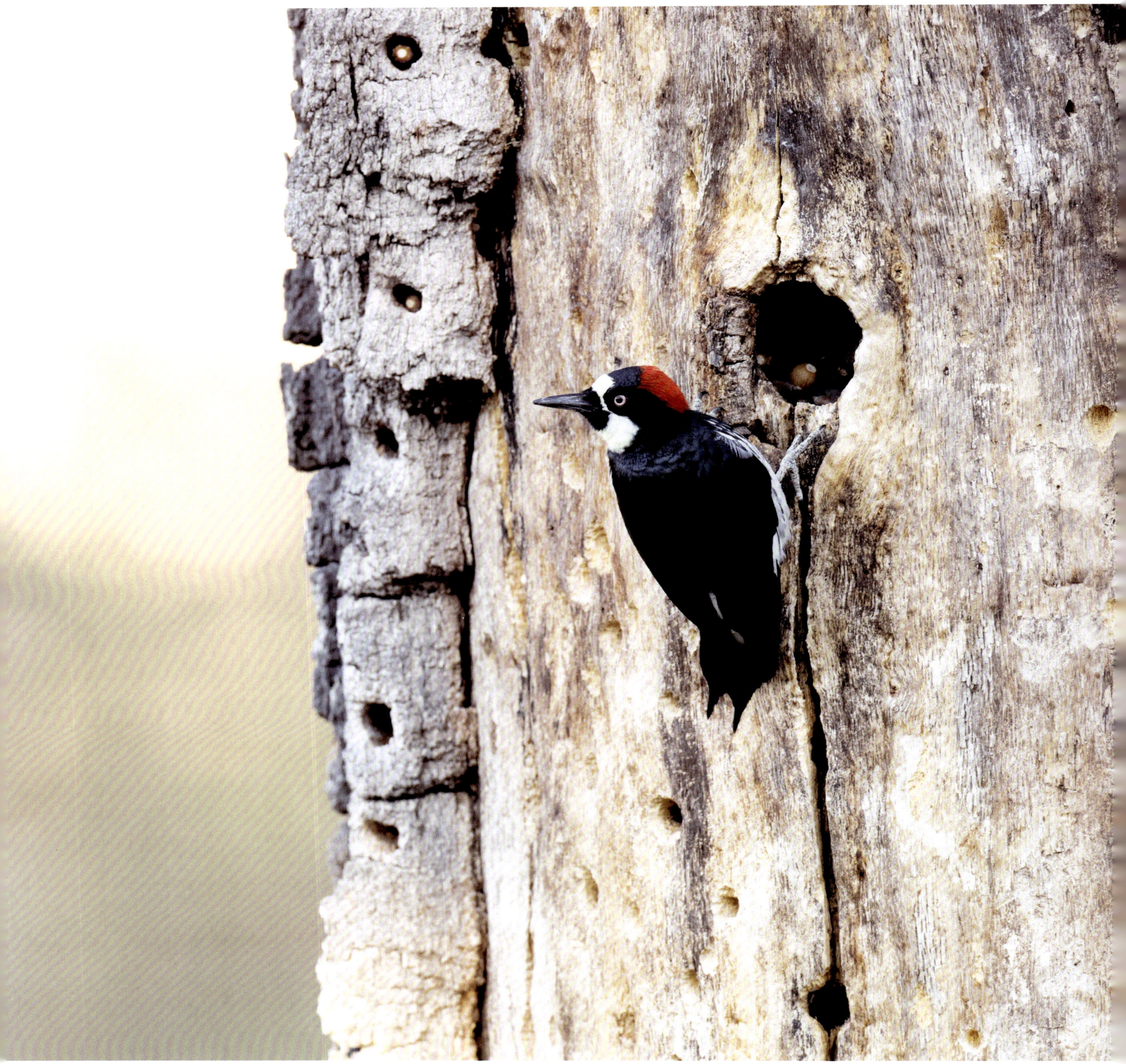

Opposite: During a mast year, when acorn crops are strong, Acorn Woodpeckers may fill existing granaries and begin filling knotholes, crevices, and even nest and roost cavities.

In Nearctic habitats, the leaders or core members of these groups are usually chickadees, nuthatches, kinglets, wood warblers, Brown Creepers, and Downy Woodpeckers. Several other woodpeckers are attendant to these groups, including but not limited to Acorn, Arizona, Golden-fronted, Hairy, Red-bellied, and Red-cockaded, as well as Yellow-bellied and Red-naped (but not Williamson's) Sapsuckers and Northern Flickers.

Neotropical mixed-species foraging flocks may form around insectivorous species such as antshrikes, woodcreepers, wood warblers, and some tanagers, and are joined at the core by Antillean Piculets or Smoky-brown Woodpeckers and attendant species such as Chestnut-colored, Black-cheeked, Golden-fronted, Golden-olive, Ladder-backed, Lineated, Puerto Rican, or Strickland's Woodpeckers.

Feeding Stations

Several woodpecker species will eagerly feed with woodpeckers of their own and other species at feeding stations, especially during the coldest days of winter or when food is scarce. Downy, Hairy, and Red-bellied Woodpeckers and occasionally Northern Flickers are common visitors. Depending on the region, habitat, food offered, and competing food available, other possible visitors include Red-headed, Golden-fronted, Acorn, Ladder-backed, Gila, Pileated, Arizona, White-headed, Golden-cheeked, and Lewis's Woodpeckers, as well as Yellow-bellied and Red-naped Sapsuckers.

Exercise caution when providing feeding stations, as feeders that are not regularly sanitized can breed and spread pathogens. Suet and nuts are normally the favored woodpecker feeder fare, but if the birds eat suet in warm weather, it can damage their feathers.

Northern Flickers like this Red-shafted female conduct most of their foraging on the ground, motivating them to migrate downslope or south to snow-free areas during the winter. Note the mud on the bill of this bird after she probed an ant mound.

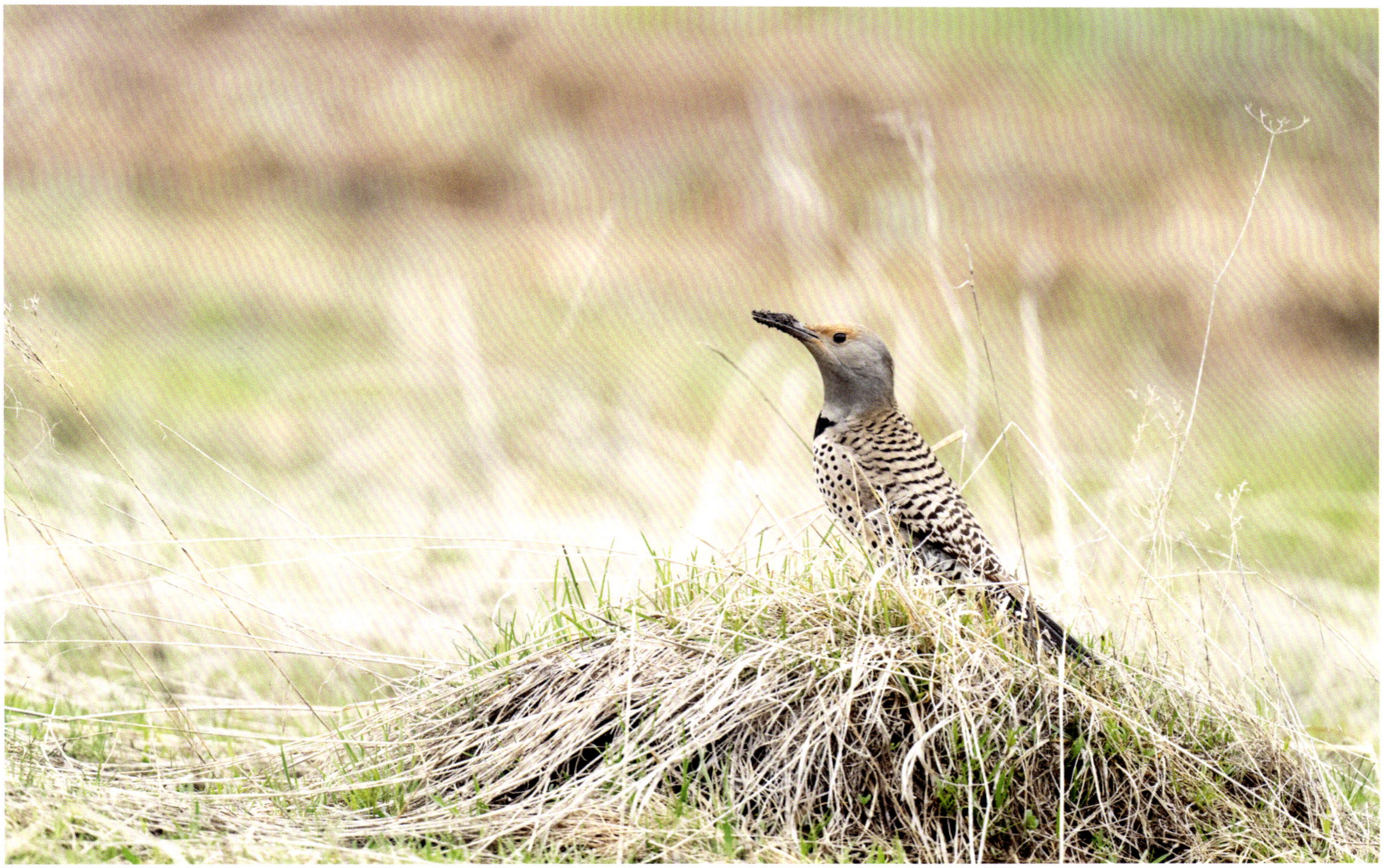

ROOST SITES

Every September I take down three Northern Flicker nest boxes in my backyard in Seattle, clean them, fill them with wood chips, and put them back up in hopes they may be chosen as nest sites during March's courtship rituals.

One year, in the middle of October, a female Northern Flicker perched at the entrances of two of the nest boxes for about five minutes each before a second female joined her. They quickly faced off. With wings and tails spread aggressively like red fans, they waved toward each other, punching at the air above each other's head with their bills. The intruder left quickly, but to my disappointment, so did the victor.

Two weeks later, temperatures dipped from 52 to 32 degrees Fahrenheit in one week. As if in response to the new chill, a female flicker flew to a box and began removing wood chips and flicking them into the cold breeze. She removed them from the outside and eventually from inside until she was able to roost against the wall of the box with her head and bill hidden from view. She returned many days before sunset and left just after sunrise until the middle of March. The second female never returned. This situation made me consider that woodpeckers must find and in some cases prepare their roost sites before the worst weather hits.

On the day the flicker began readying her roost box, I visited a mature lowland mixed-conifer forest in Seattle to see if I could witness other species preparing for winter. My eyes intuitively focused on the egg-shaped, fist-size hole about forty feet from the ground, rewarding me with a view of stiff black feathers moving up and down from within the cavity, followed by the full shape of a male Pileated Woodpecker, mouth full of wood chips. I returned over the next two days to see the male complete his roost cavity.

On the same day and in the same stand of alders, thirty yards from the Pileated, the subtle tapping of a Downy Woodpecker betrayed the location of its own roost cavity as it spat duff from the entrance. Both cavities faced away from typical winter winds. Neither of these woodpecker's cavities showed visible clues of being used unless I was nearby at the exact moment the bird entered or exited in the morning or evening.

The Northern Flicker and the Pileated and Downy Woodpeckers in the moderate lowland Pacific Northwest climate led me to understand roost cavities as essential nighttime protection once near-freezing temperatures turned summer to fall, but a winter's day in Minnesota reframed things a bit.

I arrived at a snow-blanketed oak woodland in northern Minnesota just before sunrise to an almost eerie silence. As I trudged on ice-encrusted snow among the shadows of giant trees, I heartened myself with the knowledge that Red-headed Woodpeckers, encouraged by a rich fall acorn harvest, had overwintered.

I did not see or hear anything until the golden tones of morning gave way to the gray hues of early afternoon and I focused on seeing color—any color that might hint at a woodpecker. The red looked like a spot of paint on the tree until the bird emerged from a high cavity and clung to the side of the tree, feathers fluffed, making the bird appear round. The normally hyperactive Red-headed Woodpecker moved laconically up the tree, finally settling at a spot where it pried a sliver of acorn from a crack before sluggishly climbing down the tree and back into the cavity. That experience combined with subsequent encounters with Lewis's, Acorn, and White-headed Woodpeckers, opened my mind to the way different species utilize winter cavities.

Woodpeckers, like most diurnal birds, are vulnerable in the open during cold weather, storms, or at night. They have poor vision after dark, so from the time they leave their nest they minimize their risks, first by roosting at night in dense vegetation, under a slab of bark, on the side of a trunk, underneath branches or even the eaves of buildings, under the cover of a bridge, or in a roost cavity when available.

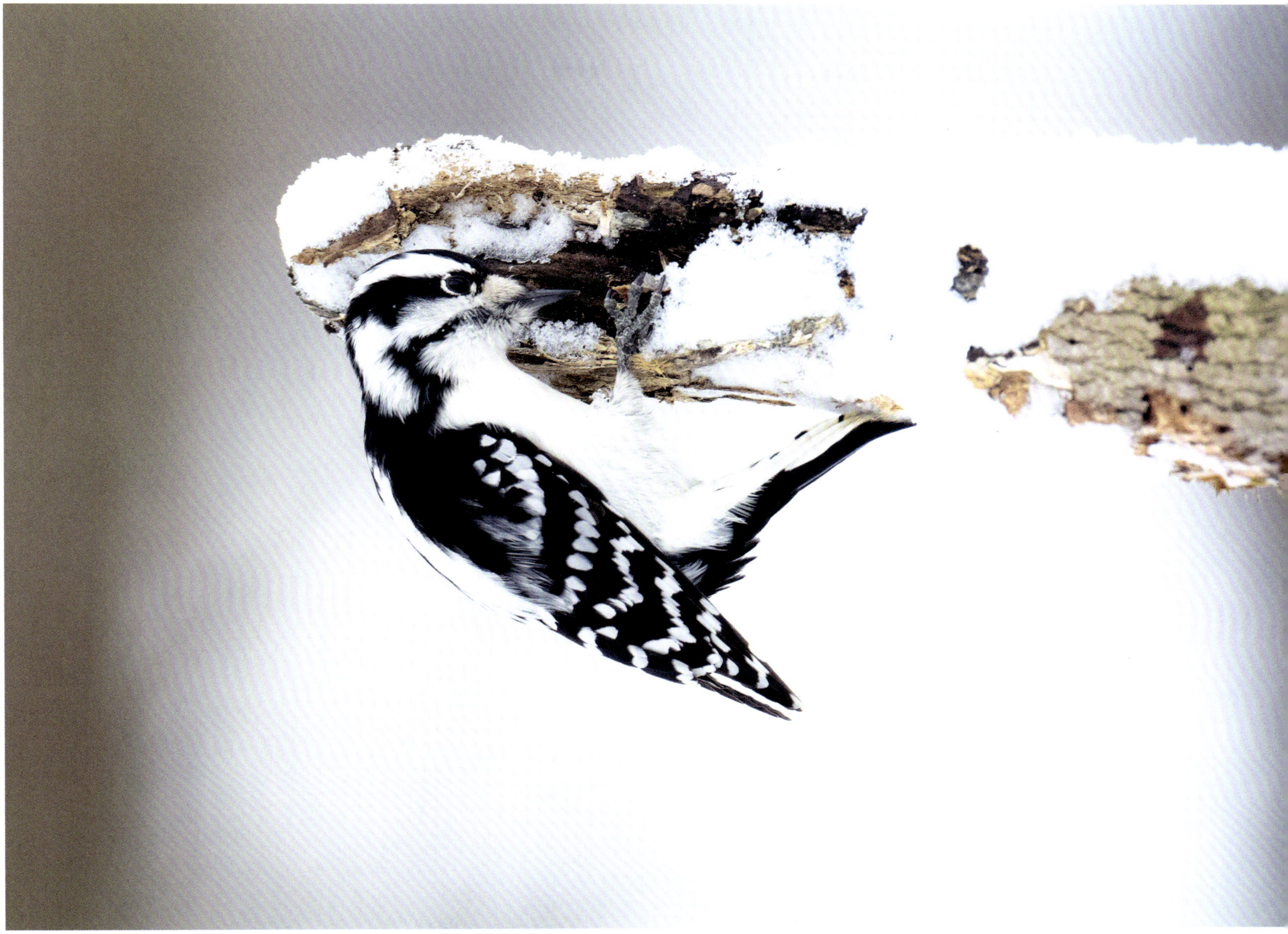

Somewhat surprisingly, the diminutive Downy Woodpecker is able to avoid migration because of its ability to probe and excavate for subsurface invertebrates.

Roost cavities are sometimes existing natural cavities, old or rejected nest cavities of their own or another species, or a hollow tree or nest box, and at other times they are intentionally created. A roost cavity does not need to be as deep or extensive as a nest cavity, so when created solely for this purpose, it may be just large enough to keep the occupant out of the elements and hidden from view, with an entrance small enough to reduce access by predators. Such cavities are created in a fraction of the time it takes to excavate a nest cavity.

Many of these preferences are not known for individual species, but we do know that the stronger the excavator, the pickier it can be, and the less likely it is to roost in a tree that might break apart, provide less insulation, have too large of an entrance, or fall. Most woodpeckers roost in their own cavities, which require time and effort to create. Juveniles typically spend their first two to five weeks out of the nest sleeping in sheltered places on trees until they locate roost cavities, although this happens more quickly in areas with a lot of abandoned holes.

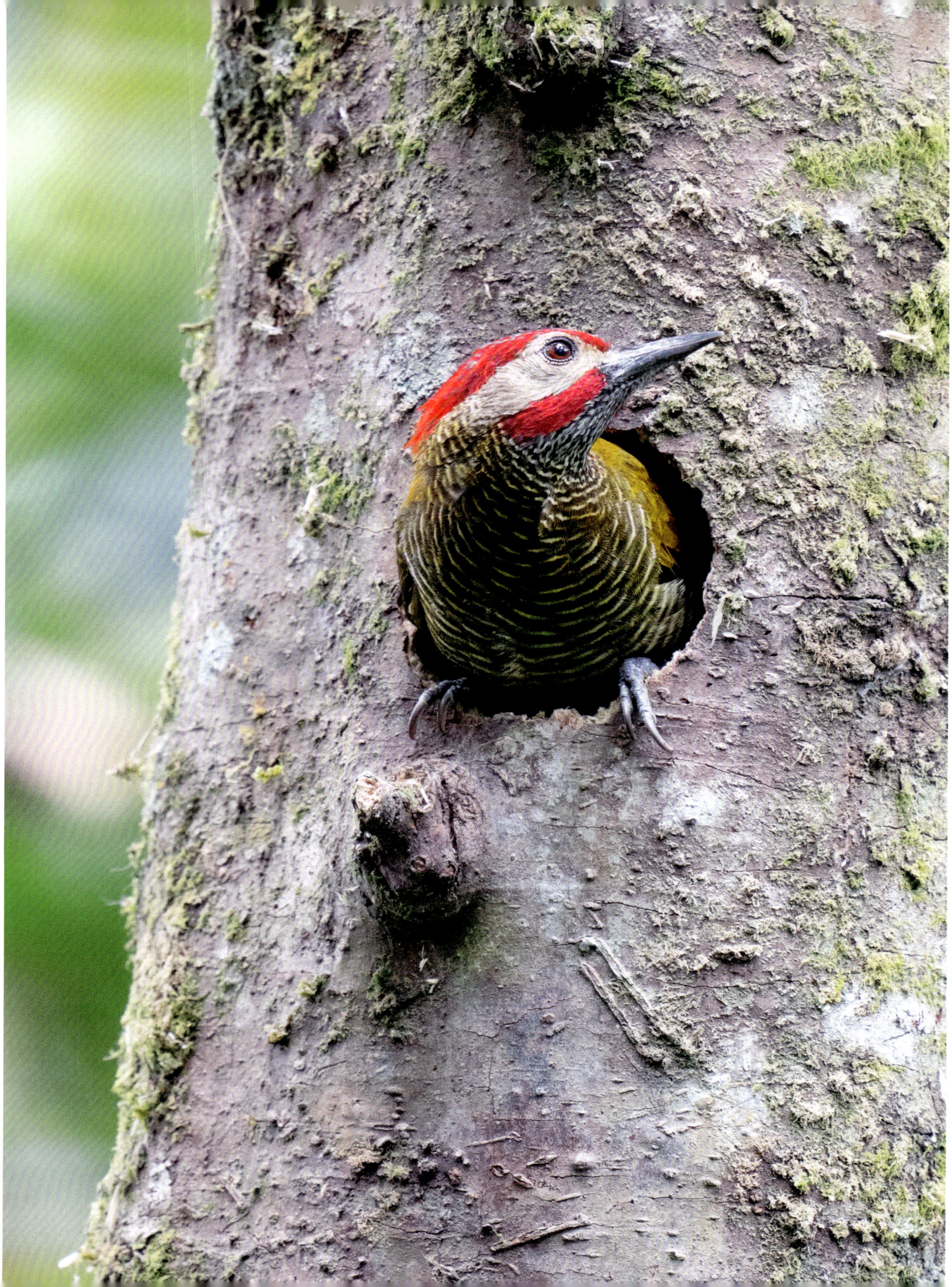

Being familiar with more roost cavities gives woodpeckers the option of alternating between them, both to reduce the chance of predation and to conserve energy when utilizing a larger home range. This helps explain why both Pileated and White-headed Woodpeckers are known to use more than ten roosting cavities outside the breeding season.

Many woodpeckers use roost cavities year-round, some rarely do, and some species use them only in winter. In the north, temperatures plummet when the sun sets, and there is no opportunity to warm through food during the extra-long nights. The inside of cavities cools more slowly and the walls break the wind, both of which reduce heat loss. The woodpecker's body raises the cavity temperature further. One study of Hairy Woodpeckers wintering in Minnesota found that roost cavities averaged 37 to 43 degrees Fahrenheit warmer than the air outside the cavity.

White-headed Woodpeckers usually roost in a single cavity at night through most of the winter, but I have also observed Red-headed, Lewis's, Acorn, and White-headed Woodpeckers regularly utilizing roost cavities in the day during snowstorms. At other times of the year, White-headed Woodpeckers might roost behind sloughed bark or in cracks and crevices in tree trunks.

Yellow-bellied, Red-naped, and Red-breasted Sapsuckers utilize roost cavities less than other woodpeckers and instead sleep clinging to a trunk, under a limb, and hidden by foliage in a vertical position with their bill under their scapular feathers. These sapsuckers never use roost cavities after the young leave the nest, although little is known about whether they do in winter. Williamson's Sapsuckers, in contrast, regularly roost in old nest cavities and also use nest starts or natural cavities but do not create roosting-specific cavities.

While sharing a roost cavity is uncommon, particularly among Nearctic species, sharing a roost tree is not, and a single tree was reported to have simultaneously hosted roosting Northern Flickers and Pileated and Hairy Woodpeckers. Pileated Woodpeckers have also been known to share the actual roost chamber with nesting Vaux's Swifts on one occasion, another time with a Hairy Woodpecker, and in a third case with a Northern Flicker. Chimney Swifts and even bats have been known to nest in hollow trees with Pileated Woodpeckers.

Sometimes roost trees are later used as nests and vice versa by an individual species, but not with Pileated Woodpeckers. The Pileated will roost in a previous nest but is not known to nest in previous roost cavities. They have more tolerance for smaller, more decayed roost cavities with more entrance holes than they would for nests. Extra holes in roost cavities provide more exits in case of invasion by a predator.

Most woodpeckers roost alone, which reduces the chance of losing a breeding pair or an entire family, although roosting in groups offers biological efficiency. One study on Acorn Woodpeckers demonstrated that four or more individuals in one cavity reduced heat loss by 17 percent.

Gray-breasted Woodpeckers, a species from warm habitats in southwestern Mexico, often roost with two to three others, and their roosts may be grouped close to several other roosts so that more than twenty-five birds roost in a tight group.

Woodpeckers use roost cavities most often at night, when the weather is wet or cold, or to guard nest sites. Golden-olive Woodpeckers begin roosting in and guarding potential nest sites while wintering woodpeckers have yet to begin their migration north.

AS WINTER TRANSITIONS . . .

The damage from a previous Florida hurricane was evident from a distance as miles of what appeared to be pale-gray sticks rose from the flat, flooded pine-oak woodland, so densely packed that they looked like quills on the back of an agitated porcupine. I was in southwest Florida to study the behavior of Red-headed Woodpeckers during the last days of winter. Descending from a hill with soaring pines and sprawling oaks at my back, I looked down across the thick snags and willow-lined ponds in front of me and could see several woodpeckers flying. Clearly this area could host a plethora of insects and the woodpeckers reliant upon them. The cackling exchanges of two Pileated Woodpeckers blended with the trill of a Northern Flicker, the occasional squeaks of Downy Woodpeckers, and the mews of wintering Yellow-bellied Sapsuckers as each species foraged. The more discordant calls came from Blue Jays and Red-bellied and Red-headed Woodpeckers. I did not need to look hard to find these last three species, which continually announced themselves as they foraged loudly near one another. I was hoping I might even see a Red-cockaded Woodpecker, as I had seen one nearby on the previous day.

After a couple of days, it was relatively easy to discern Red-headed Woodpecker territories, since several adults seemed to be anchored at clumps of specific snags. When another Red-headed or a Red-bellied crossed an invisible line, one of several behaviors occurred depending on species, how far it had ventured into another woodpecker's territory, and what it was attempting. Each individual Red-headed's territory seemed to be less than an acre and was occupied by one adult using the snags for hunting, breaking up and storing food, and roosting overnight. I imagined that one of the roost holes might serve as a future nesting cavity. In addition to the snags, the individuals seemed to defend the surrounding open areas and some of the oaks where they foraged.

Whenever multiple Red-headed Woodpeckers foraged near one another without conflict, I knew this was outside either's defended territory, although they still displaced Downy and Red-bellied Woodpeckers in these areas and raised their wings and pointed their bills in agitation when another of their kind came too close. The Red-headed appeared dominant over all woodpeckers and most other birds, with Downy and Red-bellied Woodpeckers as well as nuthatches receiving the most violent treatment.

Intruding members of their own species, even the previous year's mate, were immediately chased and supplanted. It was clear when another Red-headed had intruded, since suddenly the defending Red-headed flew swiftly to the top of the highest bleached snag; raised the brilliant red feathers on its head, giving the impression of a lion's mane; and screamed a raucous *kerr-kerr-kerr*. If the visitor did not retreat, the resident bird flew to a limb and bobbed its head toward the rival, wings slightly drooped, tail pointed up, before giving chase. Attempts to hide behind trees usually ended with a spiraling pursuit, and when the Red-headed intruder retreated, it was chased vigorously through the treetops until well out of sight.

This antagonistic response surprised me, given the defender's earlier tolerance of another Red-headed Woodpecker. It did not react until the visitor landed on one of its storage trees, and even then flew with much less energy and fewer vocalizations. When the visitor retreated, it landed fifty yards away, where it resumed peaceably working on its own acorn storage area.

The more tolerant interaction made much more sense when I returned to this same place in spring. I witnessed the conflicts between the two on these neighboring territories transform into courtship as the birds drummed at and amorously chased each other through both territories. While it can be difficult to distinguish mating rituals from aggressive interactions, the copulation gave the intent away. The territory the two birds eventually defended included the storage trees that each had protected during winter, with one later serving as the nest tree.

The coldest of winter days encourage Red-headed Woodpeckers to conserve their energy and feed upon pieces of stored acorns and other mast wedged within cracks in snags.

Page 184: Descending to the ground requires vigilance as a White-headed Woodpecker carefully prepares to take a drink.

Page 185: Lineated Woodpeckers, like this female, are a smaller relative of Pileated Woodpeckers. They are found in Mexico's coastal forests but in more open and drier habitats than their northern cousins.

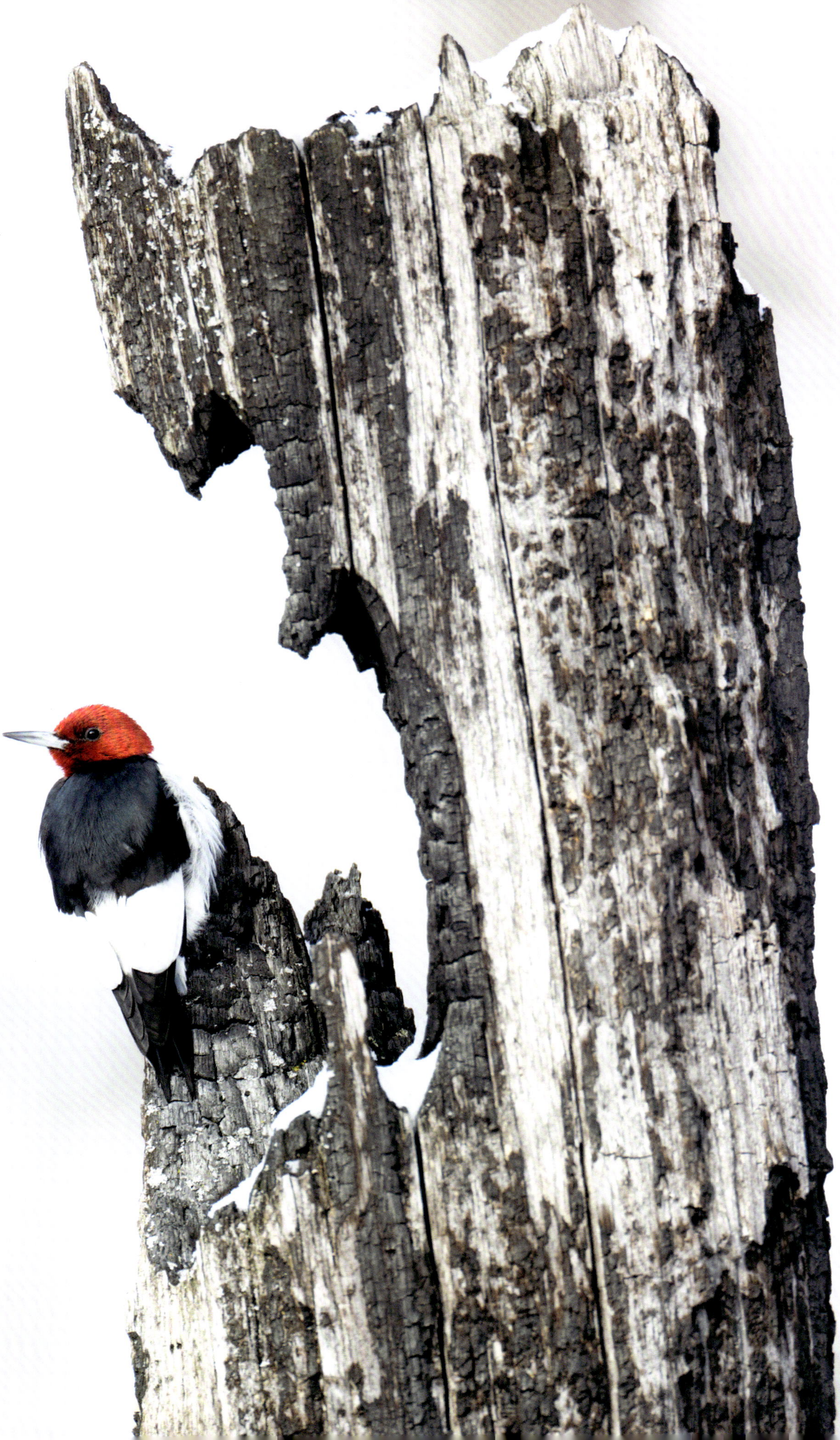

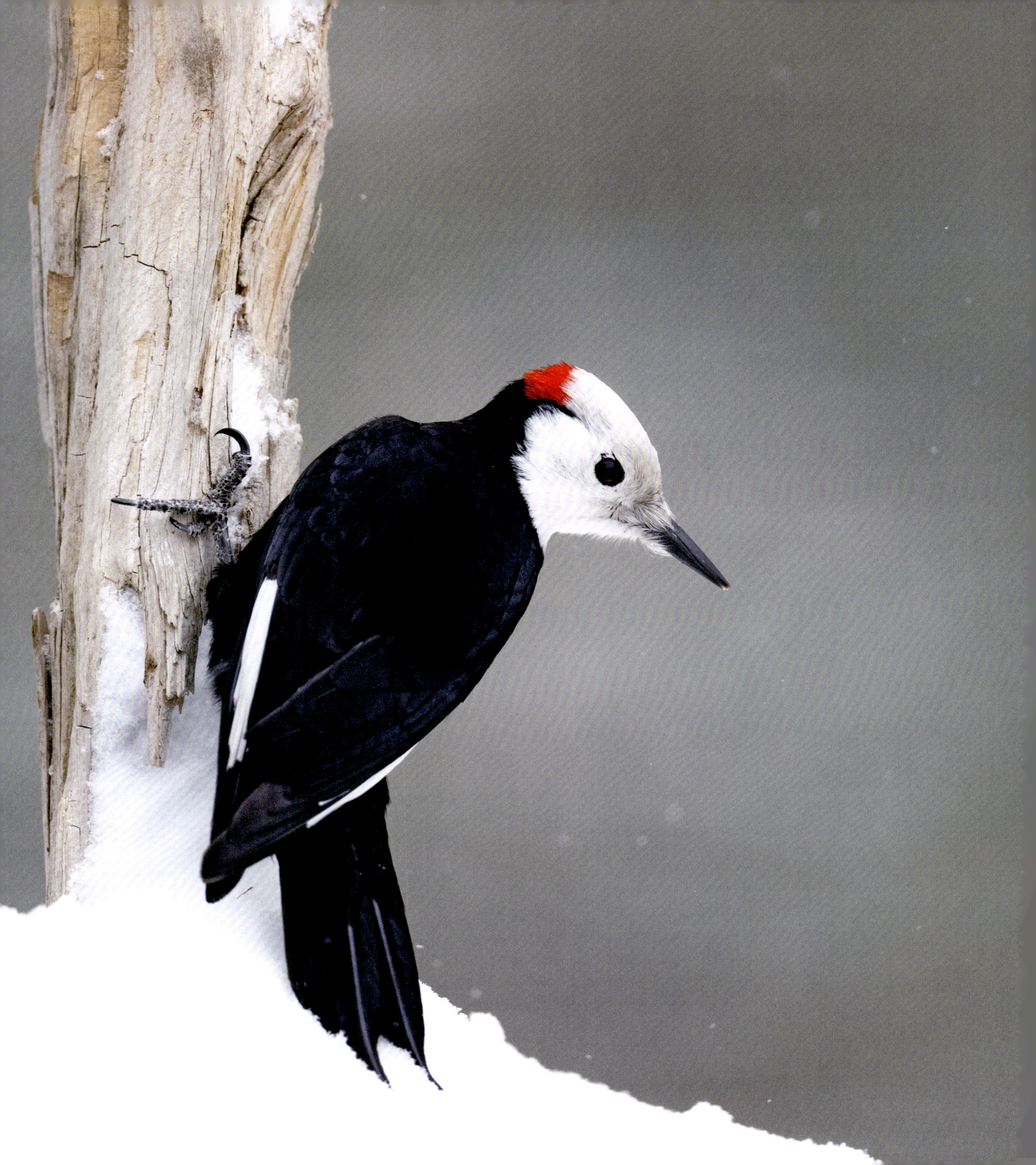

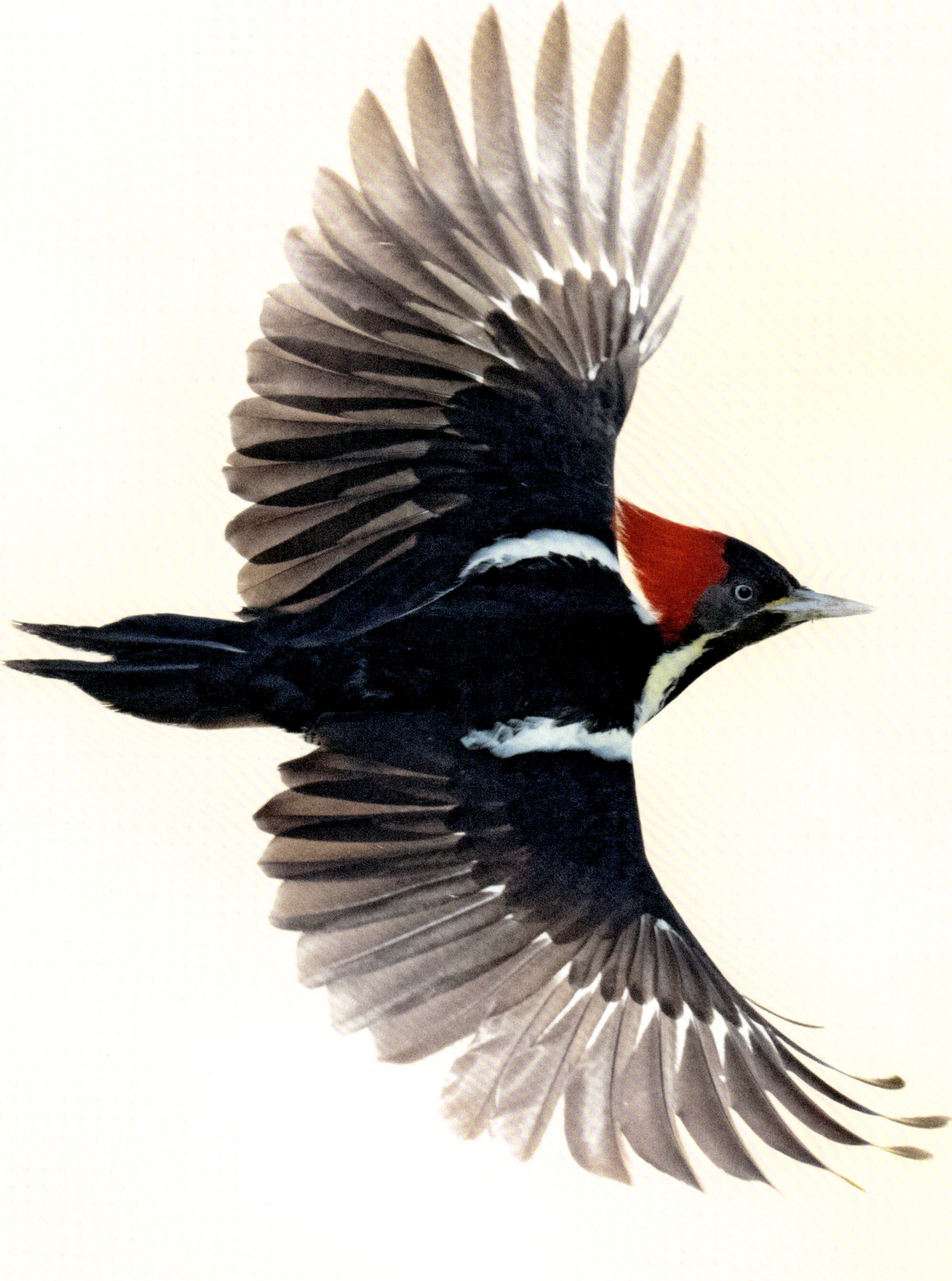

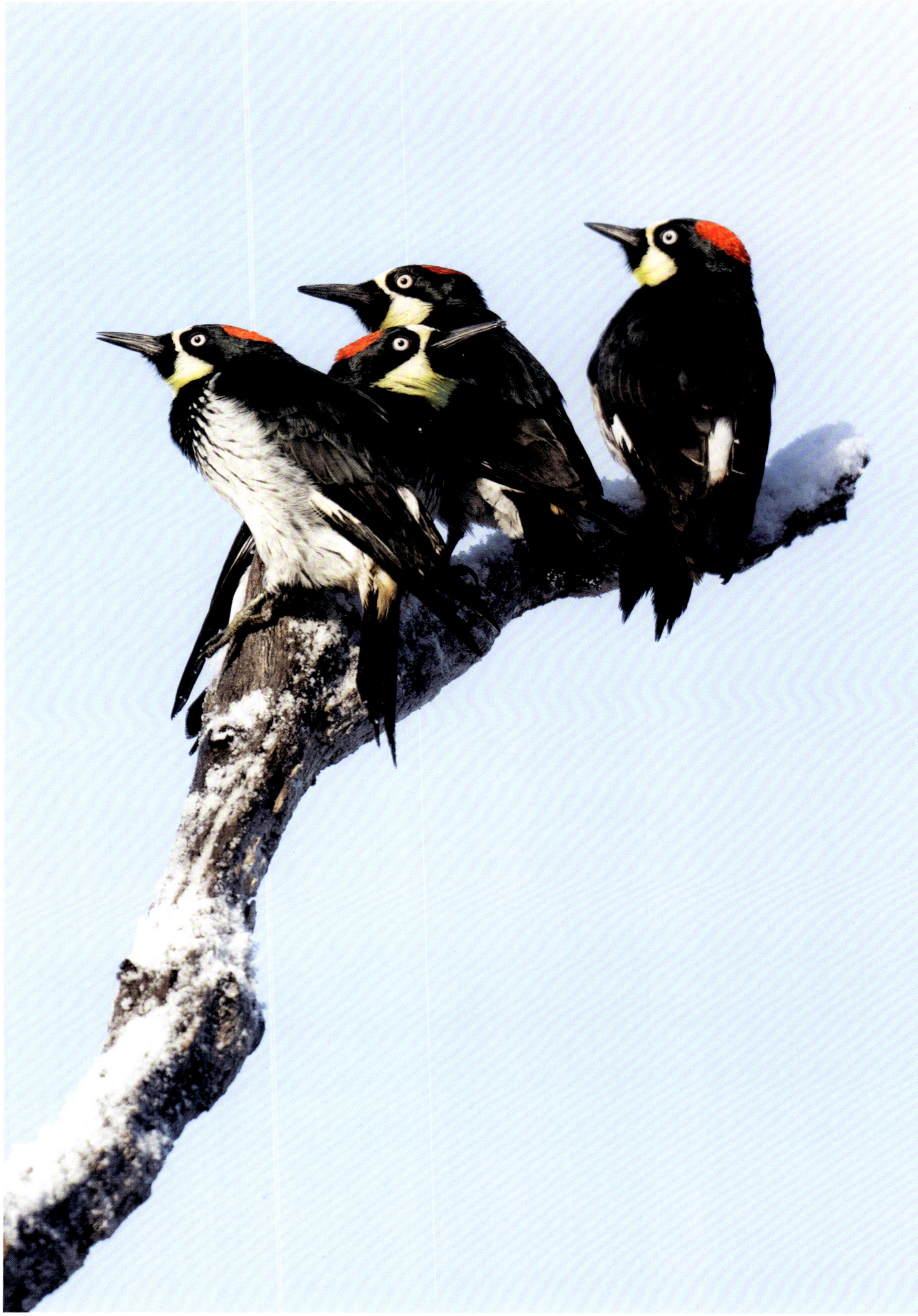

Four members of a family group of Acorn Woodpeckers perch ready to confront a neighboring group over some stored acorns.

Opposite: Red-bellied Woodpeckers are among the most adaptive woodpeckers, which has allowed them to expand their range more than any other species. While not a mast specialist, they will collect acorns and store pieces on their territories, although unlike the Red-headed, Lewis's, or Acorn Woodpeckers, they do not defend these caches. Some northern populations move south during the winter.

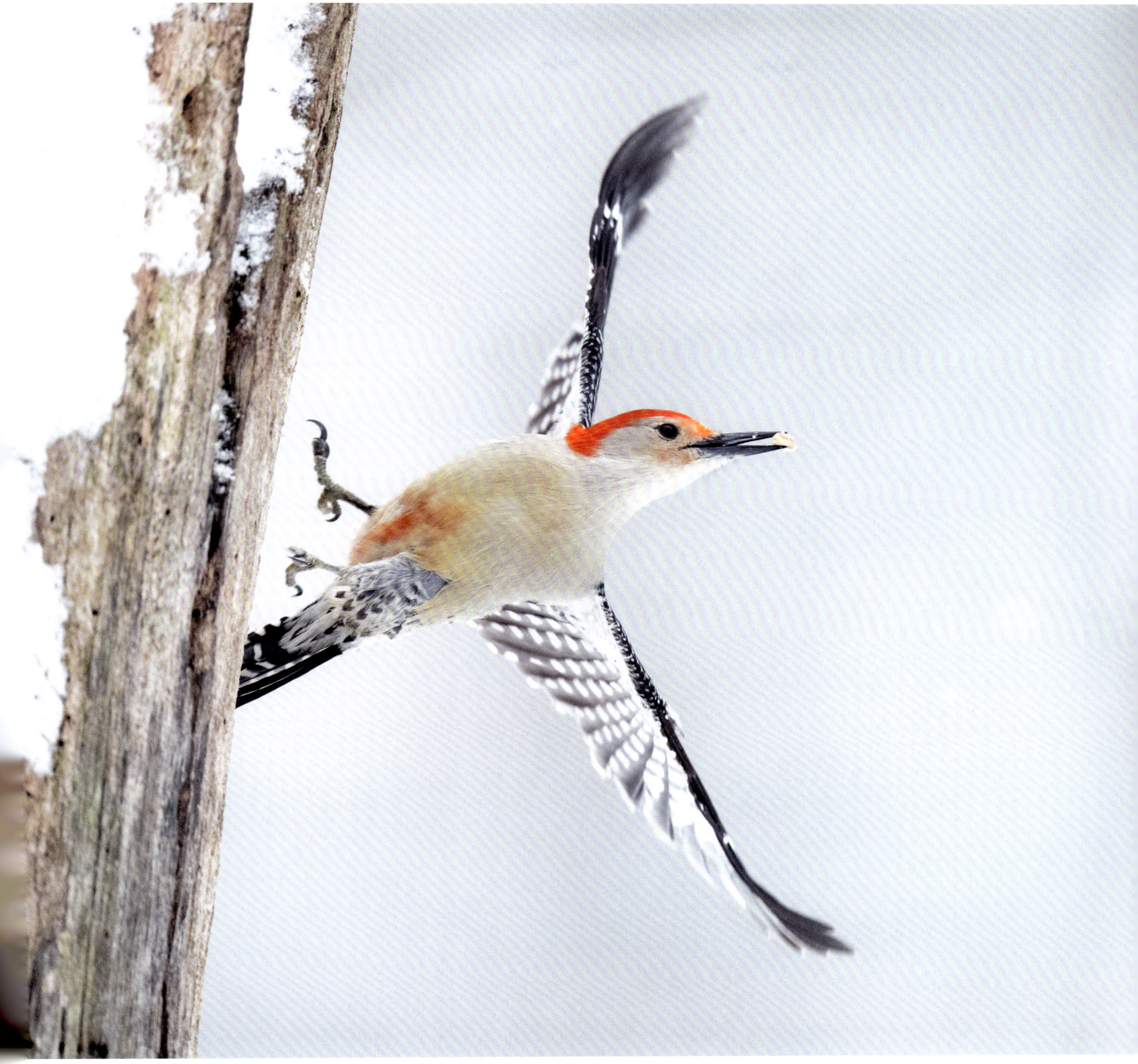

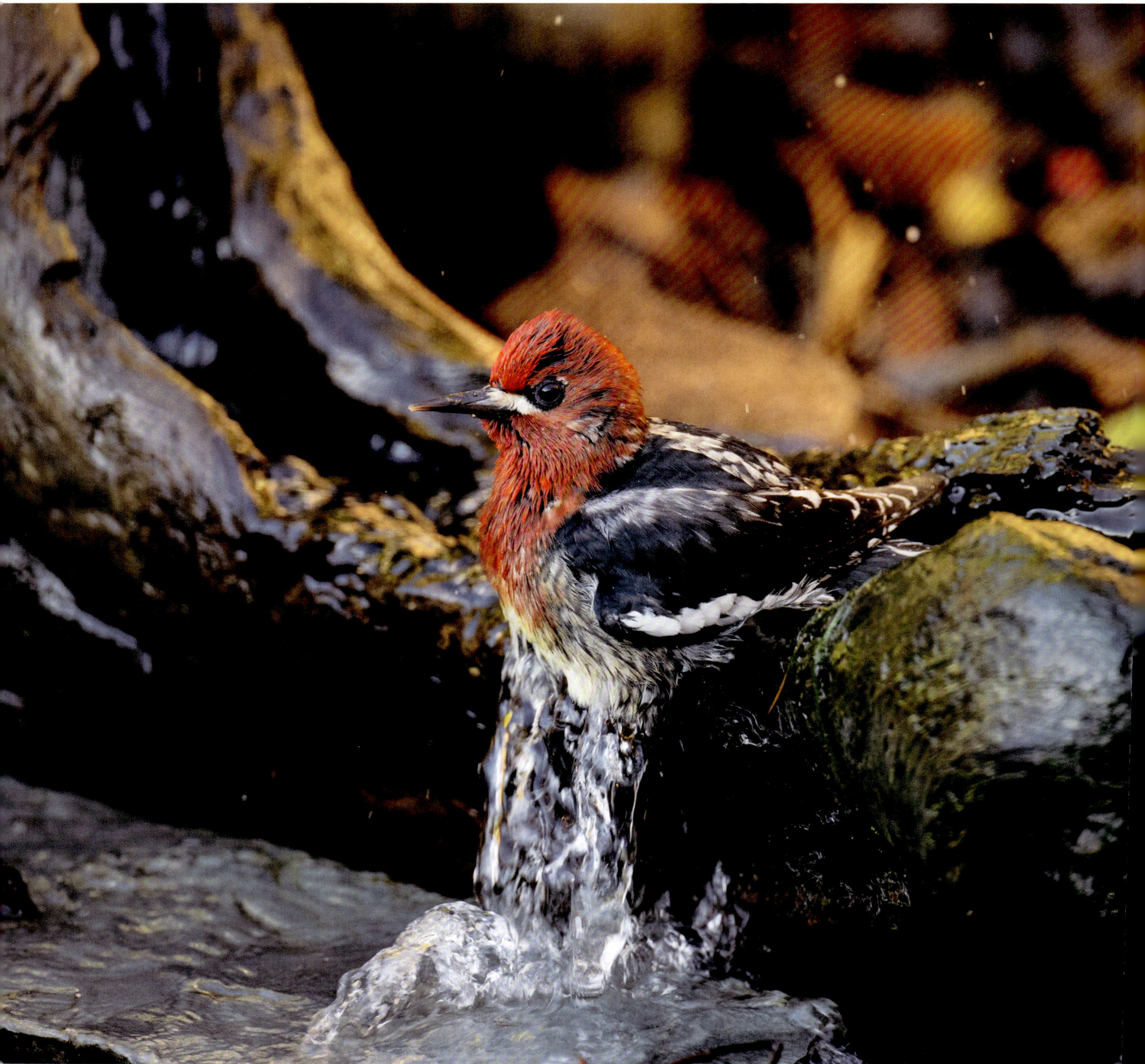

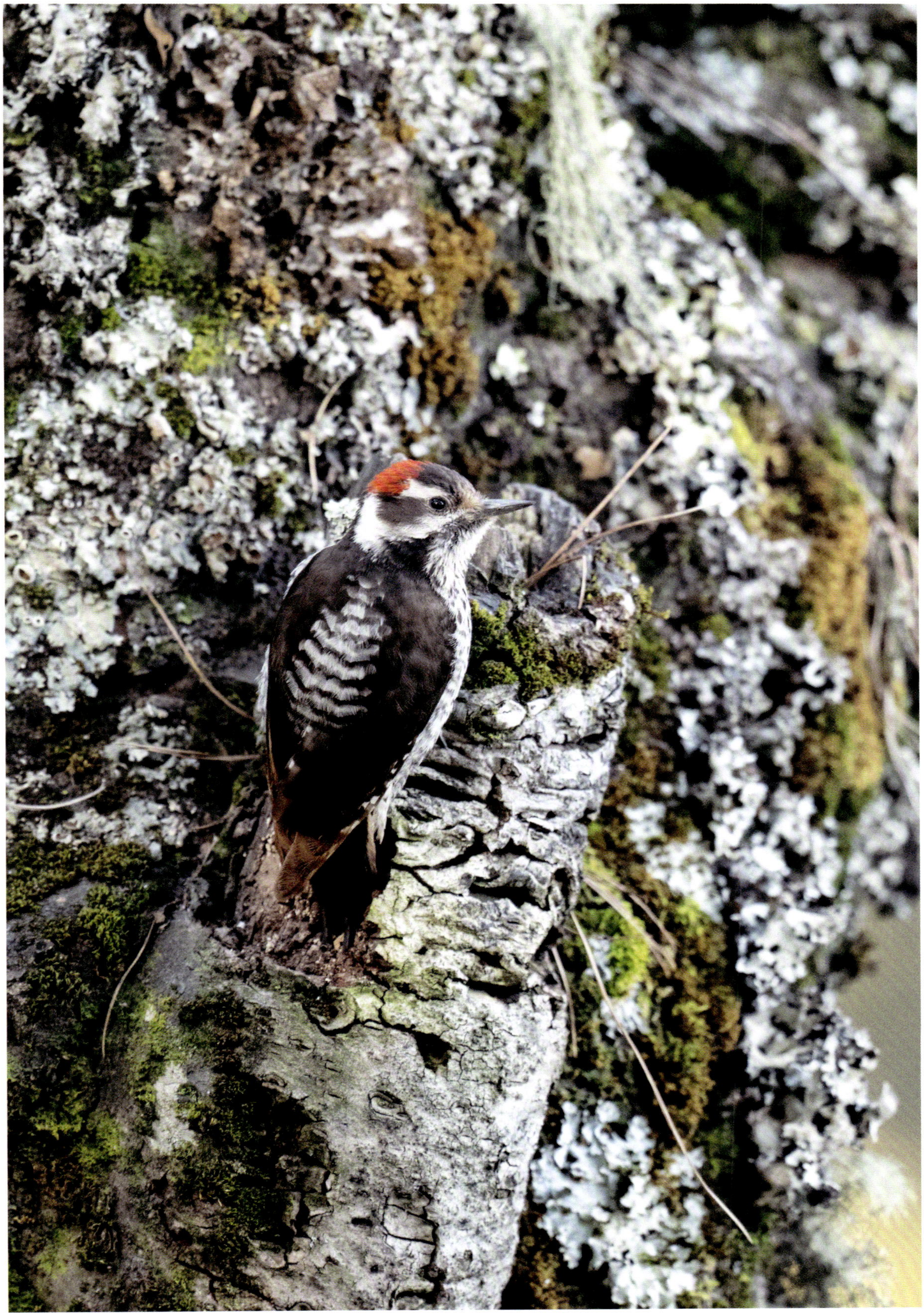

The lichen and moss on an mature alder in the trans-volcanic range of the Sierra Madre allow the Strickland's Woodpecker to be virtually invisible until it moves.

Opposite: Some woodpeckers, including this winter-bathing Red-breasted Sapsucker, utilize water year-round for bathing and drinking.

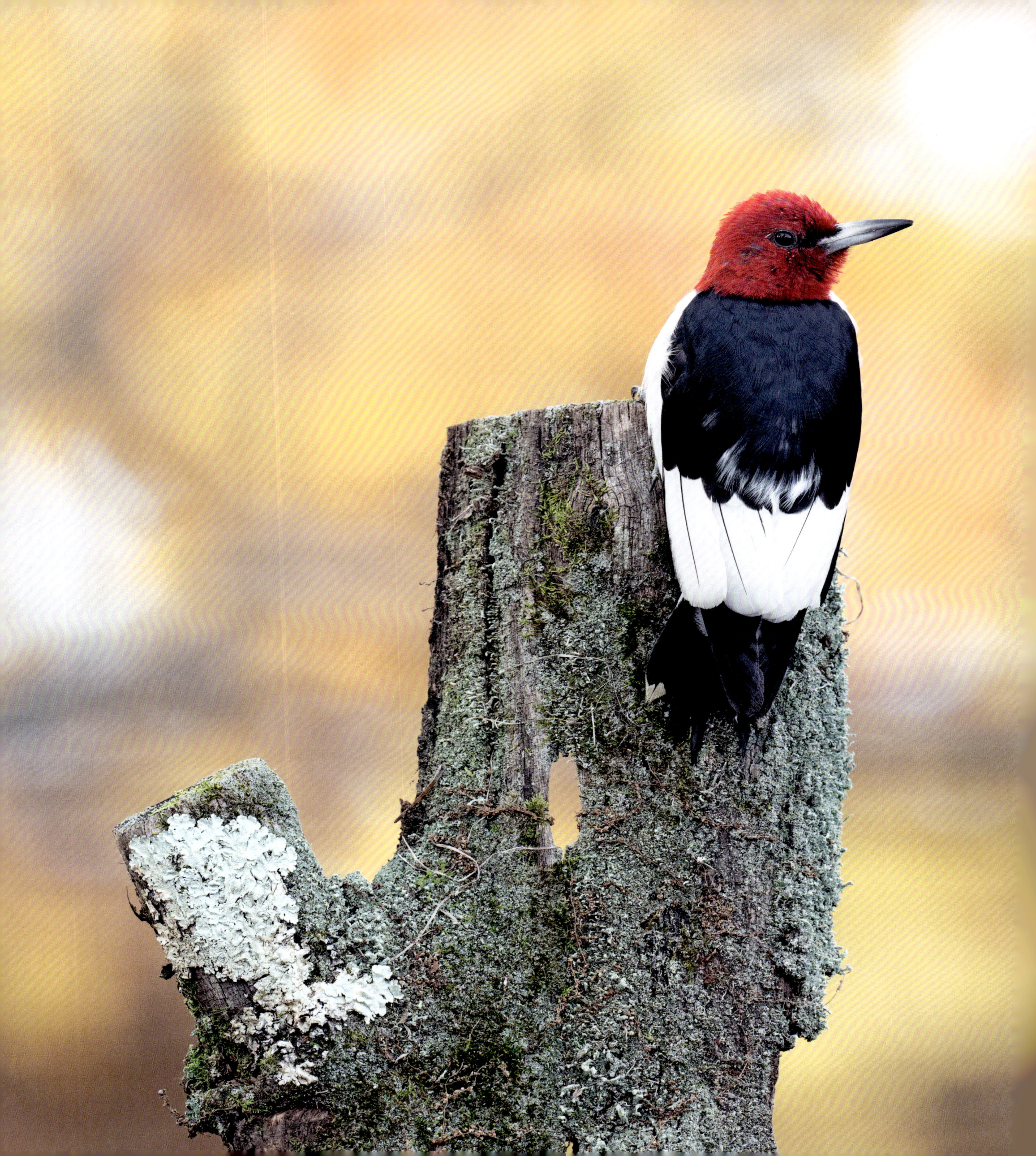

EPILOGUE

The Future of the Woodpecker

Woodpeckers must be ubiquitous for North American forest ecosystems to thrive. Fortunately, forty-one of the forty-three species that lived in North America when Europeans first arrived survive today.

Although a few, such as the Red-bellied and Golden-fronted Woodpecker, are doing relatively well compared to other wildlife, most species, especially American Three-toed, Arizona, Black-backed, Gila, Guadeloupe, Gray-crowned, Lewis's, Red-cockaded, Red-headed, and White-headed Woodpeckers, as well as Fernandina's and Gilded Flickers, Yellow-bellied and Williamson's Sapsuckers, and Antillean Piculet, are considered sensitive, vulnerable, or near threatened in parts or all of their range. Since woodpeckers can be found in every habitat with large trees or cacti, their numbers provide a tool for taking the pulse of those habitats that blanket most of our continent.

The most threatened woodpeckers are challenged by timber harvest, snag removal, fire suppression, salvage logging, alteration of riparian areas, and grazing. The impacts of climate change, including more frequent and extreme fires and storms, warming temperatures, and changes in precipitation regimes, are challenging native plants and beneficial insects, impacting both habitat and woodpecker food supply. Conversion of habitat to housing, commercial development, agriculture, and nonnative plantings, as well as use of herbicides and pesticides, fragments or eliminates habitat and reduces food or indirectly poisons birds.

Red-headed Woodpeckers are one of the more vulnerable species due to their reliance on open woodland and savanna habitat maintained by periodic fire and replete with snags.

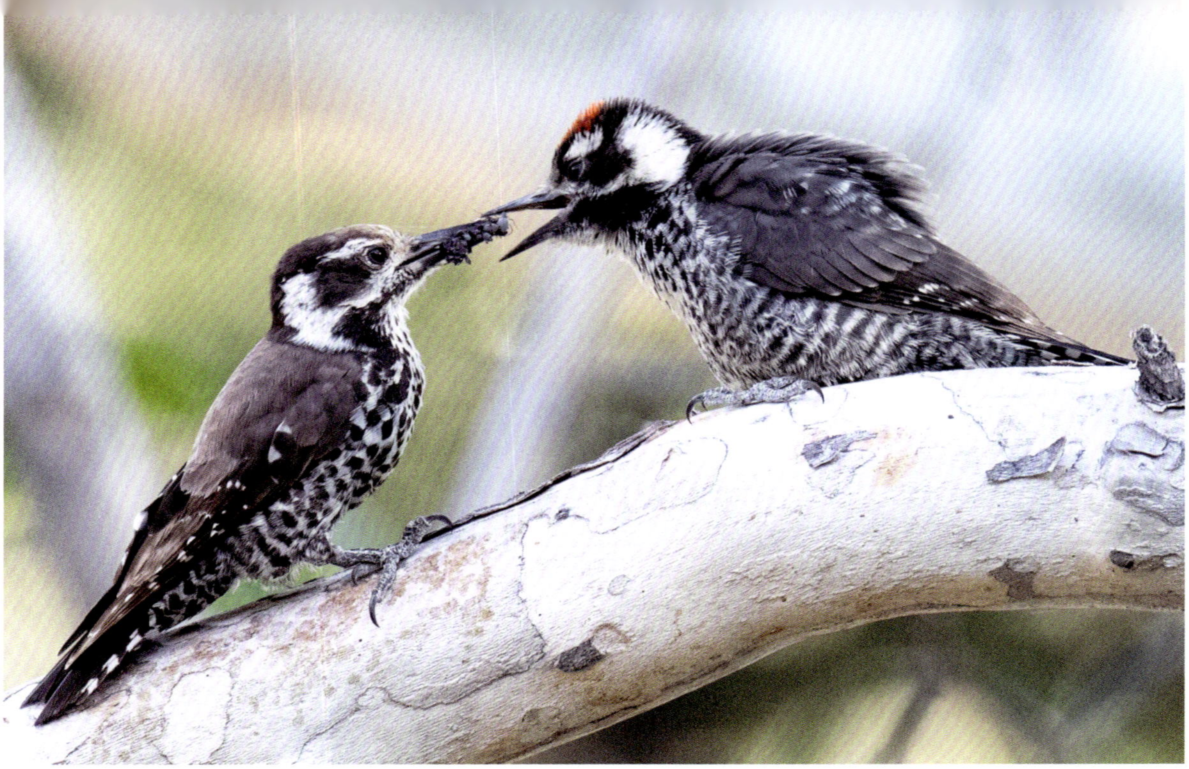

As our continent becomes more populated, the fate of woodpeckers and their habitats is in our hands. There are many things we as individuals can do to help, including landscaping with native plants for the resulting berries, fruit, nuts, acorns, flowers, and beneficial insects that feed woodpeckers. We can also consider retaining or providing nest and roost sites in the form of large mature trees, snags, or nest boxes. We can support organizations that help woodpeckers and protect and connect their habitats as well as those whose research leads us to better understand little-studied species and their needs. Since most woodpeckers are nonmigratory, and habitat is always changing, we need to think of and manage our forests on a larger scale to anticipate and meet the needs of our most specialized species.

We must vote to elect officials who support the retention of conservation laws, undeveloped public land, and sustainable forestry—including protecting old growth, retaining snags, and retaining core habitats in working landscapes—as relatively small policy changes can have large impacts. We must also recognize, listen to, and, where possible, assist those living closest to and thus being impacted the most by compromises needed to protect habitat. To protect woodpeckers and their habitats in economically challenged areas, such as in parts of Mexico and the Caribbean, we must consider how to offset the inordinate costs or offer alternative livelihoods to local people who have little to spare.

Woodpeckers are difficult to ignore in spring, when they call and drum to mates; in summer, when their young beg to be fed; and in winter, when their color adds brightness to dark days. Hopefully their appeal reaches each of us and motivates us to learn more about their lives and meet their seasonal needs. Since woodpeckers are keystone species, they allow us to assess the health of habitats, and since they are umbrella species, we know that helping them helps other birds, mammals, amphibians, and insects, and even helps the regeneration of plants, all of which creates a richer natural world.

That natural world faces ever-graver challenges, from a growing human population and demand for habitat conversion, to climate change and the resulting changing habitats, rising temperatures, and more frequent uncharacteristic fires and storms. We can assess the heartbeat of our most vulnerable forests by monitoring woodpecker populations and using what we learn to better manage these places for them and for the myriad life-forms that rely upon these habitats, including ourselves.

Days after leaving the nest, a hungry male juvenile Arizona Woodpecker is fed a mouthful of insects by his mother.

Opposite: Given how many woodpeckers occupy similar and sometimes overlapping niches in our woodlands, it is surprising how infrequently they come into conflict with each other. Here, a Black-cheeked Woodpecker quickly hitched to the top of a snag after a Pale-billed decided to forage there.

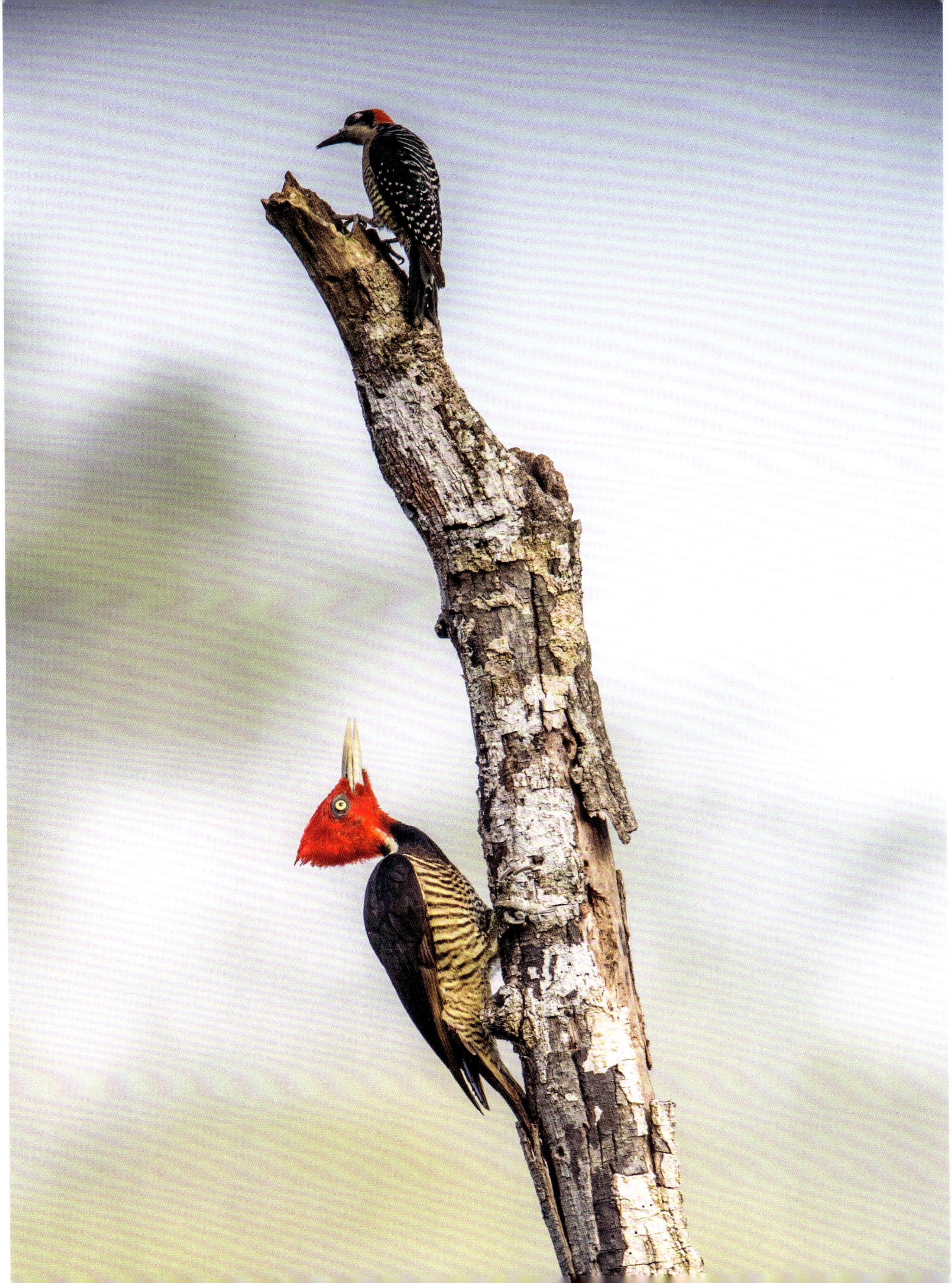

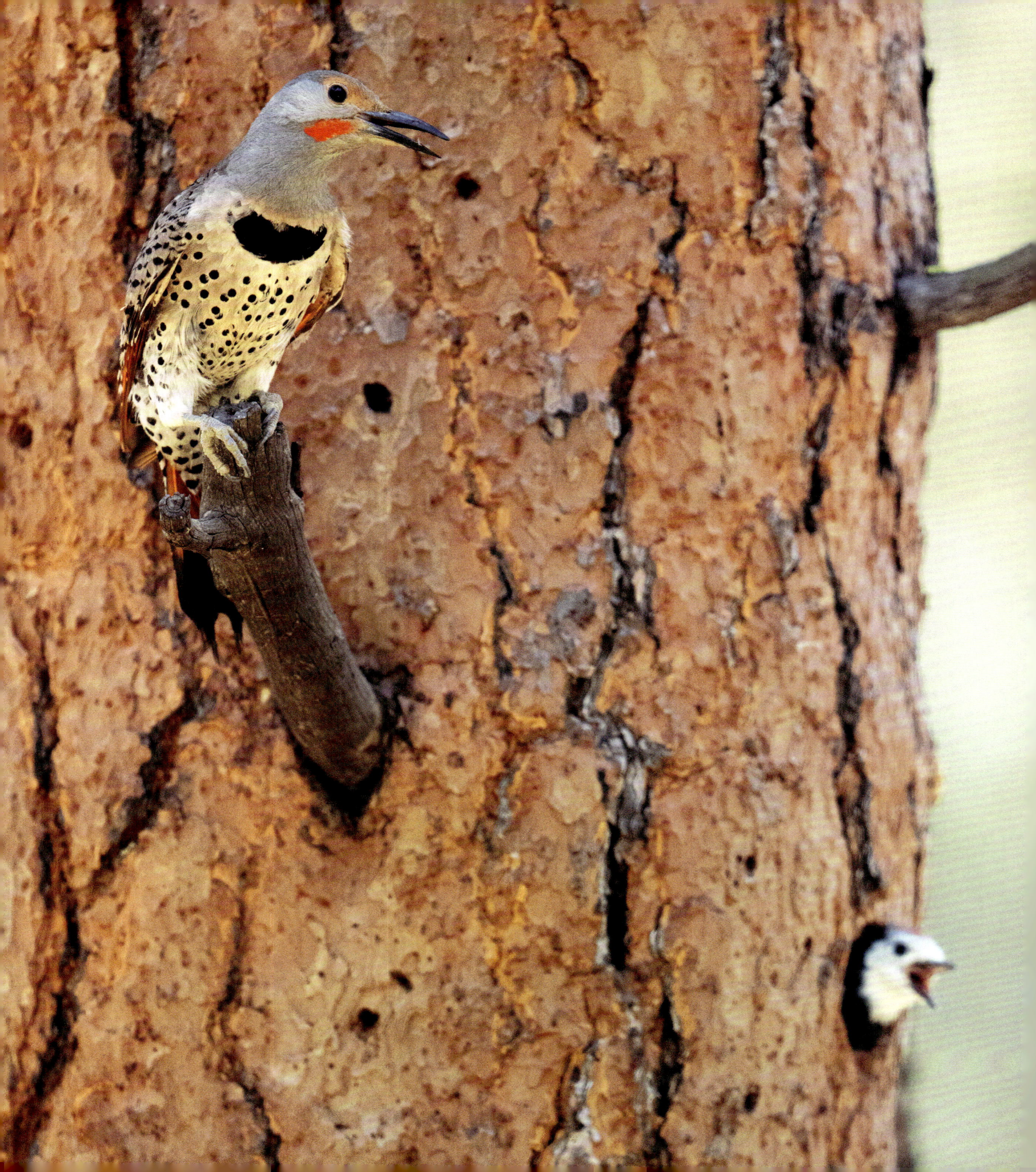

Field Guide to North American Woodpeckers

Woodpecker systematics has undergone several revisions over the past few decades, and although scientific methods have led to advancements in our understanding, there is still undoubtedly more to be clarified. The humans who categorize living things often disagree on what constitutes a species or subspecies, so different organizations will recognize different numbers of species.

The begging calls of nestling woodpeckers often draw the attention of other species. During its last week in the nest, this White-headed Woodpecker was investigated by a Northern Flicker and a Hairy Woodpecker (not pictured).

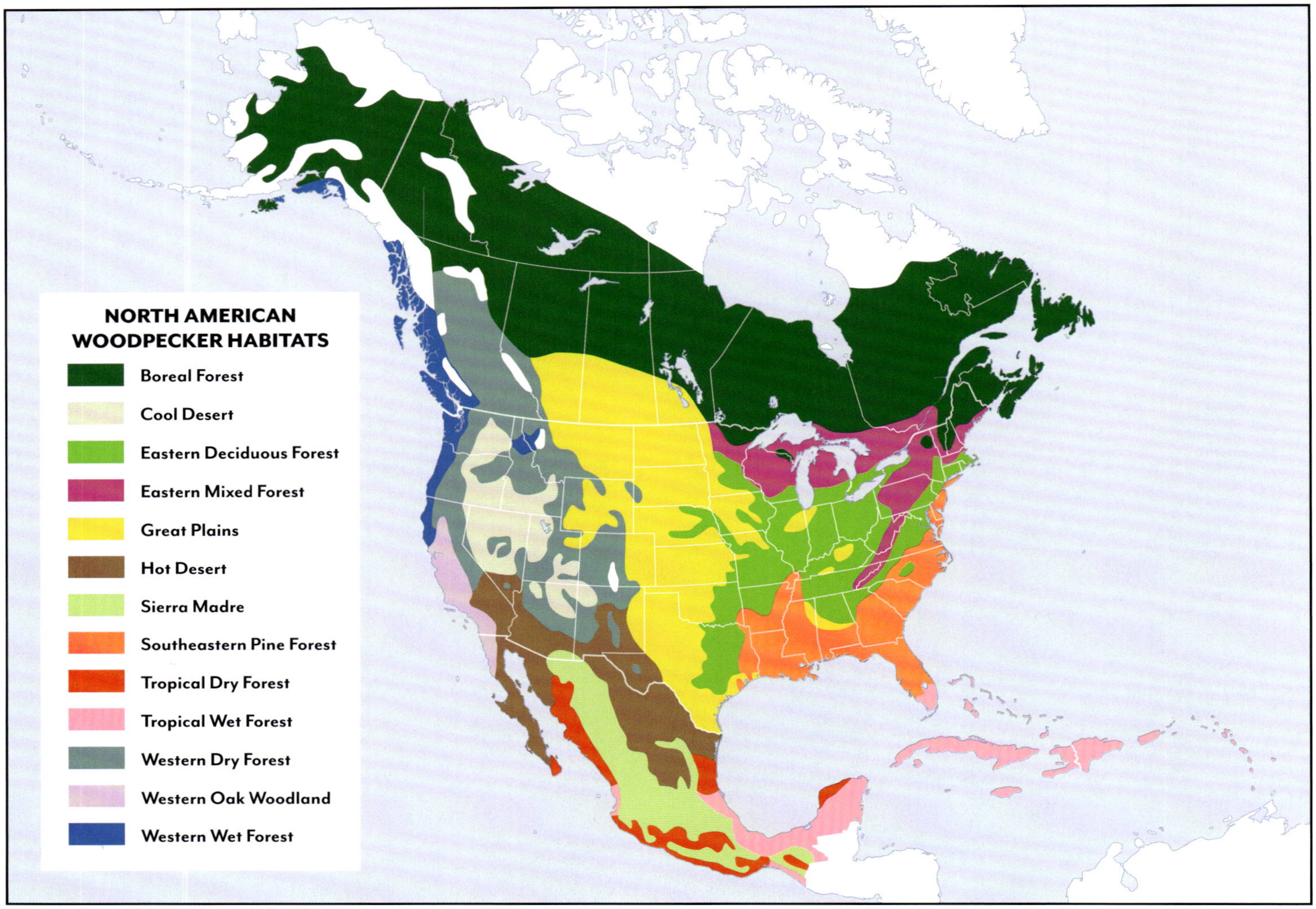

Boreal Forest: Dominated by black and white spruce, joined by birch, aspen, balsam poplar, balsam fir, tamarack and jack pine, and other trees and underlaid in many areas by permafrost, which keeps the ground moist.

Cool Desert: Also known as "steppe." Freezing winter temperatures limit vegetation to small-leaved shrubs such as sage. Lacks succulents such as cactus.

Eastern Deciduous Forest: Broadleaf forests in areas with distinct seasonal variation that cycle through warm, moist summers; cold winters; and moderate fall and spring seasons. Often dominated by oak, maples, basswood, beach, elm, and pine.

Eastern Mixed Forest: A mix of deciduous and conifer forest. Covers the transition zone between the boreal forest and the Great Plains and also covers the Appalachian Mountains between the areas of eastern deciduous forest.

Great Plains: An expanse of flat grassland, prairie, and steppe east of the Rocky Mountains from southern Canada through the United States.

Hot Desert: Sparsely vegetated, arid areas in the southwestern United States and northern Mexico, including the Sonoran, Mojave, and Chihuahuan Deserts. Features vegetation such as Joshua trees in the north, columnar cacti in the south, and shrublands in the east.

Sierra Madre: Pine-oak habitats associated with the mountain ranges from southeastern Arizona through Mexico. Features great biological diversity, particularly in pine-oak woodlands.

Southeastern Pine Forest: Mixed evergreen forests that contain the most important remaining stands of longleaf pine, often joined by loblolly and other pines and oaks.

Tropical Dry Forest: Mostly warm, arid, deciduous woodlands. Endures a long dry season when trees and shrubs lose their leaves. Many trees and shrubs develop thorns, which reflect heat and protect the plants.

Tropical Wet Forest: Year-round warm temperatures and high rainfall result in a rich variety of broadleaf evergreen trees with a complex, multilayered structure. Very little of these wet forests remain, so savannas and dry forests are important woodpecker habitats on Caribbean islands.

Western Dry Forest: Inland, mostly mountain forests typified by warm summers, cool winters, and low precipitation. Ponderosa and related pine species, as well as Douglas-fir, are major components throughout, joined by western larch, oak, and other pine species in some places. Aspen and cottonwood trees and snags are important nesting trees in wet areas.

Western Oak Woodland: Dry, open habitat dominated by oak and transitioning from chaparral, grasslands, or sage to ponderosa pine; replaced by Douglas-fir in areas with more moisture, higher elevation, or less frequent fire.

Western Wet Forest: Dense coastal conifer forests experiencing high rainfall and moderate temperatures and dominated by a variety of conifer species varying by latitude, including redwoods, western redcedar, western hemlock, and Douglas-fir.

According to the American Ornithological Society (AOS), there are 235 species in 34 genera worldwide. Between the American Arctic and Mexico's southern border, including the Caribbean islands, there are 41 living species in 10 genera all within one family. Two species within one of those genera are presumed extinct. There is the possibility that one more species, the Bermuda Flicker, known only from fossil records, may have existed when the first Europeans arrived, although I am not documenting it as such in this work.

The International Community of Ornithologists (IOC) recognizes more species, including two in Mexico, although one of those may be lumped back into another species. I include photos of these, though I identify them as subspecies. They also recognize the Arizona, Hairy, Red-cockaded, Smoky-brown, Strickland's and White-headed as members of the *Leuconotopicus* rather than *Dryobates*.

For this book, I follow the conventions of the AOS at the time of writing (2025) but acknowledge the regional differences that led some to recognize additional subspecies or species based on calls, markings, or genetic differences and recognize some of the subspecies in captions. Many, if not all, of the honorific names are subject to be changed in coming years.

Historically, bird species were distinguished by a combination of field markings, calls, and habitats. One rule suggested that if two birds could produce fertile young, they were of one species. Many birds of one species look very different in different parts of the country, while unrelated species from the same region look very similar. This evolutionary process is called *convergence*: when different species adapt to a common environment in similar ways. The best examples of convergence in North America are the Downy Woodpecker and the Hairy Woodpecker.

Downy and Hairy Woodpeckers in the Pacific Northwest tend to have less contrast between the black-and-white markings than their eastern counterparts, mountain populations are darker with more black, and eastern birds have the most white. The differences are so striking that markings of local Downy and Hairy Woodpeckers more closely resemble each other than they resemble their own species from another region.

Recent molecular science allows us to more accurately see the differences between long-separated populations, but not all species and subspecies have been analyzed in depth, and debate still exists as to how long of a separation—combined with differences in calls, markings, and habitats—distinguishes a species.

Such arguments have led the IOC to recognize two species that I regard as subspecies based on the AOS's conclusions, although that is subject to change: the Bronze-winged Woodpecker of northeastern Mexico, which the AOS includes as a subspecies of the Golden-olive Woodpecker; and the Velasquez's Woodpecker of eastern Mexico (into Guatemala), which the AOS considers a subspecies of the Golden-fronted Woodpecker.

All woodpeckers are diurnal birds that nest in cavities that they create, usually in trees. Physically, woodpeckers vary in length from 5.5 inches for the Antillean Piculet and Downy Woodpecker to 19.3 inches for the Pileated Woodpecker, from top of head to tip of tail. Woodpeckers breed in a variety of habitats with trees, large shrubs, or cacti, from high-elevation spruce forests to low-lying hot deserts to tropical coastal forests. Many are sedentary, a few are migratory, and others are irruptive.

All woodpeckers are members of family Picidae, which includes three subfamilies: woodpeckers, most piculets, and wrynecks. Species in the United States, Canada, Mexico, and the Caribbean belong to subfamily Picinae or "true woodpeckers."

What follows is a taxonomical listing of woodpeckers that provides a clearer idea of which species are most closely related.

The following abbreviations are used in this field guide for official conservation status:

BCC: Bird of Conservation Concern (US Fish and Wildlife Service). Designation mandated by law for migratory bird species that, without additional conservation actions, are likely to become candidates for listing under the Endangered Species Act.

RL: Red List of Threatened Species (International Union for the Conservation of Nature [IUCN]). Provides worldwide information about species' range, population size, habitats and ecology, use and trade, threats, and conservation actions to help inform government policy.

YWL: Yellow Watch List (Partners in Flight). Highlights species of the highest conservation concern at the range-wide scale in the US and Canada.

Nesoctites

The genus *Nesoctites* includes only the Antillean Piculet. It consumes both insects and fruit by gleaning. *Nesoctites* represents the bridge between piculets (Picumnus and Sasia) and true woodpeckers (Picinae) but are more closely related to and thus have been assigned to the latter. Their bill is relatively long and is slightly curved and pointed.

ANTILLEAN PICULET

(Nesoctites micromegas)

DESCRIPTION: A small, distinctive, atypical woodpecker with a rounded body and short, flat-ended tail. Dull green plumage above, white with dark streaks below, and a golden crown. Male's crown has a red patch. L 14–16 cm (5.5–6.2 in).

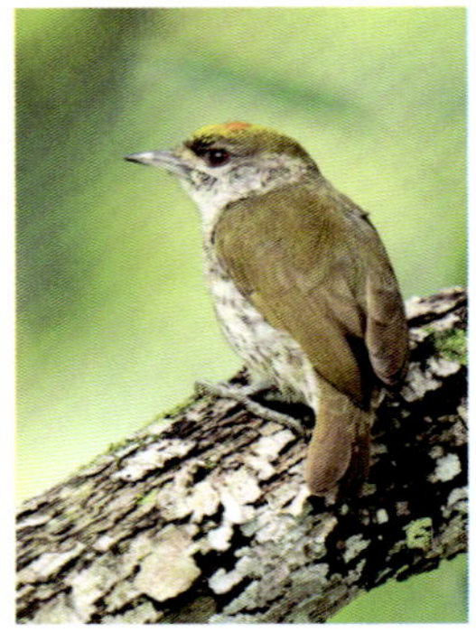

SIMILAR SPECIES: None within range but is twice the size of piculets found in Central and South America, Asia, and Africa.

INTERESTING FACT: A fossil feather attributed to *Nesoctites* shows that the ancestors of the species have been isolated on Hispaniola for at least 25 million years.

DISTRIBUTION AND HABITAT: Only on the island of Hispaniola, where it is found in many forest types including mixed desert scrub and thorn forest and in semi-arid regions and dry broadleaf forest below 1,520 meters (5,000 feet).

NESTING: February through July. Lays 2 to 4 eggs in cavities it excavates or in those created by Hispaniola Woodpeckers.

VOCALIZATION: Single *pit* notes or repeated *pew* notes when alarmed.

CONSERVATION STATUS: Least concern, but populations are declining in best locations, and isolation makes it vulnerable to impacts of climate change and deforestation, especially in Haiti.

Sphyrapicus

A genus of 4 migratory to partially migratory North American species. All 4 lick sap from wells they drill in trees, as well as feed on invertebrates, fruit, berries, and nuts. The Yellow-bellied, Red-naped, and Red-breasted are closely related, with hybridization where populations overlap and with regional variations, making species identification often difficult. Genera *Sphyrapicus*, *Melanerpes*, and *Xiphidiopicus* are believed to be closely related, with many commonalities in morphology and behavior.

WILLIAMSON'S SAPSUCKER

(Sphyrapicus thyroideus; name under review*)*

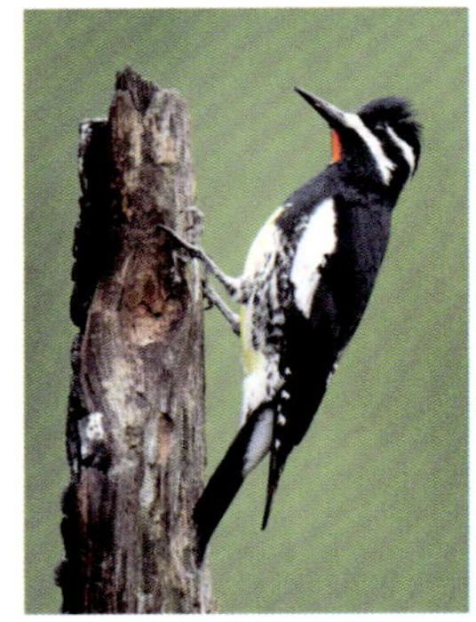

DESCRIPTION: A medium-size woodpecker with yellow belly and red throat. Male is black with more red under throat while female is mostly brown with a black barred back. L 22.5–24.8 cm (8.9–9.8 in).

SIMILAR SPECIES: Adults are unlikely to be confused with any other North American species.

INTERESTING FACT: Male and female are quite different in appearance and were once considered separate species.

DISTRIBUTION AND HABITAT: Medium- to high-elevation dry western mature conifer and mixed conifer forests featuring Douglas-fir, ponderosa pine, western larch, and pine-fir and an open understory with numerous snags, especially large aspen for nesting. Breeds in western dry forests in the US and southern Canada but winters as far south as high elevation pine-oak forests in the Sierra Madre of western Mexico.

NESTING: Late April through late July. Often nests in the same snag year to year; lays 3 to 7 eggs.

VOCALIZATION: Several consecutive *cheeur* notes.

CONSERVATION STATUS: Indicator species for western dry forests where it suffers from overharvesting, forest fragmentation, patchy breeding range, and low density. BCC for Rocky Mountain populations. Endangered in Canada due to small populations and loss of mature larch forests. Listed as "vulnerable" to "imperiled" in 7 of the 12 states and provinces where it breeds and as "apparently secure" in 5 states.

YELLOW-BELLIED SAPSUCKER

(Sphyrapicus varius)

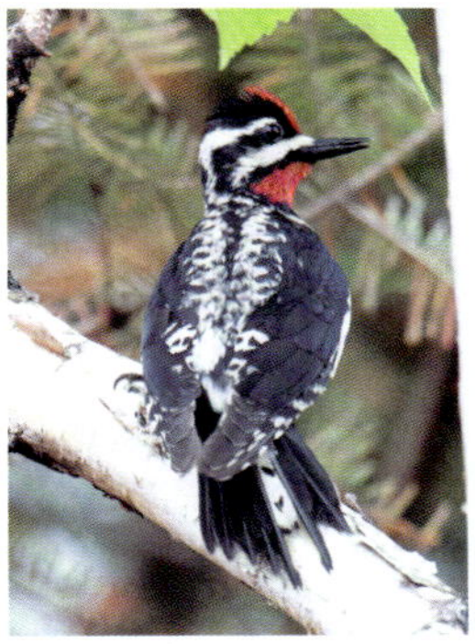

DESCRIPTION: A small to medium-size black-and-white woodpecker with red cap. Male also has red throat. L 21–22 cm (8.3–8.7 in).

SIMILAR SPECIES: Yellow-bellied, Red-breasted, and Red-naped Sapsuckers were considered regional variants of the same species until 1983. Yellow-bellied lacks red head of Red-breasted and red nape of Red-naped. Has the least red and the most white of the three and the throat of both sexes is outlined in black.

INTERESTING FACT: Like other sapsuckers, creates sapwells that are used by other animals, including Ruby-throated Hummingbirds, which may rely exclusively on the wells upon arriving on breeding grounds.

DISTRIBUTION AND HABITAT: Breeds in wooded areas throughout much of the boreal forest and eastern mixed forest and the northern Great Plains, where it requires younger forests. Winters from the southeastern deciduous forest and the Great Plains through much of Mexico and the Caribbean.

NESTING: Early March through mid-July. Often renests in the same snag or aspen and sometimes in the same cavity; lays 2 to 7 eggs.

VOCALIZATION: Catlike, mewing *keeah*.

CONSERVATION STATUS: Listed as "vulnerable" to "critically imperiled" in 14 of the 41 states, provinces, and territories where it breeds and as "apparently secure" or "secure" in 27. Southern Appalachian subspecies listed as federal "species of concern."

RED-NAPED SAPSUCKER

(Sphyrapicus nuchalis)

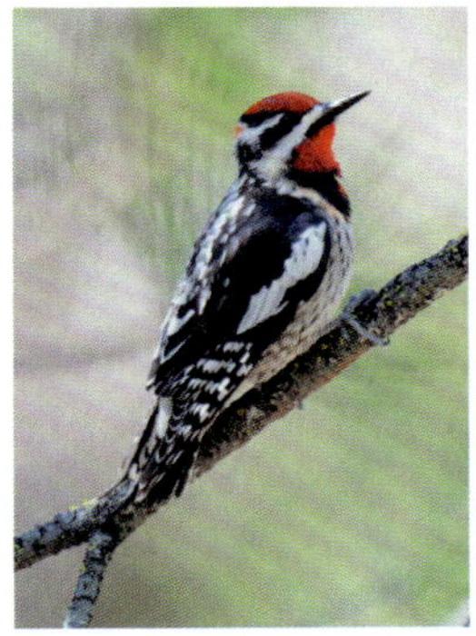

DESCRIPTION: A small to medium-size black-and-white woodpecker with red cap, and red throat. Male also has red nape. L 19–21 cm (7.5–8.3 in).

SIMILAR SPECIES: Red-naped, Red-breasted, and Yellow-bellied Sapsuckers were considered regional variants of the same species until 1983. Red-naped lacks red head and breast of Red-breasted but both sexes have some red on throat, and male has red nape absent in Yellow-bellied. Throat on females lacks white at top and has distinct black outline as in Yellow-bellied of both sexes.

INTERESTING FACT: Like other sapsuckers, creates sapwells that are used by other animals, including several birds, for food—and, in particular, are used by several species of hummingbirds during migration.

DISTRIBUTION AND HABITAT: Breeds in western dry forest and winters in a variety of habitats from the southwestern United States through northwest Mexico including western oak woodlands, western dry forest, the Sierra Madre, and riparian areas where it prefers to forage in willows and other deciduous trees.

NESTING: Mid-March through early August. Often renests in the same snag or aspen and sometimes in the same cavity; lays 4 to 7 eggs.

VOCALIZATION: Catlike, mewing *keeah*.

CONSERVATION STATUS: No special status. Protected under Migratory Bird Treaty Act in the US and Canada.

RED-BREASTED SAPSUCKER

(Sphyrapicus ruber)

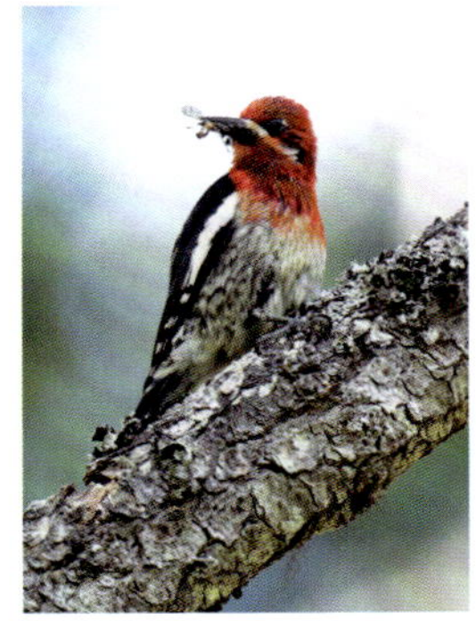

DESCRIPTION: A small to medium-size black, white, and red woodpecker. L 20–22 cm (7.9–8.7 in).

SIMILAR SPECIES: Red-breasted, Red-naped, and Yellow-bellied Sapsuckers were considered regional variants of the same species until 1983. Red-breasted distinguished by mostly red head and red breast, particularly in northern part of range. Two subspecies recognized.

INTERESTING FACT: Like other sapsuckers, creates sapwells that are used by other animals, including several birds, for food—especially by the Rufous and other species of hummingbirds during migration. Usually selects trees previously damaged or with previously drilled sapwells.

DISTRIBUTION AND HABITAT: Breeds in old-growth western wet forests and westernmost areas of western dry forest from southeast Alaska through Southern California. Winters in westernmost areas of western wet forests and in western oak woodland as far south as northwestern Baja California.

NESTING: Early March through July. Often renests in the same snag or tree; lays 4 to 7 eggs.

VOCALIZATION: Catlike, mewing *keeah*.

CONSERVATION STATUS: No special status. Protected under Migratory Bird Treaty Act in the US and Canada.

Xiphidiopicus

A genus with one species found only in Cuba and adjacent small islands. Often found in pairs. Distinct family believed to be most closely related to *Melanerpes* and *Sphyrapicus* woodpeckers.

CUBAN GREEN WOODPECKER

(*Xiphidiopicus percussus*)

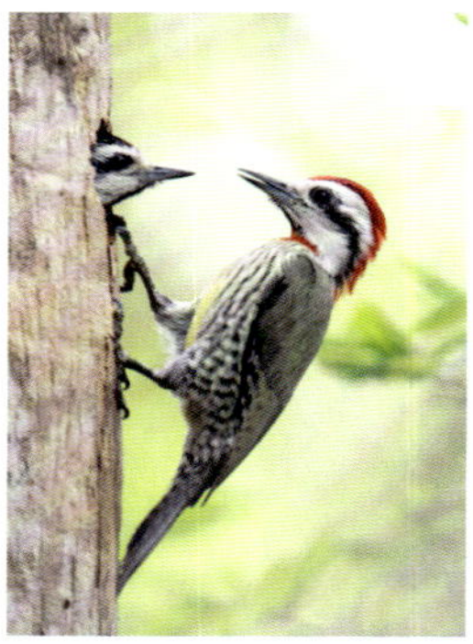

DESCRIPTION: A small to medium-size green-backed woodpecker with red crown (more extensive on male), nape, and bib; white face; and yellow breast and belly. L 21–25 cm (8.3–9.8 in).

SIMILAR SPECIES: No similar species. Wintering Yellow-bellied Sapsucker is similar in size and facial pattern but lacks green and has broad black band across breast.

INTERESTING FACT: Adults make whirring sound with their wings as they arrive at the nest to replace their mate.

DISTRIBUTION AND HABITAT: Endemic to Cuba, where it is widespread in a diverse range of forest types.

NESTING: February to August. Nests in cavity it excavates in live or dead trees or in boxes; lays 3 to 4 eggs.

VOCALIZATION: Brief, low *jhorrr* repeated several times; repetitive *yea-yea-yea*.

CONSERVATION STATUS: Least concern, but because it lives only in Cuba, populations could be impacted by climate change.

Melanerpes

A genus of 23 small to medium-size new-world woodpeckers, 15 of which live in the geography covered by this book. Includes our most colorful woodpeckers and 3 of our 4 monomorphic (genders look alike) species. This is our most social genus and includes several colonial species. Most are sedentary, but the northernmost species are at least partially migratory. This is also the most omnivorous group with the most specialized fly-catchers, but while insects comprise a large part of their diet, they also consume large quantities of fruit, nuts, and acorns, and some even eat sap or small animals and snails. They lack the bounding flight typical of other woodpeckers. *Melanerpes* are closely related to both *Sphyrapicus* and *Xiphidiopicus*.

LEWIS'S WOODPECKER

(*Melanerpes lewis*; name under review)

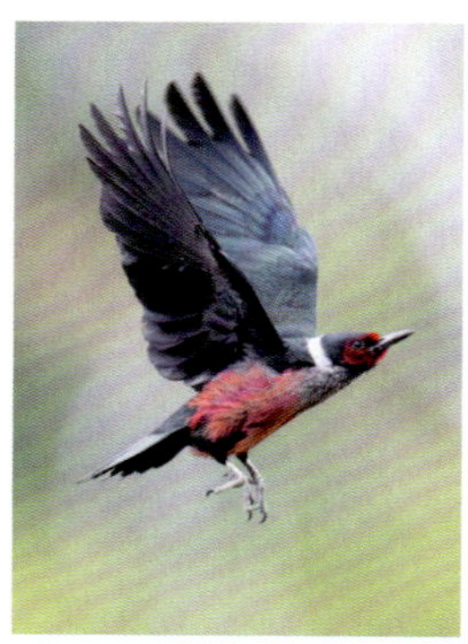

DESCRIPTION: A medium to large long-winged and long-tailed dark green woodpecker with a green back and head, crimson face, silver collar, and pink belly. L 26–28 cm (10.2–11 in).

SIMILAR SPECIES: Similar looking to the darker and less colorful Guadeloupe Woodpecker, limited to the island of Guadeloupe.

INTERESTING FACT: Often seen fly-catching from elevated perches.

DISTRIBUTION AND HABITAT: Breeds in open habitats within western dry forest, western Great Plains, and western oak woodland from south-central British Columbia into northwestern Baja California that include oak, ponderosa pine, juniper, mature aspen, cottonwood riparian areas, and especially recent stand-replacing burns. Winters in similar habitats from south-central Washington to Baja California in the West and from Colorado through much of Nevada and Arizona in the interior, although the abundance of acorns, nuts, or grains becomes more important.

NESTING: Late April through September. Weak excavator. Sometimes returns to same cavity in consecutive years. Uses natural cavities, those created by flickers, or one that it excavates in soft wood; lays 5 to 9 eggs.

VOCALIZATION: *Churr* lasting roughly a half a second and given a few to several times in succession.

CONSERVATION STATUS: BCC, YWL, "threatened" in Canada. Broad population declines along with loss of habitat and nesting snags have motivated its listing as "vulnerable" to "critically imperiled" by 15 of the 17 states and provinces where it breeds; extirpated from previous areas such as western Washington; apparently secure in 2 states.

GUADELOUPE WOODPECKER

(*Melanerpes herminieri*)

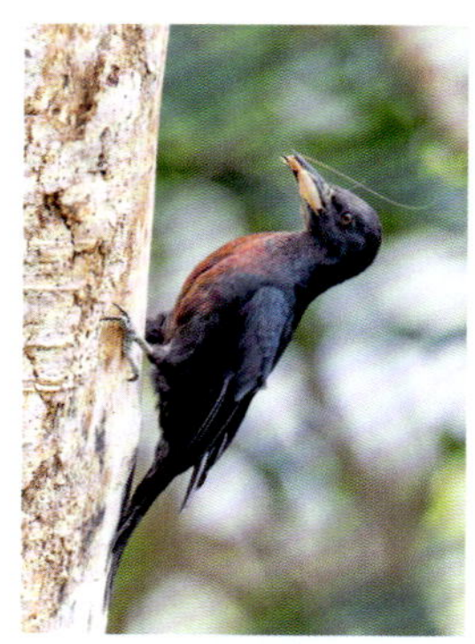

DESCRIPTION: A distinctive dark-colored medium-size woodpecker with a red-to-purple wash on belly. L 24–29 cm (9.4–11.4 in).

SIMILAR SPECIES: Not likely to be confused as it is the only woodpecker on the Caribbean island of Guadeloupe. Appears similar to but darker than Lewis's Woodpecker.

INTERESTING FACT: Unique enough to originally have been placed in its own genus, *Linneopicus*.

DISTRIBUTION AND HABITAT: Only on Guadeloupe, where it takes advantage of all wooded habitat but prefers humid semi-deciduous and evergreen forests.

NESTING: January through August. Creates cavities in dead coconut palms, other trees, or utility poles; lays up to 5 eggs.

VOCALIZATION: Raspy single or repeated *kwa*.

CONSERVATION STATUS: Currently considered "near threatened" based on restricted range and vulnerability to habitat loss from human activity and hurricanes.

PUERTO RICAN WOODPECKER

(*Melanerpes portoricensis*)

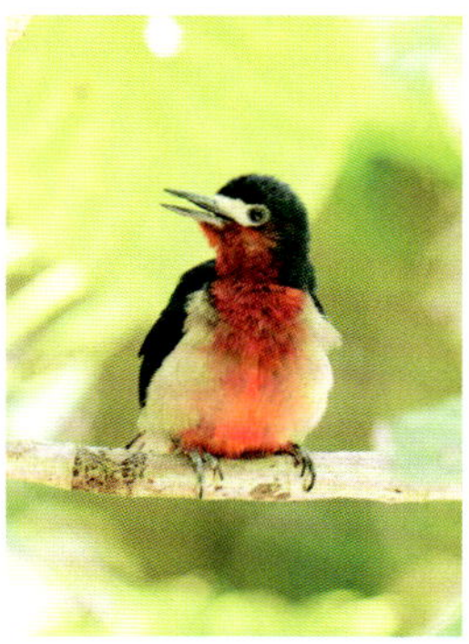

DESCRIPTION: A distinctive medium-size steely-blue-backed, red-bellied, white-masked woodpecker. L 23–27 cm (9–10.6 in).

SIMILAR SPECIES: Endemic to Puerto Rico where it is the only woodpecker, except when the very different-looking Yellow-bellied Sapsucker appears.

INTERESTING FACT: One of 5 *Melanerpes* woodpeckers that live only on Caribbean islands.

DISTRIBUTION AND HABITAT: Most wooded areas of Puerto Rico and adjacent Vieques but prefers dry forest. Formerly found on Saint Thomas.

NESTING: January through June. Constructs nests in live or dead trees and utility poles; lays 4 to 5 eggs.

VOCALIZATION: High-pitched *pep* given singly or in a series and a *churr* call typical of *Melanerpes*.

CONSERVATION STATUS: Least concern, but climate change is altering the moisture cycle and will likely impact primary food (arthropods and fruits).

RED-HEADED WOODPECKER

(*Melanerpes erythrocephalus*)

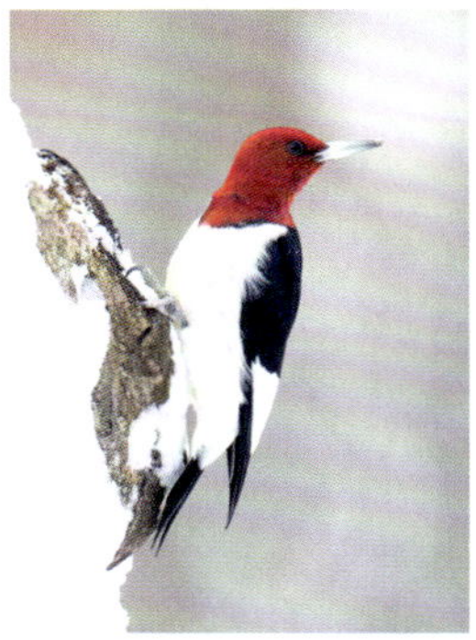

DESCRIPTION: A distinctive medium-size red-headed, black-and-white-bodied woodpecker. L 19.4–23.5 cm (7.6–9.2 in).

SIMILAR SPECIES: Unlikely to be confused with any other North American species, although Red-bellied Woodpeckers and Red-breasted Sapsuckers are often referred to as Red-headed Woodpeckers by novice bird-watchers.

INTERESTING FACT: Sometimes resident, sometimes migratory, and sometimes nomadic during the winter when it wanders in search of nuts and acorns.

DISTRIBUTION AND HABITAT: Breeds throughout much of the eastern US (outside of New England and the southern tip of Florida) and southern Canada, and winters from the southernmost eastern mixed forest south through the rest of the breeding range.

NESTING: Early May until early September. Often returns to same tree in consecutive years. Lays 3 to 10 eggs; 4 to 7 most common. A second and, less commonly, a third brood might be raised in a single year.

VOCALIZATION: *Tchurr, tchurr* with each vocalization lasting less than half a second.

CONSERVATION STATUS: BCC, RL, YWL. Listed as "vulnerable" to "critically imperiled" in 18 of the 39 states and provinces where it breeds and "apparently secure" or "secure" in 21. Considered a high conservation priority in more states and provinces than any other woodpecker due to loss of habitat and snags.

ACORN WOODPECKER

(*Melanerpes formicivorus*)

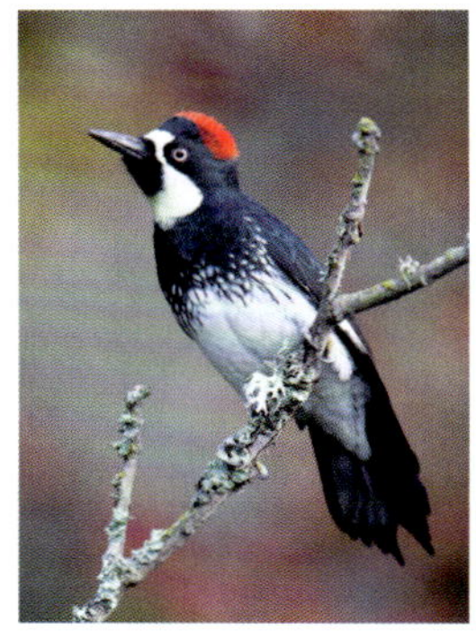

DESCRIPTION: A medium-size clown-faced woodpecker with black back and red cap. Male's red extends farther toward bill. L 13–15 cm (5.1–5.9 in).

SIMILAR SPECIES: Unlikely to be confused with any other North American species.

INTERESTING FACT: They are cooperative breeders that live in multimember social groups year-round.

DISTRIBUTION AND HABITAT: Occurs with oaks in western North America in two disjunct populations: a small, isolated population in the western oak woodland of extreme south-central Washington along the West Coast to Southern California and one in western dry forest and the Sierra Madre from central Arizona to western New Mexico and south through Mexico.

NESTING: Mid-March through September. Cooperative group broods 2 to 8 eggs. Excavates cavities in whatever large trees are available, including telephone poles.

VOCALIZATION: *Waka, waka, waka.*

CONSERVATION STATUS: Listed as least concern by BirdLife International. Fairly common throughout its range and even increasing range in some areas. However, listed as "vulnerable" or "critically imperiled" in 3 of the 5 states where it breeds and "apparently secure" or "secure" in 2. The loss of oak woodlands in Oregon has contributed to its "vulnerable" status there.

BLACK-CHEEKED WOODPECKER

(Melanerpes pucherani)

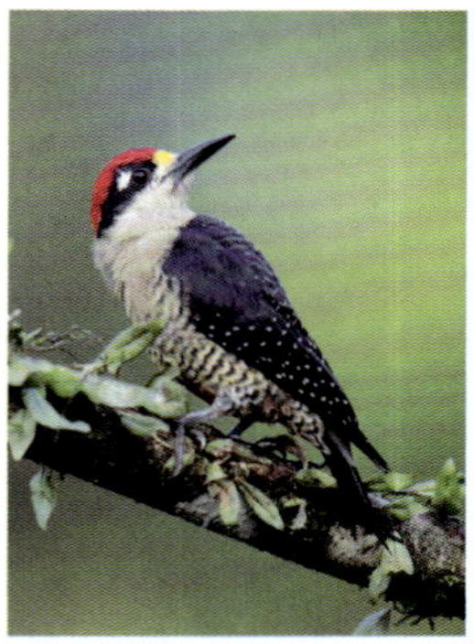

DESCRIPTION: A small zebra-backed woodpecker with barred belly, black mask, yellow forehead, and red nape, which extends to a red cap on the male. L 17–19 cm (6.7–7.5 in).

SIMILAR SPECIES: Golden-fronted Woodpecker is larger and lacks black mask and barred belly.

INTERESTING FACT: Fledglings sometimes return to their nests to roost, and adults may start a second brood.

DISTRIBUTION AND HABITAT: Tropical wet forest in parts of southern Mexico through parts of South America.

NESTING: March through June in holes in dead trees; lays 2 to 4 eggs.

VOCALIZATION: Very similar to other *Melanerpes* woodpeckers, including rattling trills, *churr* calls.

CONSERVATION STATUS: Common to fairly common throughout its range.

HISPANIOLAN WOODPECKER

(Melanerpes striatus)

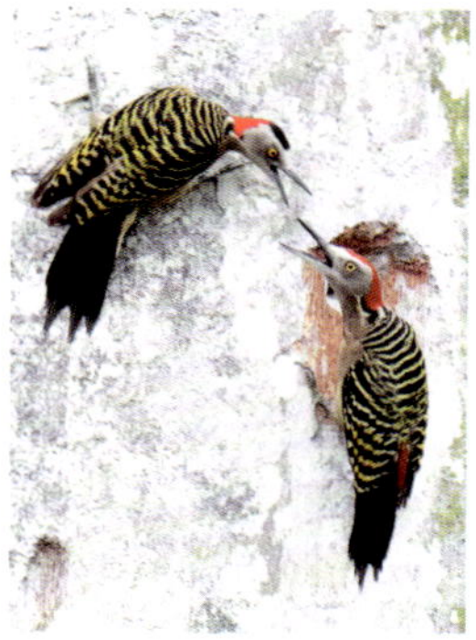

DESCRIPTION: A medium-size black-and-yellow barred woodpecker with red crown, nape, and rump. Female has black cap. L 22–28 cm (8.7–11 in).

SIMILAR SPECIES: None within range.

INTERESTING FACT: Has been known to breed in colonies of up to 20 pairs.

DISTRIBUTION AND HABITAT: Only on the island of Hispaniola, where it occupies all treed areas.

NESTING: Year-round but especially February through July. Carves cavities in live or dead palm, cactus, or utility poles where it lays 4 to 6 eggs.

VOCALIZATION: A lot of variation, but a long series of *waaa* notes is common.

CONSERVATION STATUS: Common in all wooded habitats, but urbanization is a potential threat because its range is restricted to a single island.

JAMAICAN WOODPECKER

(Melanerpes radiolatus)

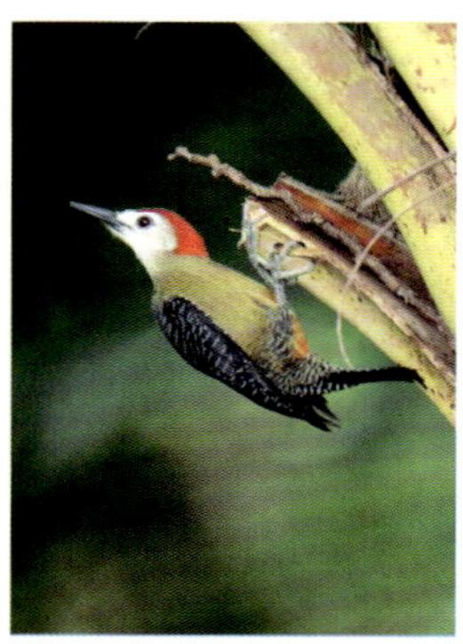

DESCRIPTION: A medium-size woodpecker with black back with fine white bars, olive belly. Male has red crown and nape. Female has gray cap. L 24–26 cm (9.4–10.2 in).

SIMILAR SPECIES: None. It is the only woodpecker on the island of Jamaica, although red cap and nape are reminiscent of Red-bellied Woodpecker.

INTERESTING FACT: Rarely forages on trunks more than 10 cm (3.9 in) in diameter.

DISTRIBUTION AND HABITAT: Only on the island of Jamaica in most habitats.

NESTING: December through August; may have 2 or even 3 broods in a single season. Excavates in dead trees, dead parts of live trees, or in utility poles; lays 3 to 5 eggs.

VOCALIZATION: Loud *kaaa* repeated 2 or 3 times is most common call.

CONSERVATION STATUS: Least concern. Occurs in most treed habitats, although habitat loss is a threat.

GOLDEN-CHEEKED WOODPECKER

(Melanerpes chrysogenys)

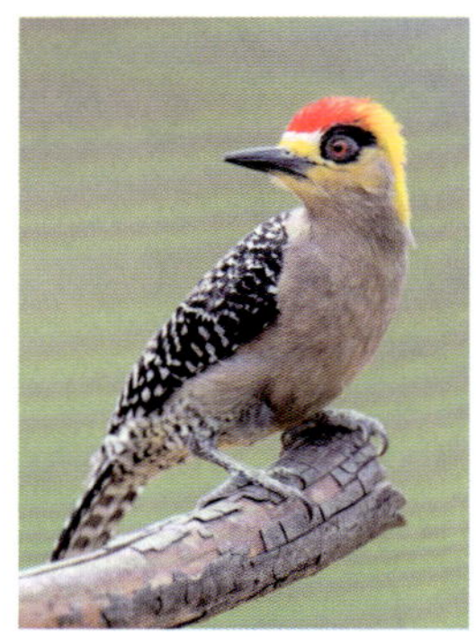

DESCRIPTION: A small to medium-size zebra-backed, brown-bellied woodpecker with black eye patch and distinctive red cap transitioning to orange at the nape. Female cap is brown. L 19–22 cm (7.5–8.7 in).

SIMILAR SPECIES: Golden-fronted has less color on head, with gap between yellow and red, and lacks mask. Gila Woodpecker lacks yellow and mask. Gray-breasted Woodpecker is grayish and lacks yellow on nape.

INTERESTING FACT: Forages for insects and fruit in the middle and upper parts of the forest.

DISTRIBUTION AND HABITAT: Endemic to Mexico's Pacific slope and plain in arid or semihumid treed areas of tropical wet and tropical dry forest below 1,500 m (4,920 ft).

NESTING: May through July. Excavates in trees or cacti; lays unknown number of eggs.

VOCALIZATION: *Churr* calls and brief nasal *ki-di-dick*, among others.

CONSERVATION STATUS: Generally common throughout range.

GRAY-BREASTED WOODPECKER

(Melanerpes hypopolius)

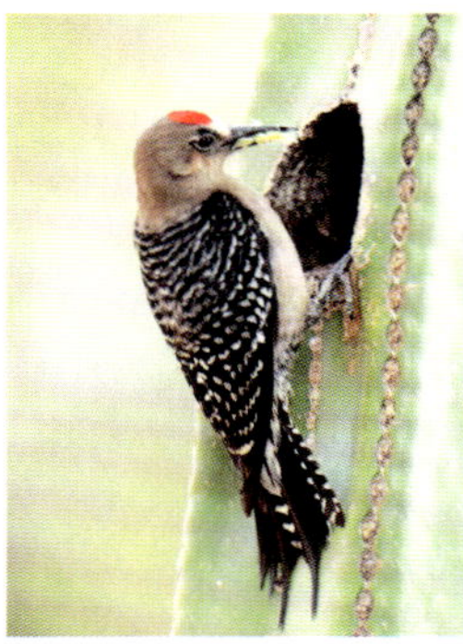

DESCRIPTION: A small to medium-size zebra-backed woodpecker, gray with small red cap, buff-white forehead, black around eyes, and reddish below eye. L 19–21 cm (7.5–8.3 in).

SIMILAR SPECIES: Golden-fronted and Golden-cheeked Woodpeckers are more colorful overall with brownish underparts and golden napes.

INTERESTING FACT: Cooperative breeder that also forages in groups of up to 10 for fruit and insects.

DISTRIBUTION AND HABITAT: Only in southwestern Mexico in open and semi-open arid areas of tropical dry forest with scattered organ pipe cacti and other vegetation where they forage at all levels.

NESTING: Late April through July. Cooperative and colonial breeder. Builds nests in trees or cacti; lays unknown number of eggs.

VOCALIZATION: Nasal *yekka-yekka* and a series of rattling *chiir* calls.

CONSERVATION STATUS: Range restricted but is common to fairly common through its range.

YUCATAN WOODPECKER

(Melanerpes pygmaeus)

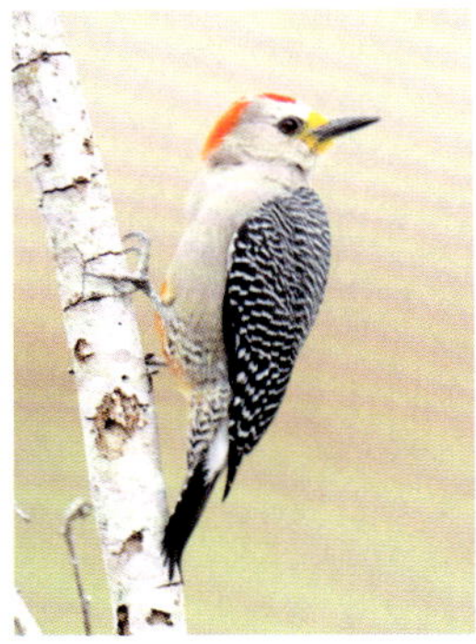

DESCRIPTION: Small zebra-backed woodpecker with buff-brown underparts, golden-orange forehead and nape, and yellow nasal tufts. No red cap on female. L 16–18 cm (6.3–7.1 in).

SIMILAR SPECIES: Golden-fronted Woodpecker larger with larger bill, reddish nasal tufts, and larger red nape patch.

INTERESTING FACT: Very closely related to Black-cheeked, Golden-cheeked, Gray-breasted, Golden-fronted, Gila, West Indian, and Red-bellied Woodpeckers with which they were once grouped under the genus *Centrus*.

DISTRIBUTION AND HABITAT: Only on and near the Yucatan Peninsula, at edges of both tropical dry forest and tropical wet forest and in coastal scrub at low elevations.

NESTING: April through May in cavities it excavates; lays unknown number of eggs.

VOCALIZATION: Soft rolling *churr*. Also *pyurrr* repeated in a series.

CONSERVATION STATUS: Fairly common within small range.

GILA WOODPECKER

(Melanerpes uropygialis)

DESCRIPTION: A medium-size zebra-backed tan woodpecker. Male has red cap. L 22–25 cm (8.7–9.8 in).

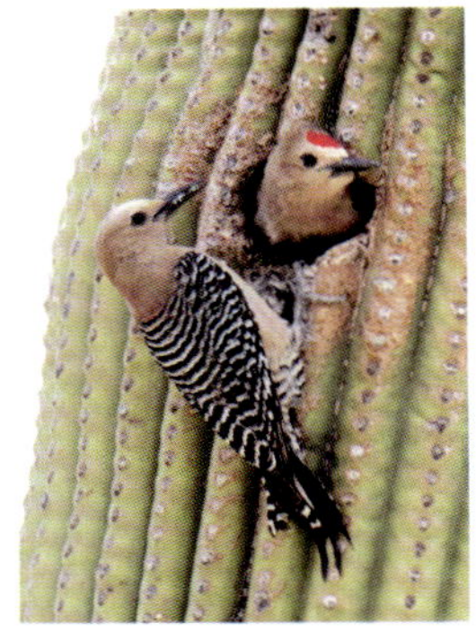

SIMILAR SPECIES: Closely related to both the Golden-Fronted Woodpecker and the Red-bellied Woodpecker. Red-bellied has colored nape, not crown. Female Williamson's Sapsucker looks superficially similar to female Gila but is unlikely to be in same habitat type except rarely in winter.

INTERESTING FACT: Eats cactus fruits, flowers, small animals, insects, and bird eggs.

DISTRIBUTION AND HABITAT: Arid parts of hot desert and tropical dry forest from extreme southeastern California, southern Arizona, and extreme southwestern New Mexico through northwest Mexico.

NESTING: April through August. Excavates in saguaro cacti; lays 3 to 6 eggs. Occasionally a second and, less frequently, a third clutch.

VOCALIZATION: Double *churr* with each note lasting roughly one-quarter of a second.

CONSERVATION STATUS: BCC. Listed as "vulnerable" or "imperiled" in all 3 states where it breeds. Threatened by competition with European Starlings and development of Sonoran Desert.

GOLDEN-FRONTED WOODPECKER

(Melanerpes aurifrons)

DESCRIPTION: A medium- to large-size zebra-backed woodpecker with golden nape (red in Velasquez's). Male has red cap. L 22–26 cm (8.7–10.2 in).

SIMILAR SPECIES: Closely related to Gila Woodpecker and Red-bellied Woodpecker and especially the much smaller Yucatan Woodpecker.

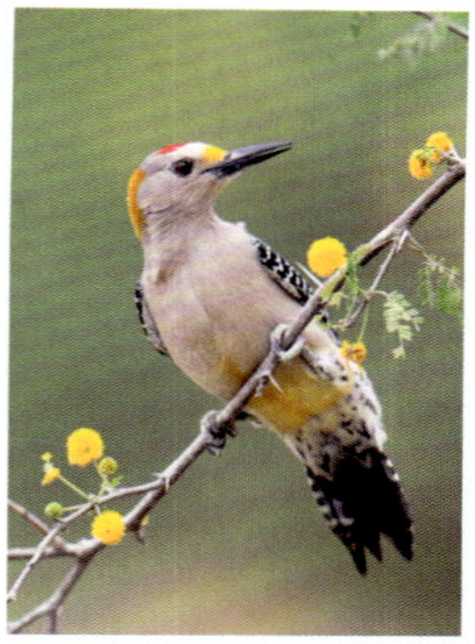

INTERESTING FACT: One of two woodpeckers in the United States that is expanding its range. A subspecies in Belize and southeast Mexico with a red rather than yellow nape is considered by some to be a separate species called Velasquez's Woodpecker (name under review).

DISTRIBUTION AND HABITAT: Breeds in arid brushlands and semi-open woodlands within the Great Plains, tropical wet, and tropical dry forest habitats from southwestern Okla-homa through southern Texas and throughout much of eastern and southern Mexico.

NESTING: Mid-March through July. Often reuses cavities; lays 4 to 7 eggs. Second broods are common.

VOCALIZATION: Loud, harsh *krrrr* lasting less than half a second.

CONSERVATION STATUS: Least concern overall. Secure in Texas but "imperiled" in Oklahoma due to small population size.

RED-BELLIED WOODPECKER

(Melanerpes carolinus)

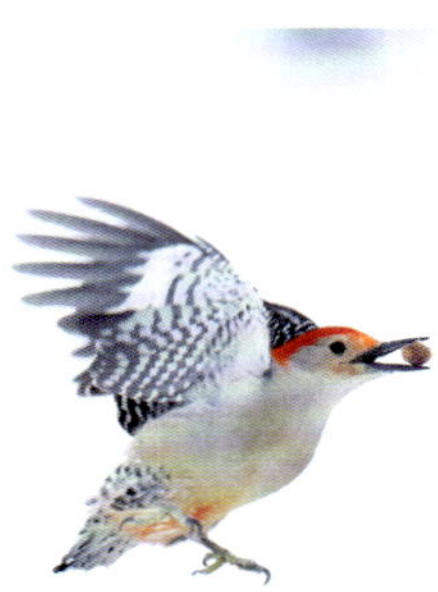

DESCRIPTION: A medium-size zebra-backed woodpecker with red nape. Male has red cap. L 22–26.7 cm (8.7–10.5 in).

SIMILAR SPECIES: Closely related to Gila, Gray-breasted, and Golden-fronted Woodpeckers.

INTERESTING FACT: Extremely omnivorous and eats a wide variety of foods, including lizards and even frogs. Expanding its range at a faster rate than any other woodpecker.

DISTRIBUTION AND HABITAT: Breeds in all forest types in eastern North American south of the boreal forest and north of Mexico. It is the most common woodpecker throughout much of the southern part of its range.

NESTING: Mid-April through mid-September. Commonly returns to same snag or dead limb to excavate new cavity; lays 3 to 6 eggs. Might raise 2 or even 3 broods in southern part of range.

VOCALIZATION: A nearly half-second-long *kwirr*.

CONSERVATION STATUS: This species is highly adaptable and is expanding its range, although populations are less secure in newly inhabited northeastern states and northern provinces.

WEST INDIAN WOODPECKER

(Melanerpes superciliaris)

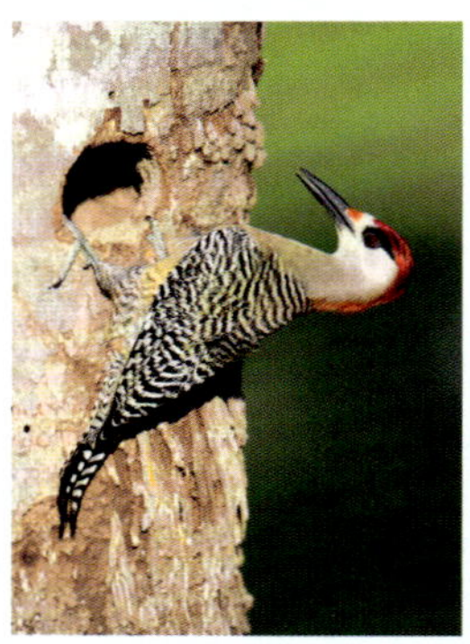

DESCRIPTION: A medium to large zebra-backed woodpecker with red nape. Male has red cap. Distinct black above eye in some populations. L 26–30 cm (10.2–11.8 in).

SIMILAR SPECIES: Some populations look very much like Red-bellied Woodpecker but with less red on the head. In other populations, also distinguished by black above eye. Very closely related to Red-bellied, Black-cheeked, Golden-cheeked, Gray-breasted, Golden-fronted, Gila, and Yucatan Woodpeckers with which they were once grouped under the genus *Centrus*.

INTERESTING FACT: Diverse diet is more than half animal matter, including spiders, insects, frogs, and reptiles.

DISTRIBUTION AND HABITAT: All wooded habitats of the Bahamas, Cuba, and Grand Cayman.

NESTING: February through August in dead trees or utility poles; lays 5 to 6 eggs. Sometimes 2 broods.

VOCALIZATION: *Kwirr* and a trilling call similar to Red-bellied.

CONSERVATION STATUS: Abundant and of least concern on Cuba, Isle of Pines, and Grand Cayman. Populations on other islands are vulnerable to habitat destruction, more frequent storms, and climate change. Extirpated from Grand Bahama.

Picoides

Picoides is a group of 3 species of the northern latitudes and high elevations of North America and Eurasia that include 2 species covered in this text. They are commonly known as three-toed woodpeckers as they lack the hallux (first digit) found in all other North American species. These are the only woodpeckers in our range without red in any plumage. Males are distinguished by bright yellow caps both in juvenile and adult stages. The bills are broader at the base, but most other physical and behavioral features are similar to those found in *Dryobates* woodpeckers, which were formerly part of the genus. *Picoides* species obtain more than 75 percent of their insect prey through subsurface excavation, in contrast to *Dryobates* species, such as the Hairy Woodpecker, that consume less than half of their prey in this manner.

AMERICAN THREE-TOED WOODPECKER

(Picoides dorsalis)

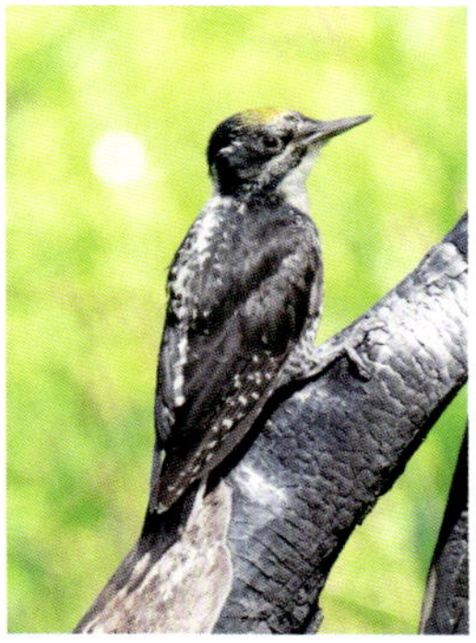

DESCRIPTION: A medium-size black-and-white woodpecker with white mustache and irregular white markings down center of back. Male has yellow cap. L 20–22 cm (7.9–8.7 in).

SIMILAR SPECIES: Black-backed Woodpecker is larger with a solid black back.

INTERESTING FACT: Nests farther north than any other woodpecker species.

DISTRIBUTION AND HABITAT: Medium- to high-elevation spruce and lodgepole pine forests of boreal forest, western dry forest, and, uncommonly, in northern parts of the eastern mixed forest. Often associated with recent (<3 years old) burns, and insect-killed or flooded forest where food and nest snags are plentiful.

NESTING: Mid-May through July. Nests in live or recently dead trees or snags; lays 3 to 7 eggs.

VOCALIZATION: *Kik* call given on its own or in a series, similar to Downy Woodpecker's call.

CONSERVATION STATUS: Globally ranked as secure, but status varies by region. Listed as "vulnerable" to "critically imperiled" in 16 of the 27 states, provinces, and territories where it breeds and as "apparently secure" or "secure" in 11.

BLACK-BACKED WOODPECKER

(Picoides arcticus)

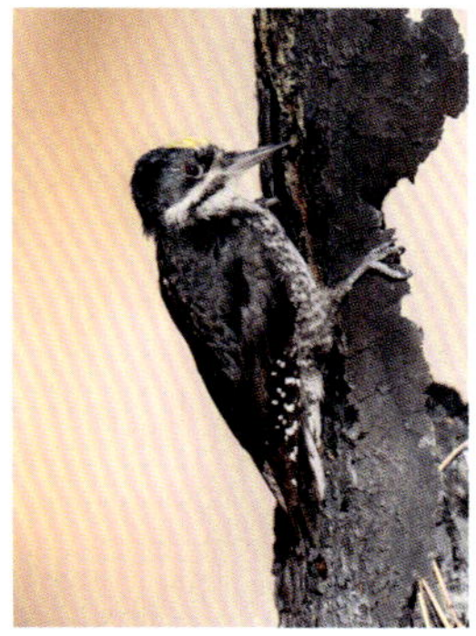

DESCRIPTION: A medium-size, stocky black woodpecker with white limited to the mustache, throat, and belly. Male has yellow cap. L 23–25 cm (9–9.8 in).

SIMILAR SPECIES: American Three-toed Woodpecker is smaller with more white, usually most noticeable on the back.

INTERESTING FACT: Specialist of recently burned areas, particularly in mixed conifers.

DISTRIBUTION AND HABITAT: Primarily boreal but can be found in mid- to sub-alpine habitats in western and northern parts of western dry forest and northern parts of eastern mixed forest. Prefers burned areas adjacent to unburned stands.

NESTING: Late May through late August. Nests in live trees or snags; lays 2 to 6 eggs.

VOCALIZATION: Among the most complex woodpecker calls, although its *kyik* call sounds very similar to Hairy Woodpecker's call. Also uses a distinctive three-part scream-rattle-snarl call.

CONSERVATION STATUS: Globally ranked as secure but status varies by region. The US Forest Service Northern Region and the Bureau of Land Management consider it a "sensitive species." Considered "at risk" by the US Forest Service and Bureau of Land Management. Listed as "vulnerable" to "critically imperiled" in 16 of the 28 states, provinces, and territories where it breeds and "apparently secure" or "secure" in 12. In 2012, two populations of the Black-backed Woodpecker, Oregon Cascades-California and Black Hills, were petitioned for listing under the US federal Endangered Species Act. In California listed as "candidate" species under the California Endangered Species Act.

Dryobates

A genus comprising 25 small to medium-size Eurasian and American woodpeckers, 9 of which are found in the range covered by this book. Only one species (Smoky-brown) ranges into the Southern Hemisphere. These are arboreal species that spend little if any time on the ground, and many of them have a hallux (first toe) that is smaller than in other species and is little used when climbing. They are mostly black-and-white or brown-and-white woodpeckers with the presence of red on the nape or cap differentiating males from females. This group primarily feeds upon insects through a mixture of subsurface excavation, gleaning from surfaces, probing cavities with their bills and tongues, and fly-catching. They will also feed upon sap, and some species will take pollen. Females of most species are generally smaller and feed upon tinier branches and higher up than males. All *Dryobates* were once considered part of *Picoides*.

DOWNY WOODPECKER

(Dryobates pubescens)

DESCRIPTION: A small black-and-white woodpecker. Male has red nape patch. L 14.5–17 cm (5.7–6.7 in).

SIMILAR SPECIES: Hairy Woodpecker is larger with a chisel-shaped bill that is longer than the length of its head.

INTERESTING FACT: Smallest woodpecker in North America.

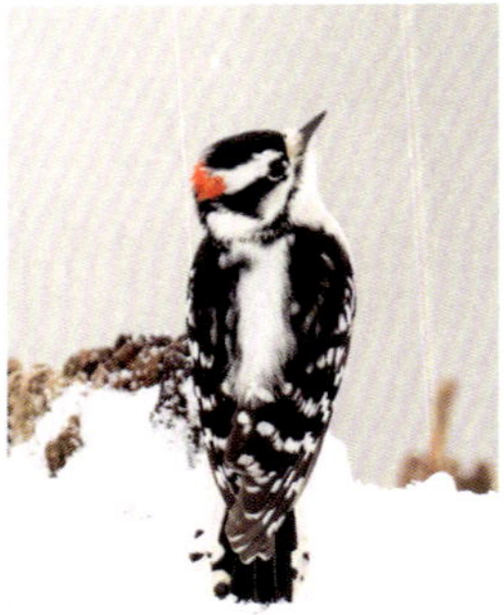

DISTRIBUTION AND HABITAT: Widespread. Breeds in all deciduous and mixed habitats from near tree line in Canada to the southern tip of Florida with patchy distribution in the southwest.

NESTING: Early April through late July. Usually nests in a dead tree or dead part of a living tree; lays 3 to 8 eggs.

VOCALIZATION: Simple *pik* calls, alone or in a series, at a higher pitch than similar call of Hairy Woodpecker.

CONSERVATION STATUS: Species of least concern across its range.

NUTTALL'S WOODPECKER

(Dryobates nuttalli; name under review)

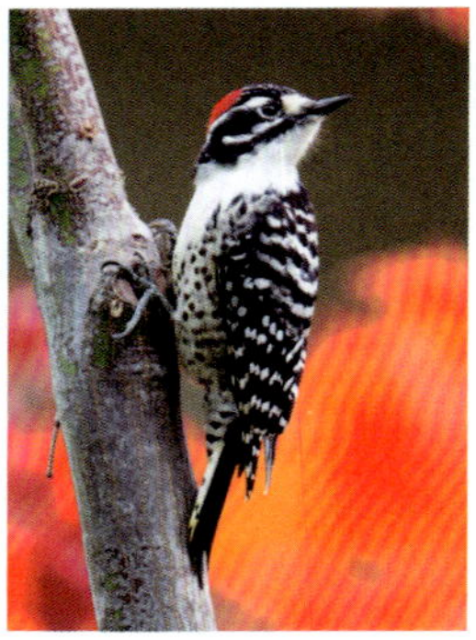

DESCRIPTION: A small black-and-white woodpecker with barred back and wings. Male has red crown. L 17–19 cm (6.7–7.5 in).

SIMILAR SPECIES: Ladder-backed Woodpecker is slightly smaller with more red on male's head and barring on the upper part of the back.

INTERESTING FACT: Acrobatic; frequently feeds while hanging from small twigs or berry clusters.

DISTRIBUTION AND HABITAT: Oak and riparian habitat throughout western oak woodland in California south to northwestern Baja California. The importance of riparian areas increases southward as oaks decline and Downy Woodpeckers disappear.

NESTING: Early April through mid-July. Excavates in extremely soft wood of decayed trunk or limb; lays 3 to 6 eggs.

VOCALIZATION: High-pitched *pit* or *pitick*.

CONSERVATION STATUS: BCC given limited range and dependence on oak woodlands and riparian areas, which themselves are threatened by urban, suburban, and agricultural development.

LADDER-BACKED WOODPECKER

(Dryobates scalaris)

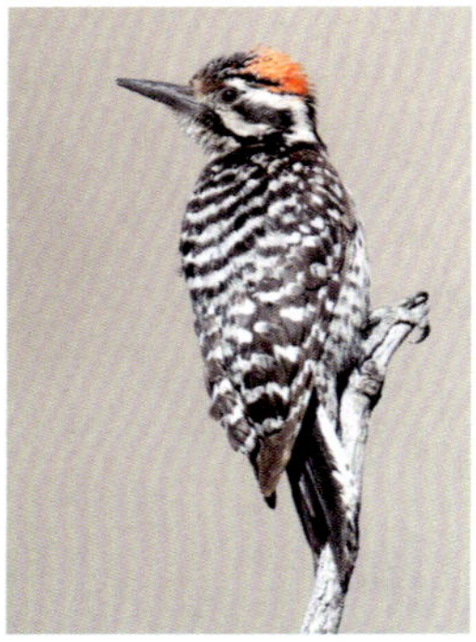

DESCRIPTION: A small black-and-white woodpecker with barred back and wings. Male has red crown. L 16–18 cm (6.3–7.1 in).

SIMILAR SPECIES: Nuttall's Woodpecker is slightly larger with less red on the head and barring that does not extend to the upper part of the back.

INTERESTING FACT: Nimble; frequently feeds from the surface of cacti, trees, and shrubs.

DISTRIBUTION AND HABITAT: Arid landscapes across all habitat types in southeastern California, southern Nevada, much of Arizona, New Mexico, Texas, and much of Mexico.

NESTING: Mid-April through late July. Excavates in trees, utility poles, or fence posts; lays 4 to 7 eggs.

VOCALIZATION: *Peek* call similar to Nuttall's but louder and higher in pitch.

CONSERVATION STATUS: Considered least concern globally and "apparently secure" or "secure" in its core range in Arizona, New Mexico, Texas, and Nevada, but "vulnerable" to "critically imperiled" in the 3 states at the edges of its range.

RED-COCKADED WOODPECKER

(Dryobates borealis)

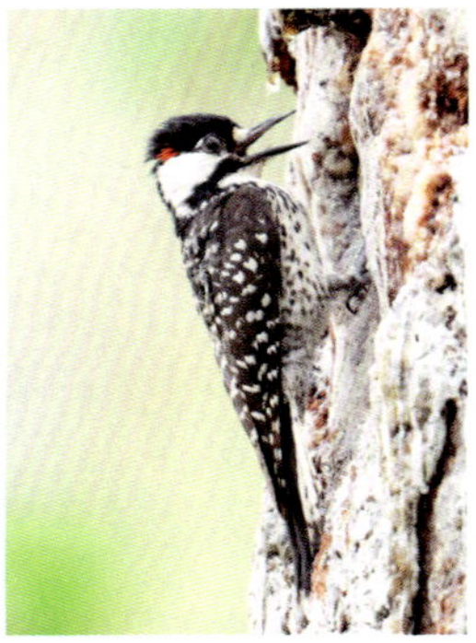

DESCRIPTION: A medium-size black-and-white woodpecker with barred back and wings. Male's few red feathers (cockade) usually hidden. L 20–22 cm (7.9–8.7 in).

SIMILAR SPECIES: Ladder-backed and Nuttall's Woodpeckers, although Red-cockaded has completely white cheeks compared to their black eyestripes.

INTERESTING FACT: Cooperative breeder and lives in family groups.

DISTRIBUTION AND HABITAT: Sporadic in mature southeastern pine forest, particularly longleaf pine.

NESTING: Early April through July. Nests in mature live pines; lays 2 to 5 eggs.

VOCALIZATION: *Chirt* is primary call note given at intervals of 2 to 4 seconds.

CONSERVATION STATUS: Listed as Threatened under the Endangered Species Act and Near Threatened by Global IUCN. Listed as "imperiled" or "critically imperiled" in all 11 states where it currently breeds. It is "presumed extirpated" from 4 additional states.

HAIRY WOODPECKER

(Dryobates villosus)

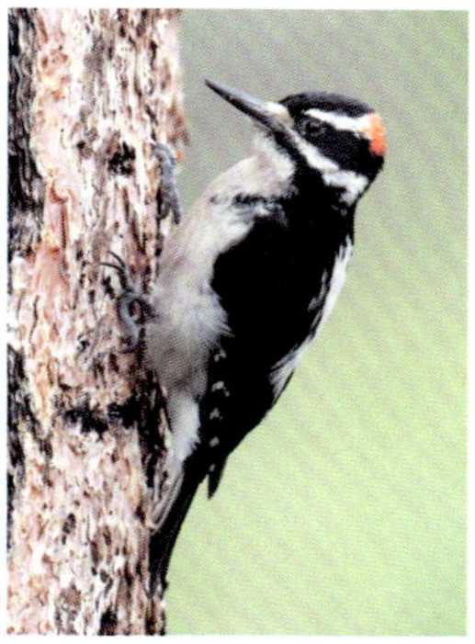

DESCRIPTION: A medium-size black-and-white woodpecker. Male has red nape patch. L 17–26 cm (6.7–10.2 in).

SIMILAR SPECIES: Downy Woodpecker is smaller with a bill that is shorter than the length of its head.

INTERESTING FACT: Plumage varies regionally, with the whitest individuals on the East Coast and sootier ones in the Pacific Northwest.

DISTRIBUTION AND HABITAT: Widespread. Breeds in all types of coniferous forests from near tree line in Canada to the southern tip of Florida and throughout the mountains of Mexico.

NESTING: Mid-March through July. Usually nests in a live tree with fungal rot; lays 3 to 7 eggs.

VOCALIZATION: Monosyllabic *peek* at lower pitch than similar call of Downy Woodpecker.

CONSERVATION STATUS: Listed as "apparently secure" or "secure" in 52 of the 55 states, provinces, and territories where it breeds and as "vulnerable" in 3.

WHITE-HEADED WOODPECKER

(Dryobates albolarvatus)

DESCRIPTION: A medium-size white-headed woodpecker. White on wing primaries and red on nape of male. L 21–23 cm (8.3–9 in).

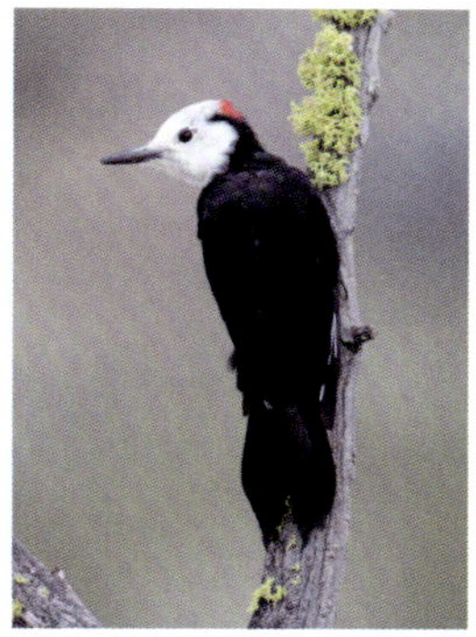

SIMILAR SPECIES: None within range. North America's only black-bodied woodpecker with a white head.

INTERESTING FACT: Specializes in eating pine nuts.

DISTRIBUTION AND HABITAT: Western dry forest from southeastern British Columbia through Oregon, Idaho, California, and Nevada.

NESTING: Late April through mid-August. Excavates in large conifers (usually dead); lays 4 to 5 eggs.

VOCALIZATION: *Peek-ik*.

CONSERVATION STATUS: Considered a "sensitive species" in the Pacific Northwest, intermountain, and northern regions of the US Forest Service; "endangered" in Canada. Listed as "vulnerable" to "critically imperiled" in all 5 states and provinces where it breeds.

SMOKY-BROWN WOODPECKER

(Dryobates fumigatus)

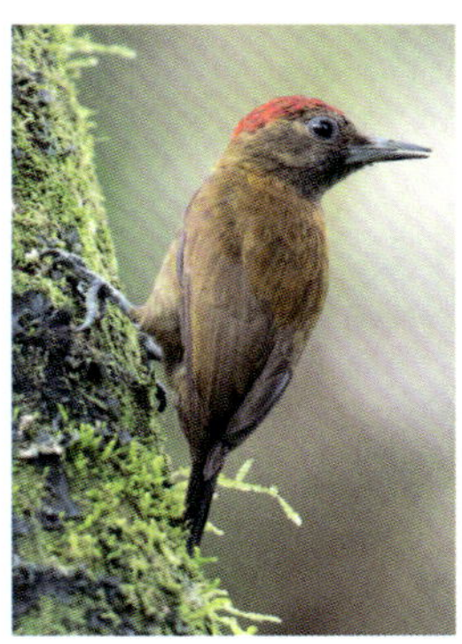

DESCRIPTION: A small brown woodpecker. Male has red cap and nape. L 17–18.5 cm (6.7–7.3 in).

SIMILAR SPECIES: None within range.

INTERESTING FACT: The only *Dryobates* woodpecker found in the Southern Hemisphere and the only one with no white feathers.

DISTRIBUTION AND HABITAT: Tropical wet forest and moist Sierra Madre of Mexico up to 1,500 m (4,920 ft). Often feeds in tangles of vines, slender branches, and foliage excavating for small wood-boring beetles and their larvae.

NESTING: February through June. Excavates in dead limb, trunk, or fence post; lays 4 eggs.

VOCALIZATION: Short *wik*, *pwik*, and *chuck* notes similar but softer and less crisp than Hairy Woodpecker; also high-pitched rattle call and squeaky *wicka* call.

CONSERVATION STATUS: Least concern but is not common anywhere and in some places populations are thin.

STRICKLAND'S WOODPECKER

(Dryobates stricklandi; name under review)

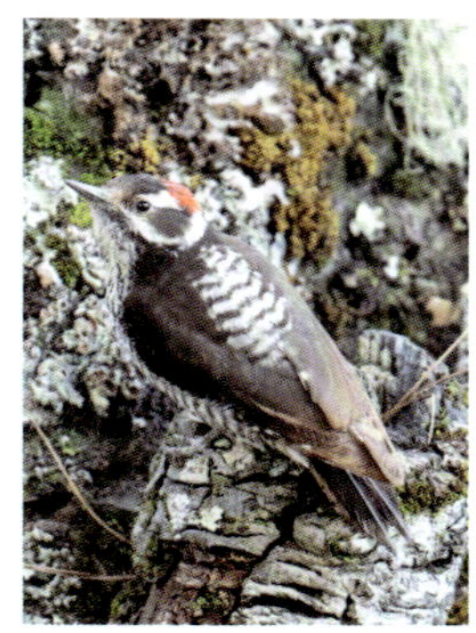

DESCRIPTION: A small to medium-size dark brown woodpecker with a white (usually barred) patch on the back. Male has red nape. L 18–19 cm (7.1–7.5 in).

SIMILAR SPECIES: Arizona Woodpecker is lighter brown, has larger body and bill, no white on the back, and is more spotted on the belly.

INTERESTING FACT: Arizona Woodpecker used to be considered Strickland's Woodpecker.

DISTRIBUTION AND HABITAT: Temperate coniferous and mixed forests dominated by firs, oaks, and alders within the highest and southernmost part of the Sierra Madre between 2,500 and 4,100 m (8,200–13,450 ft).

NESTING: March through July. Excavates in dead trees, dead parts of live trees, or agave stems; lays 2 to 4 eggs.

VOCALIZATION: Single *peep*, distinctive rattle *peep chree-chree-chree-chree-chree*.

CONSERVATION STATUS: Least concern but generally uncommon and found in isolated pockets where it faces significant threat from habitat fragmentation, logging, and heavy grazing, which can reduce regeneration of nest trees and the water table upon which important trees depend.

ARIZONA WOODPECKER

(Dryobates arizonae)

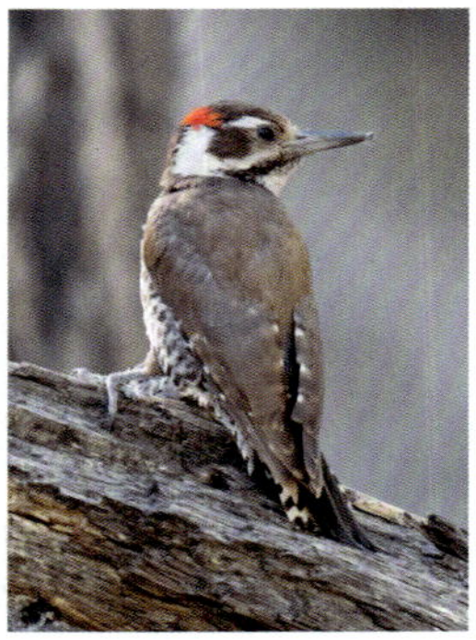

DESCRIPTION: A medium-size brown-backed woodpecker. Male has red nape. L 18–20 cm (7.1–7.9 in).

SIMILAR SPECIES: Strickland's Woodpecker is dark brown with white-marked back and shorter bill.

INTERESTING FACT: The only *Dryobates* woodpecker in the United States and Canada that is brown and white rather than black and white.

DISTRIBUTION AND HABITAT: Sierra Madre from southwest New Mexico and southeastern Arizona through the Sierra Madre Occidental of Mexico in Madrean oak and pine-oak woodland and adjacent riparian woodland at 1,220–2,130 m (4,000–6,990 ft) just below Hairy and just above Ladder-backed where their ranges overlap.

NESTING: Mid-April through late June. Excavates in a tree (usually dead) or dead branch; lays 2 to 4 eggs.

VOCALIZATION: A long *peep*.

CONSERVATION STATUS: BCC, YWL. US Forest Service Region 3 (Arizona and New Mexico) lists Arizona Woodpecker as a "sensitive species." "Vulnerable" in Arizona and "imperiled" in New Mexico. Overdraft of groundwater from heavy grazing can impact regeneration of important sycamores (nesting) in the US Southwest and Mexico.

Campephilus

Campephilus is a distinctive-looking group of 11 new-world species with 3 within our range, although, sadly, 2 are likely extinct. They are all large, mostly black-and-white birds, often with red heads and impressive crests. They have a long fourth toe often held roughly perpendicular to the surface to help pull in their large weight while they are climbing or while excavating wood for insect larvae. Most of these species have a distinctive 2- to 3-beat drum that sounds like a hard strike on a drum followed by a pause and softer echo. They are not close relatives to the superficially similar *Dryocopus*.

PALE-BILLED WOODPECKER

(Campephilus guatemalensis)

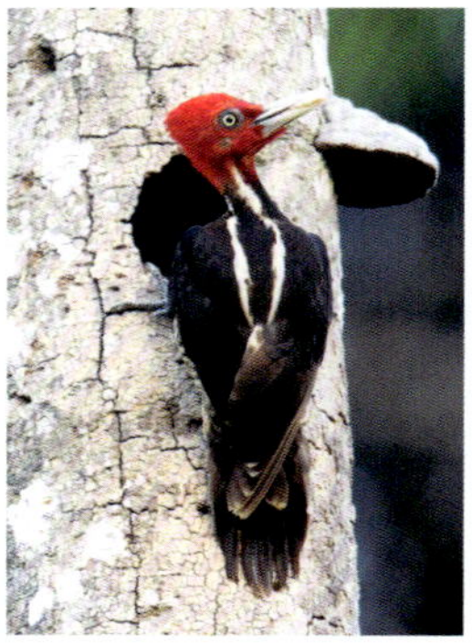

DESCRIPTION: Large black-and-white woodpecker with red crested head (female has black forecrown). Vertical white lines on each side of neck meet in distinctive V on upper back. L 35.5–38 cm (14–15 in).

SIMILAR SPECIES: Lineated Woodpecker overlaps in habitat and range but has white stripes on its face and only a red crest and red malar (male).

INTERESTING FACT: This is the only living *Campephilus* woodpecker in our range since the Ivory-billed and Imperial Woodpeckers are likely extinct.

DISTRIBUTION AND HABITAT: Lowland tropical dry and tropical wet forests of Mexico from Sonora south on the Pacific coast and from Tamaulipas south on the Atlantic coast up to 2,000 m (6,560 ft).

NESTING: December through June in Mexico. Excavates in large, often dead trees; lays 2 eggs.

VOCALIZATION: A variable *ka-ka-ka-ka-Kay-nuk*; also a chatter or sputter call.

CONSERVATION STATUS: Least concern. Requires large trees for foraging, however, and disappears from areas that are deforested or heavily logged.

IVORY-BILLED WOODPECKER

(Campephilus principalis)

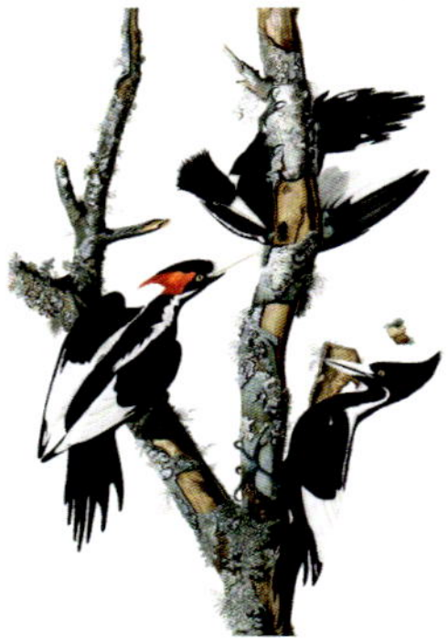

DESCRIPTION: A very large black-and-white woodpecker with an impressive curved crest (red on male, black on female); ivory-colored, chisel-shaped bill; and white stripes extending up back to back of face. L 48–53 cm (18.9–20.9 in).

SIMILAR SPECIES: None within range. Pileated Woodpecker is large, crested, and black and white but much smaller, and white on underwings limited to lining and small primary patch.

INTERESTING FACT: While many believe the Ivory-billed is extinct, several sightings have been reported by credible observers since 1999, although at this point no photographs, videos, or recordings have been widely seen as unequivocal.

DISTRIBUTION AND HABITAT: Historically found in southeastern pine forest region in extensive contiguous mature lowland hardwood-forested flooded plains, alluvial swamps, hummocks, and upland pinewoods from the US to Cuba.

NESTING: January through May. Created cavities in live or partially dead trees; laid 3 to 6 eggs.

VOCALIZATION: Most widely described call was simple, often repeated *kent* singly or in pairs.

CONSERVATION STATUS: Critically endangered and possibly extinct due to hunting, and cutting and fragmentation of habitat.

IMPERIAL WOODPECKER

(Campephilus imperialis)

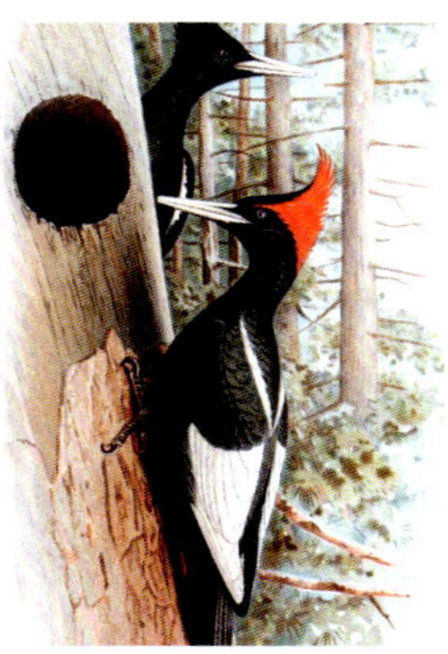

DESCRIPTION: A huge black-and-white woodpecker with a distinctive crest: pointed and crimson red on the male; long, curled, and black on the female. L 53.5–60 cm (21.1–23.6 in).

SIMILAR SPECIES: Within its range the Pale-billed Woodpecker was most similar but at least 30 percent smaller, and both sexes have much more red on the head.

INTERESTING FACT: The Imperial was the largest woodpecker in the world and up to 20 percent larger than the Ivory-billed Woodpecker.

DISTRIBUTION AND HABITAT: Lived in mature western Sierra Madre pine forests 1,920 m (6,300 feet) in elevation.

NESTING: Likely January through June.

VOCALIZATION: Various accounts mention "nasal penny-trumpet-like" *keent-keent* notes and tinny nuthatch-like calls that sounded very much like the Ivory-billed.

CONSERVATION STATUS: Critically endangered and likely extinct due to subsistence hunting, logging, and clearing for agriculture.

Dryocopus

A genus of 6 species including 4 from the New World, 2 of which are in our range. These are very large, mostly black woodpeckers with white markings, red crests, and color badges. Most specialize in ants and their larvae. They will also take fruits, berries, and nuts. Their loud drumming and expressive cackling calls are heard over great distances. Their annual nest cavities provide homes for many birds and other animals, making them a keystone species in the large-treed forested habitats in which they live year-round. Their flights are strong and direct. The most closely related genera in our range are *Colaptes* and *Celeus*.

LINEATED WOODPECKER

(Dryocopus lineatus)

DESCRIPTION: A large black-and-white woodpecker with red crest, barred underparts, and white line extending from the breast up to the bill. Male has red malar. L 30–36 cm (11.8–14.2 in).

SIMILAR SPECIES: Pale-billed Woodpecker has a red head rather than just red crest.

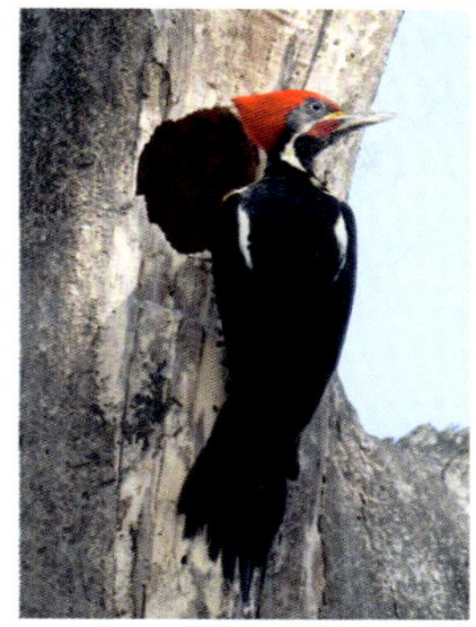

INTERESTING FACT: Very closely related to the Pileated Woodpecker of the US and Canada. Will forage or even nest near (even in the same tree) the Pale-billed Woodpecker.

DISTRIBUTION AND HABITAT: Lowland tropical wet and tropical dry forests but prefers relatively open areas such as forest edges on both coasts of Mexico southward through much of South America.

NESTING: April through May. Excavates in dead trees; lays 2 to 4 eggs.

VOCALIZATION: Rapid, intensifying laughing series of notes lasting a few to several seconds, rising at first and then falling, described as *yihh-yihh-yihh*. Other calls reminiscent of Pileated Woodpecker in length and structure.

CONSERVATION STATUS: Least concern.

PILEATED WOODPECKER

(Dryocopus pileatus)

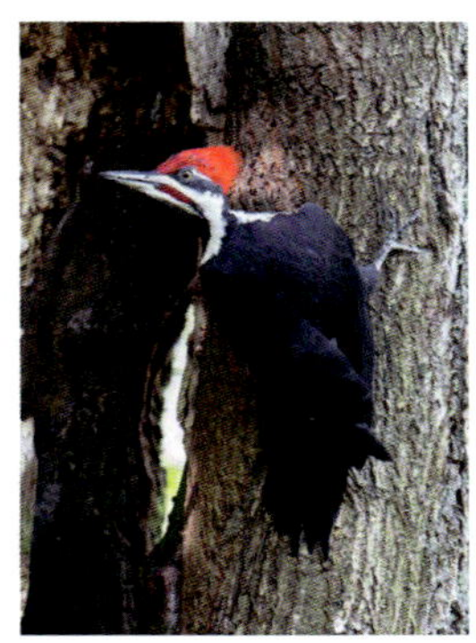

DESCRIPTION: A large black woodpecker with a white neck and distinctive red crest. Female's crest does not reach bill. Male has red malar. L 40–49 cm (15.7–19.3 in).

SIMILAR SPECIES: Unlikely to be confused with any other woodpecker within its range.

INTERESTING FACT: Largest woodpecker in North America.

DISTRIBUTION AND HABITAT: Throughout all forest types in the US, including most east of the Great Plains, plus the southern half of the boreal forest, the northern Rockies, and much of the West Coast, including the Cascades, Olympics, and Sierras.

NESTING: Early May through mid-July. Excavates in dead trees; lays 1 to 6 eggs. Nests often reused as roost cavities but not as nests.

VOCALIZATION: Most well-known call is extended *kuk-kuk, kuk-kuk* that can be heard in jungle scenes of some old movies.

CONSERVATION STATUS: Not currently listed as Threatened or Endangered. Listed as "apparently secure" or "secure" in 38 of the 51 states, territories, and provinces where it breeds and "vulnerable" to "critically imperiled" in 13.

Celeus

Celeus is a genus of 12 tropical and subtropical species with only one in our range. These medium-size to large new-world woodpeckers are rufous or cinnamon-colored with bushy crests and pale bills. Red malar distinguishes male from female. They eat mostly ants and termites near the ground or in the canopy where they also forage for fruit and seeds. Their morphology reflects their penchant for gleaning, probing, and collecting over excavating with their curved bill, which lacks protective nostril feathers. They capture food with their long tongues.

CHESTNUT-COLORED WOODPECKER

(Celeus castaneus)

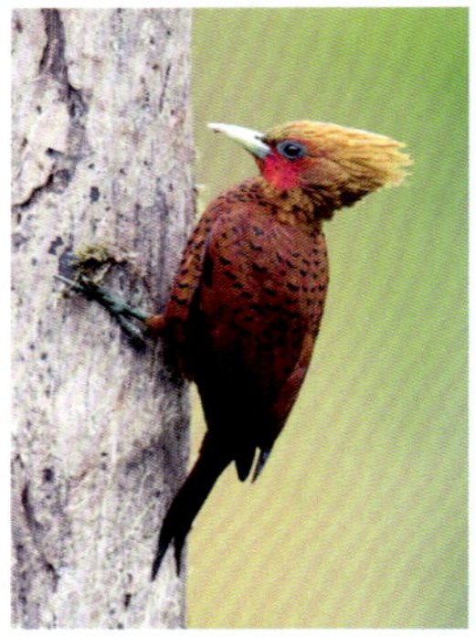

DESCRIPTION: A medium- to large-size chestnut-brown woodpecker with a lighter-colored head featuring a distinctive large shaggy crest. Male has red malar. L 21.5–24 cm (8.5–9.4 in).

SIMILAR SPECIES: None within range.

INTERESTING FACT: The northernmost member of genus *Celeus* and the only one in our range.

DISTRIBUTION AND HABITAT: Humid evergreen and semideciduous tropical wet forest in southeastern Mexico.

NESTING: March through July. Excavates in dead trees and lays approximately 4 eggs.

VOCALIZATION: Descending, whistling *skew* or *peew* trailed by a few piercing, nasal *keew, wet-wet-wet*.

CONSERVATION STATUS: Least concern but vulnerable to habitat loss or degradation.

Colaptes

This new-world genus of medium to large woodpeckers comprises 12 species with 5 in our area. Each pursues a diet of ants, termites, beetles and their larvae, and, less frequently, fruit and nuts. Flickers and their relatives have decurved bills and long tongues to assist them in seizing their target prey. Underwings range from beige to bright yellow and red. The species fall into two groups whose color schemes better camouflage them while foraging; the flickers that spend more time on the ground are generally brown with black on their backs and wings, while the more arboreal species are green in coloration. The gender-distinguishing badges include darker, often more colorful malar feathers in the flickers and additional red on the head in the others.

GOLDEN-OLIVE WOODPECKER

(Colaptes rubiginosus)

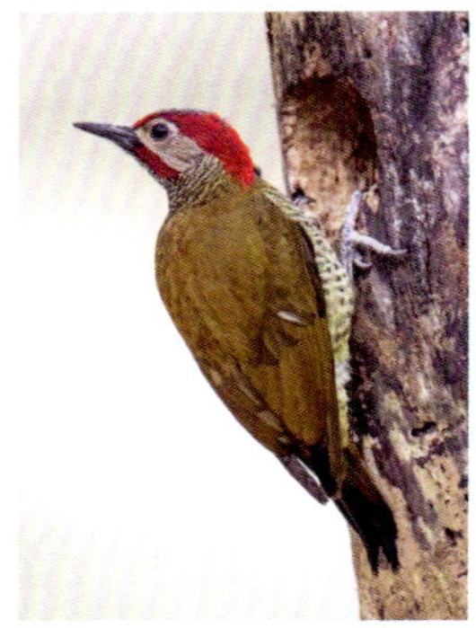

DESCRIPTION: A medium-size olive-green woodpecker with a solid back tending toward bronze green on the wings, barred breast, gray face, dark gray crown (edged in red on male) red nape, and red mustache (gray in female). L 18–23 cm (7.1–9.5 in).

SIMILAR SPECIES: None within range. Gray-crowned Woodpecker from western Mexico similar but with much less red on the head, which is lighter overall.

INTERESTING FACT: Northern populations are considered by the International Ornithological Committee and others to be a separate species, the Bronze-winged Woodpecker, based on different call, more bronze-colored wings, and scalloped rather than barred pattern on breast. Also, red on sides of male's crown does not extend to eye.

DISTRIBUTION AND HABITAT: Tropical wet and tropical dry forests and Sierra Madre from northeastern Mexico through southern Mexico from near sea level to 2,100 m (6,890 ft).

NESTING: January through May. Excavates in dead or live trees or palms; lays 2 to 4 eggs.

VOCALIZATION: Long rattling trill. Sound of northern (Bronze-winged species or subspecies) is slow-paced series of high-pitched calls.

CONSERVATION STATUS: Least concern and common throughout range.

GRAY-CROWNED WOODPECKER

(Colaptes auricularis)

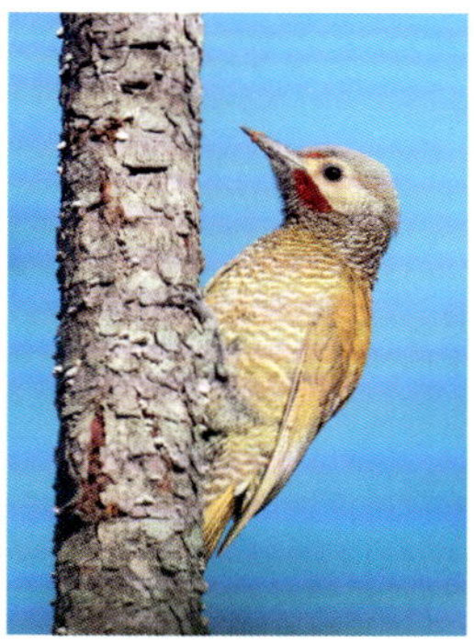

DESCRIPTION: A medium-size bronze-green woodpecker with a gray crown, cream-colored face, and green barred underparts. Male has red malar; female has gray malar. L 19.5–21 cm (7.7–8.3 in).

SIMILAR SPECIES: Smoky-brown Woodpecker is smaller and browner.

INTERESTING FACT: Once considered a subspecies of the Golden-olive Woodpecker.

DISTRIBUTION AND HABITAT: Endemic to the Pacific slope of western Mexico in tropical dry and tropical wet forests and lower Sierra Madre from near sea level to 2,400 m (7,870 ft) and also in pine-oak forests.

NESTING: February through April. Excavates in dead trees; number of eggs unknown.

VOCALIZATION: Rapid rattling and trilling notes.

CONSERVATION STATUS: Least concern by the IUCN but high concern by Partners in Flight, which considers it of "high tri-national concern." Endemic to northwest Mexico, where it is threatened.

NORTHERN FLICKER

(Colaptes auratus)

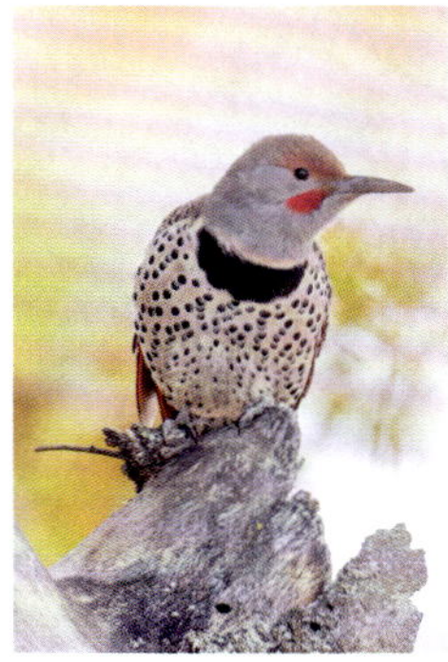
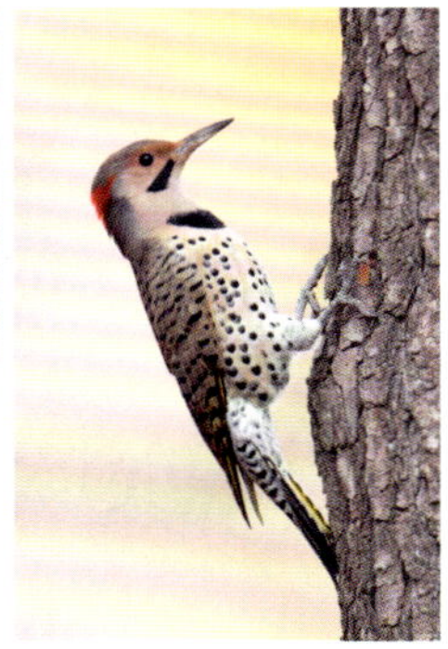

DESCRIPTION: A large, distinctive brown woodpecker with polka-dotted belly, barred back, black chest patch, and bright color under wings and tail. Eastern Yellow-shafted subspecies has yellow underwings and tail and a red crescent on nape. Western Red-shafted subspecies has red underwings and tail and no marking on nape. L 28–31 cm (11–12.2 in).

SIMILAR SPECIES: Gilded Flicker is smaller with brown nape. Neither Red-shafted nor Yellow-shafted subspecies have brown nape.

INTERESTING FACT: Often forages on the ground, where it specializes in ants.

DISTRIBUTION AND HABITAT: Breeds throughout all treed landscapes of North America from tree line to southern Florida to central Mexico, the Florida Keys, Cuba, and Grand Cayman Island. Red-shafted flickers found to the south and west of the Rocky Mountains, with an intergrade zone to the east and north of the Rockies.

NESTING: Late April through early August. Excavates in dead or diseased trees; lays 3 to 12 eggs.

VOCALIZATION: Long series of repeated *wicka, wicka, wicka*, similar to but lower pitched than Gilded Flicker.

CONSERVATION STATUS: Considered least concern by the IUCN, but has lost more than 50 percent of its population in the past 40 years, which is alarming considering their ecological importance. "Apparently secure" or "secure" in 52 of the states, provinces, and territories where it breeds but "vulnerable" in 3.

NOTE: At the time of writing, the Guatemalan Flicker was recognized only as a subspecies, but just as this book went to press it was elevated to a species by AOS.

GILDED FLICKER

(Colaptes chrysoides)

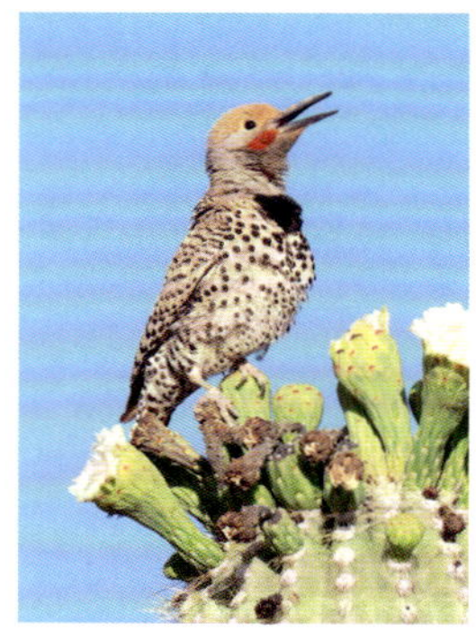

DESCRIPTION: A medium-large grayish-brown woodpecker with dark bars above, black spots on whitish breast, large black crescent mark on upper breast, and bright yellow ventral surface of tail and wings. Males have red malar. L 25.4–30.5 cm (10–12 in).

SIMILAR SPECIES: Northern Flicker larger with gray nape (both Red-shafted and Yellow-shafted).

INTERESTING FACT: One of two North American woodpeckers that excavate in saguaro cacti.

DISTRIBUTION AND HABITAT: Hot desert habitat of the Sonoran Desert of southeast California and southern Arizona.

NESTING: Early March through mid-August. Excavates in saguaro cacti; lays 3 to 5 eggs, with 2 and sometimes 3 broods per year.

VOCALIZATION: Long series of repeated *wicka, wicka, wicka*, similar to but higher pitched than Northern Flicker.

CONSERVATION STATUS: BCC, YWL. Listed as "vulnerable" to "critically imperiled" in the 3 states where it breeds.

FERNANDINA'S FLICKER

(*Colaptes fernandinae;* name under review)

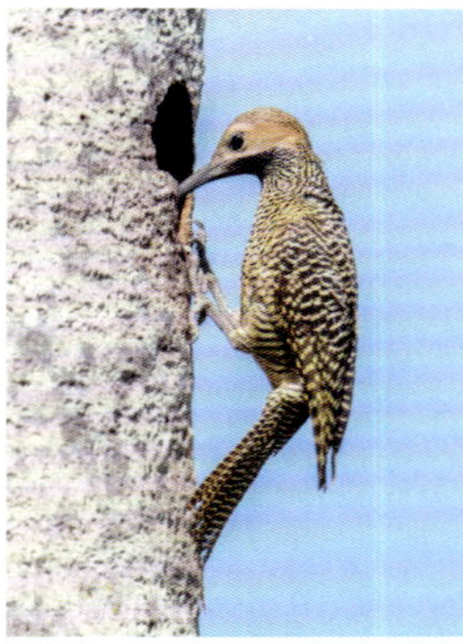

DESCRIPTION: A large woodpecker with tan or yellowish plumage with bars of black on the back, tail, and belly, tan head with black mustache on male (streaked on female). L 30–34 cm (11.8–13.4 in).

SIMILAR SPECIES: Cuban race of Northern Flicker has red nape and black chest patch.

INTERESTING FACT: Often nests in cavities initiated or completed by West Indian Woodpecker.

DISTRIBUTION AND HABITAT: Endemic to Cuba where it prefers open woodland and pasture with palms but can nest in denser woodland.

NESTING: March through June. Excavates or inherits cavities in dead or (especially) live palms; lays 3 to 5 eggs.

VOCALIZATION: Loud series of *pic* calls and typical flicker *wicka* series.

CONSERVATION STATUS: "Vulnerable" or "endangered" due to small, fragmented, and declining population (300–440 pairs remain). Parrot hunters push over palms containing cavities, and storms inundate prime habitat with saltwater.

The absence of a fourth toe allows the American Three-toed Woodpecker (pictured here) and Black-backed Woodpecker to apply more force while excavating but makes for more awkward cllmbing.

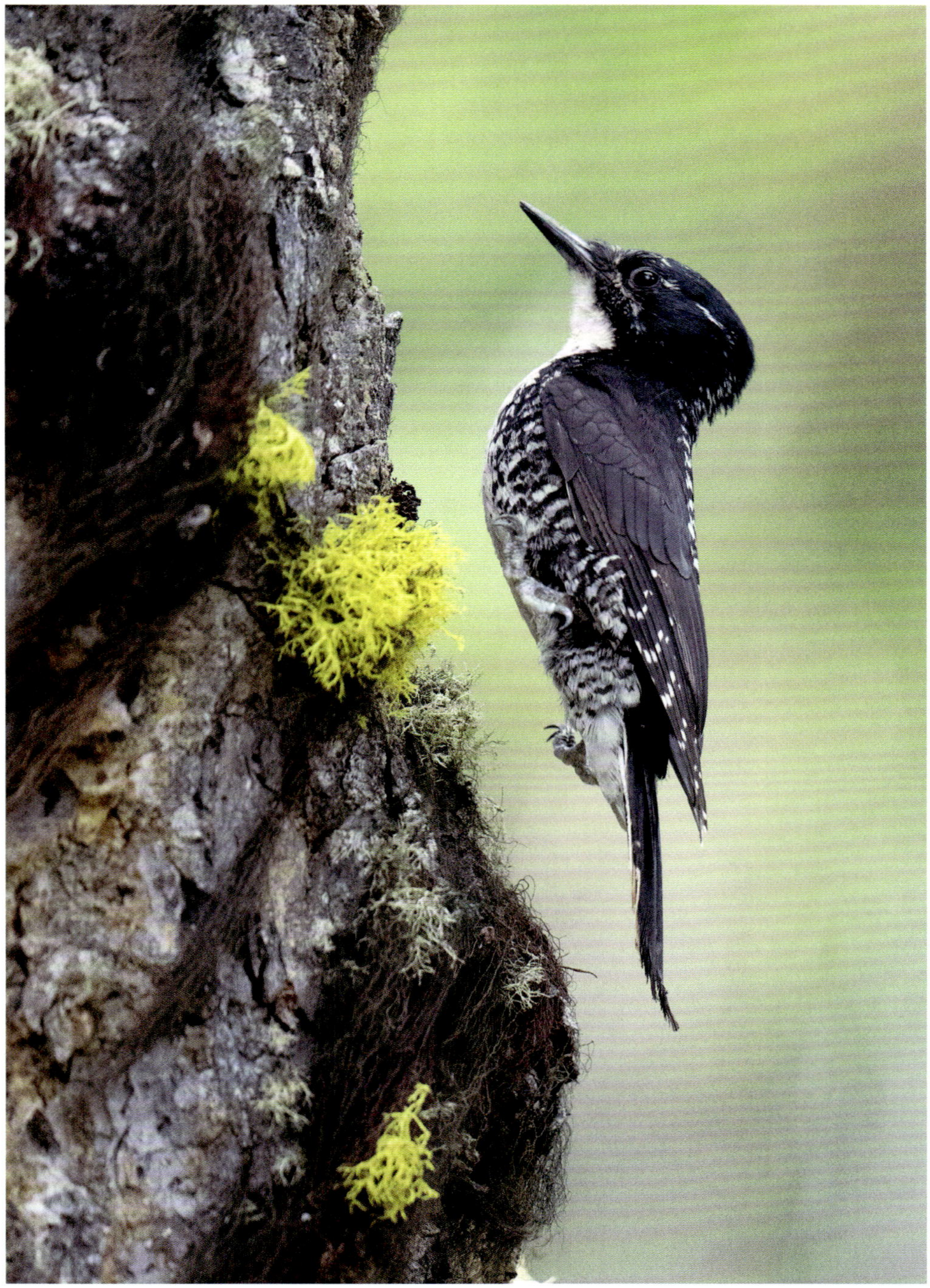

Acknowledgments

Thanks to all of those who helped in my efforts to photograph and study woodpeckers, whether that be by joining me in the field, helping me gain access to closed locations, or providing me a place to stay: Jamie Acker, Dr. Phil Allman, Brady Beck, Mark Christiansen, Steve Godwin, Donna Gould, Jim Howitz, Keith Kingdon, Walter D. Koenig, Linda Nichols, Jud Parson, Dr. Caitlin Barale Potter, Megan Sandri, Sean and Anneke Sober, Siah St. Clair, Peter Thiemann, and Jean Tuomi.

A special thanks to Michelle Privat Obermeyer for her invaluable review of my drafts and photos throughout the process, and to Matthew Danielson, Katy Love, and Matt Bannick for valuable feedback on particular sections of the manuscript.

This work would not have been possible without a talented team including Kate Rogers, whose deft stewardship and creative contributions were critical to the book's success; Janet Kimball, who was a dependable partner throughout every step of the process and ensured that everything and everyone was on track; Linda Gunnarson, who reviewed and edited the first draft; and Laura Whittemore, who helped guide and refine the edits of the book. I am also grateful to designer Cat Grishaver and page proofer Jenn Kepler for the addition of their talents to the book. Finally, I would like to thank Teresa Lorenz for reviewing the manuscript for scientific accuracy.

With Gratitude

Thank you to the following members, volunteers, and donors whose financial contributions help keep our independent nonprofit publishing thriving. With your support, our publications can reach more people and help them to enjoy, respect, and sustain the natural world. To learn more about how you can elevate the work and impact of Mountaineers Books, please visit mountaineersbooks.org/donate.

$5,000

Gary Rygmyr

$2,500

M. Martha Ries & David W. Day

$1,000

Don & Marci Heck
Rachel & Doug McCall
Madeline Nguyen, in honor of Tracey Ash
Bri Vanderlinden & Anvesh Reddy Yedugundla
Auntie B, in honor of Trae Cates Whitted, Rhoen Jack Whitted, Jere Porter Mancha, & Claire Angel Mancha

$500

Dan & Heather Irwin
Mike & Maria Ruth
The Sanders / Zanini Family
James Pierce Whinston
LaVerne Woods, in memory of Erin F. & W. William Woods

$250

Helen & Arnie Cherullo
Jen Walsh Fisher & Richard Fisher
Mr. & Mrs. D. Gelinas
Danielle Graham, in honor of Andrew Pedersen
Peter Hendrickson & Nancy Temkin
Mary Hiney, in honor of Aspen S. Adams
Danielle Lail
Ken Lans
Andrea Ma
Mandi Maycumber, in honor of Arthur Maycumber
Susan Pavlansky
Peggy J. Printz
Krestine Reed
Caryn Schutzler
Alan Stewart
Tom Vogl & Katie Koepke
Meimin Xie, in honor of Rob March

Generous support from Tom and Sonya Campion funds the Braided River imprint and Paul Bannick's events and outreach on behalf of woodpeckers and their diverse natural habitats.

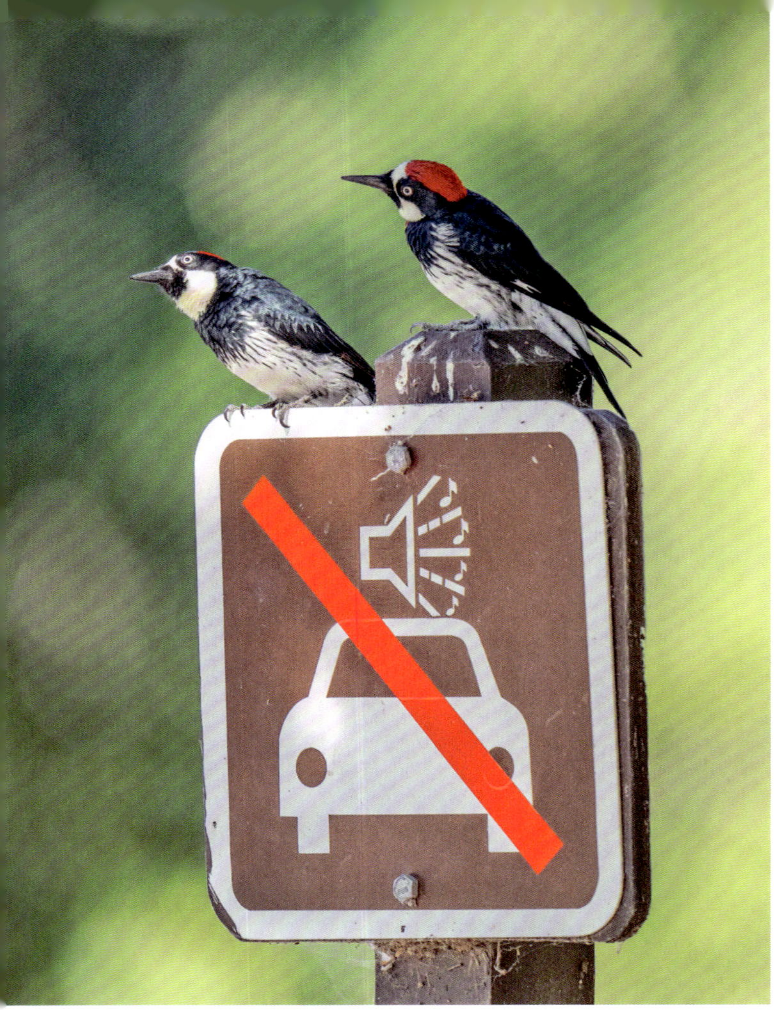

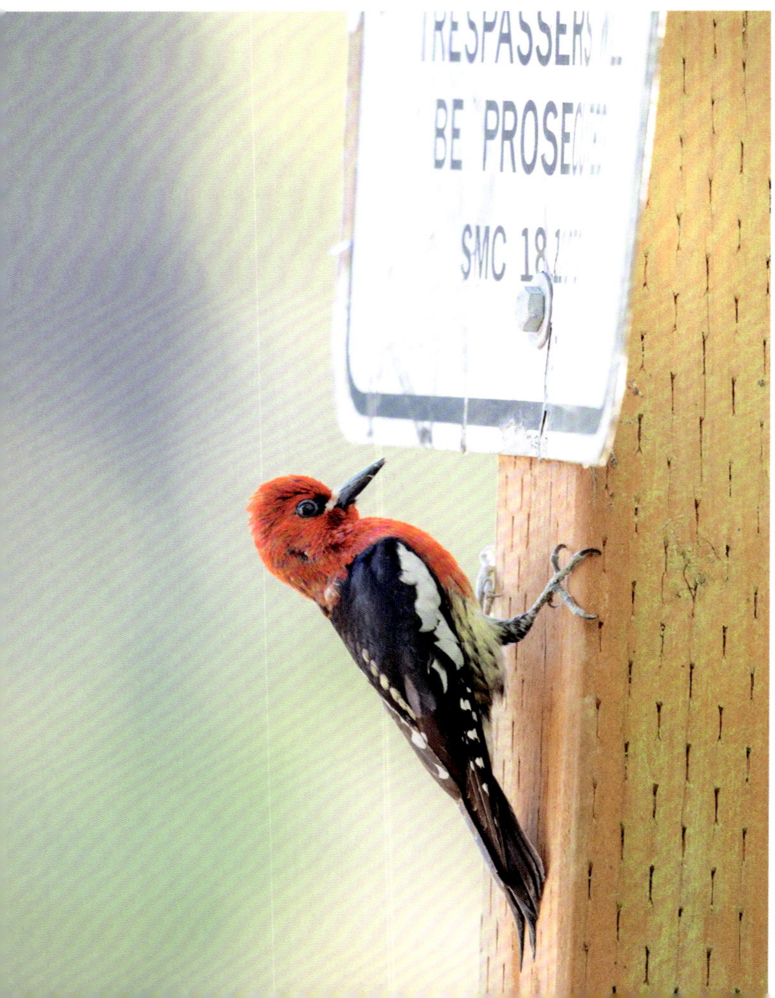

Selected Bibliography

Books and Journal Articles

Aitken, K. E., H. L. Wiebe, and K. Martin. "Nest-Site Reuse Patterns for a Cavity-Nesting Bird Community in Interior British Columbia." *The Auk* 119, no. 2 (April 2002): 391–402.

Aubry, K. B., and C. M. Raley. "Selection of Nest and Roost Trees by Pileated Woodpeckers in Coastal Forests of Washington." *Journal of Wildlife Management* 66, no. 2 (April 2002): 392.

Austin, G. T., and E. L. Smith. "Winter Foraging Ecology of Mixed Insectivorous Bird Flocks in Oak Woodland in Southern Arizona." *The Condor* 74 (1972): 17–24.

Bock, C. E., H. H. Hadow, and P. Somers. "Relations Between Lewis' and Red-headed Woodpeckers in Southeastern Colorado." *Wilson Bulletin* 83, no. 3 (1971): 237–48.

Bock, C. E., and D. L. Larson. "Winter Habitats of Sapsuckers in Southeastern Arizona." *The Condor* 88 (1986): 246–47.

Bock, W. J., and W. D. Miller. "The Scansorial Foot of Woodpeckers with Comments on the Evolution of Climbing and Feet in Birds." *American Museum Novitates*, no. 1931 (1959).

Bull, Evelyn L., Roy C. Beckwith, and Richard S. Holthausen. "Arthropod Diet of Pileated Woodpeckers in Northeastern Oregon." *Northwestern Naturalist* 73, no. 2 (Autumn 1992): 42–45.

Bull, E. L., and E. C. Meslow. "Breeding Biology of the Pileated Woodpecker—Management Implications." Research Note PNW-RN-474 (1988). Portland, OR: US Department of Agriculture, Forest Service, Pacific Northwest Research Station.

Bull, E. L., S. R. Peterson, and J. W. Thomas. "Resource Partitioning Among Woodpeckers in Northeastern Oregon." Research Note PNW-444 (1986). Portland, OR: US Department of Agriculture, Forest Service, Pacific Northwest Research Station.

Crockett, A. B., and H. H. Hadow. "Nest Site Selection by Williamson's and Red-naped Sapsuckers." *The Condor* 77, no. 3 (1975): 365–68.

Crockett, A. B., and P. L. Hansley. "Coition, Nesting, and Post Fledging Behavior of Williamson's Sapsuckers in Colorado." *Living Bird*, no. 16 (1977): 7–19.

Daniels, Susan J., and Jeffrey R. Walters. "Between-Year Breeding Dispersal in Red-cockaded Woodpeckers: Multiple Causes and Estimated Cost." *Ecology* 81, no. 9 (September 2000): 2473–84.

del Hoyo, Josep, Andrew Elliott, and Jordi Sargatal. *Handbook of the Birds of the World*. Vol. 7, *Jacamars to Woodpeckers*. Lynx Editions, 2002.

du Plessis, M. A., W. W. Weathers, and W. D. Koenig. "Energetic Benefits of Communal Roosting by Acorn Woodpeckers During the Nonbreeding Season." *The Condor* 96 (1994): 631–37.

Gorman, Gerald. *Woodpeckers of the World: The Complete Guide*. Christopher Helm, 2014.

Hammond, Ruby L., and Tad C. Theimer. "A Review of Tree-Scale Foraging Ecology of Insectivorous Bark-Foraging Woodpeckers in North America." *Forest Ecology and Management* 478 (December 2020). doi:10.1016/j.foreco.2020.118516.

Top: Acorn Woodpeckers are among the species that become quite accustomed to people when valuable food sources are in parks or backyards.

Bottom: Woodpeckers, like this Red-breasted Sapsucker, amplify their courtship and territorial advertisements by drumming on the most resonant objects they can find.

Harding, Sergio R. "The Dynamics of Cavity Excavation and Use by the Red-Cockaded Woodpecker (*Picoides borealis*)." Master's thesis, Virginia Polytechnic Institute and State University, 1997.

Howell, Steve, and Sophie Webb. *A Guide to the Birds of Mexico and Northern Central America*. Oxford University Press, 1995.

Hoyt, J. S., and S. J. Hannon. "Habitat Associations of Black-backed and Three-toed Woodpeckers in the Boreal Forest of Alberta." *Canadian Journal of Forest Research* 32, no. 10 (2002): 1881–88.

Ingold, D. J. "Influence of Nest-Site Competition Between European Starlings and Woodpeckers." *Wilson Bulletin* 106, no. 2 (1994): 227–41.

Inouye, D. W. "Nonrandom Orientation of Entrance Holes to Woodpecker Nests in Aspen Trees." *The Condor* 78, no. 1 (January 1976): 101–2.

Kirwan, Guy M., Anthony Levesque, Mark Oberle, and Christopher J. Sharpe. *Birds of the West Indies*. Lynx Editions, 2019.

Koenig, Walter D., Frank A. Pitelka, William J. Carmen, Ronald L. Mumme, and Mark T. Stanback. "The Evolution of Delayed Dispersal in Cooperative Breeders." *Quarterly Review of Biology* 67, no. 2 (June 1992): 111–50.

Koplin, James R. "Measuring Predator Impact of Woodpeckers on Spruce Beetles." *Journal of Wildlife Management* 36, no. 2 (April 1972): 308–320.

Kozma, Jeffrey M. "Composition of Forest Stands Used by White-Headed Woodpeckers for Nesting in Washington." *Western North American Naturalist* 71, no. 1 (2011): 1–9.

Kozma, Jeffrey M. "Nest-Site Characteristics of Three Woodpecker Species in Managed Ponderosa Pine Forests of the Eastern Cascade Range." *Northwestern Naturalist* 93 (Autumn 2012): 111–19.

Kozma, Jeffrey M., and Andrew Kroll. "Woodpecker Nest Survival in Burned and Unburned Managed Ponderosa Pine Forests of Northwestern United States." *The Condor* 114, no. 1 (February 2012): 173–84.

Li, Pingjun, and Thomas E. Martin. "Nest Site Selection and Nesting Success of Cavity-Nesting Birds in High Elevation Forest Drainages." *The Auk* 108, no. 2 (April 1991): 405–18.

Lorenz, Teresa J. "Between-Year Breeding Dispersal by White-headed Woodpeckers: A Caution About Using Color Bands to Estimate Survival." *Northwestern Naturalist* 97, no. 3 (Winter 2016): 252–56.

Lorenz, T. J., P. C. Fischer, A. Stillman, M. Tingley, and R. Siegel. "Survival and Dispersal in Juvenile White-headed and Black-backed Woodpecker." 135th Annual Meeting of the American Ornithological Society, East Lansing, MI, 2017.

Lorenz, T. J., K. T. Vierling, T. R. Johnson, and P. C. Fischer. "The Role of Wood Hardness in Limiting Nest Site Selection in Avian Cavity Excavators." *Ecological Applications* 25, no. 4 (June 2015): 1016–33.

Lorenz, Teresa J., Kerri T. Vierling, Jeffrey M. Kozma, Janet E. Millard, and Martin G. Raphael. "Space Use by White-headed Woodpeckers and Selection for Recent Forest Disturbances." *Journal of Wildlife Management* 79, no. 8 (August 2015): 1286–97.

Mastretta-Yanes, Alicia, Alejandra Moreno-Letelier, Daniel Piñero, Tove H. Jorgensen, and Brent C. Emerson. "Biodiversity in the Mexican Highlands and the Interaction of Geology, Geography and Climate Within the Trans-Mexican Volcanic Belt." *Journal of Biogeography* 42, no. 9 (June 2015): 1586–1600.

Mumme, R. L., W. D. Koenig, and F. A. Pitelka. "Individual Contributions to Cooperative Nest Care in the Acorn Woodpecker." *The Condor* 92, no. 2 (1990): 360–68.

Murphy, Edward C., and William A. Lehnhausen. "Density and Foraging Ecology of Woodpeckers Following a Stand-Replacement Fire." *Journal of Wildlife Management* 62, no. 4 (October 1998): 1359–72.

Newlon, Karen R., and Victoria. A. Saab. "Nest-Site Selection and Nest Survival of Lewis's Woodpecker in Aspen Riparian Woodlands." *The Condor* 117, no. 2 (2011): 183–93.

Saab, V. A., and K. T. Vierling. "Reproductive Success of Lewis's Woodpecker in Burned Pine and Cottonwood Riparian Forests." *The Condor* 103, no. 3 (2001): 491–501.

Schepps, J., S. Lohr, and T. E. Martin. "Does Tree Hardness Influence Nest-Tree Selection by Primary Cavity Nesters?" *The Auk* 116, no. 3 (1999): 658–65.

Short, Lester L. "Habits and Interactions of North American Three-toed Woodpeckers (*Picoides arcticus* and *Picoides tridactylus*)." *American Museum Novitates* 547 (1974): 1–42.

Short, Lester L. *Woodpeckers of the World*. Delaware Museum of Natural History, 1982.

Shunk, Stephen A. *Peterson Reference Guide to Woodpeckers of North America*. Mariner Books, 2016.

Smith, K. G. "On Habitat Selection of Williamson's and 'Red-naped' Yellow-bellied Sapsuckers." *Southwestern Naturalist* 27, no. 4 (1982): 464–66.

Speights, Jason R., and Warren C. Conway. "Habitat and Host Tree Selection by Wintering Yellow-bellied Sapsuckers in East Texas." *American Midland Naturalist* 164, no. 1 (July 2010): 107–18.

St-Amand, Julien, Junior A. Tremblay, and Kathy Martin. "Foraging Ecology of the Williamson's Sapsucker." *The Condor* 120, no. 3 (August 2018): 680–702.

Stillman, Andrew N., Rodney B. Siegel, Robert L. Wilkerson, Matthew Johnson, Christine A. Howell, and Morgan W. Tingley. "Nest Site Selection and Nest Survival of Black-backed Woodpeckers after Wildfire." *The Condor* 121, no. 3 (August 2019): 1–13.

Stillman, Andrew N., Rodney B. Siegel, Robert L. Wilkerson, Matthew Johnson, and Morgan W. Tingley. "Age-Dependent Habitat Relationships of a Burned Forest Specialist Emphasize the Role of Pyrodiversity in Fire Management." *Journal of Applied Ecology* 56, no. 4 (December 2018): 880–90.

Vierling, K. T. "Interactions Between European Starlings and Lewis' Woodpeckers at Nest Cavities." *Journal of Field Ornithology* 69, no. 3 (1998): 376–79.

Vierling, K. T., D. J. Gentry, and A. M. Haines. "Nest Niche Partitioning of Lewis's and Red-headed Woodpeckers in Burned Pine Forests." *Wilson Journal of Ornithology* 121, no. 1 (2009): 89–96.

Walters, Jeffrey R., Phillip D. Doerr, and J. H. Carter III. "Delayed Dispersal and Reproduction as a Life-History Tactic in Cooperative Breeders: Fitness Calculations from Red-cockaded Woodpeckers." *American Naturalist* 139, no. 3 (March 1992): 623–43.

Winkler, Hans, David A. Christie, and David Nurney. *Woodpeckers: An Identification Guide to the Woodpeckers of the World*. Houghton Mifflin, 1995.

Websites

Cornell Lab, Birds of the World, https://birdsoftheworld.org/bow/content/source-content. A significant resource for in-depth information of all aspects of the life histories and behaviors of all species covered in this book.

Natureserve Explorer, https://explorer.natureserve.org. Provides a vast array of biodiversity data for the US and Canada.

North American Bird Conservation Initiative, https://nabci-us.org/resources/bird-conservation-regions-map/#bcr43. Provides an array of tools including a bird conservation regions map.

Index

About the Author

Paul Bannick is an award-winning author and photographer who captures authentic images to inspire wildlife education and conservation. He is the author and photographer of four books: the award-winning *Owl: A Year in the Lives of North American Owls*, *The Owl and the Woodpecker: Encounters with North America's Most Iconic Birds*, *Snowy Owl: A Visual Natural History*, and *Great Gray Owl: A Visual Natural History*. Paul's photography has won awards from several prestigious contests, including those hosted by Audubon magazine and the International Conservation Photography Awards. His work is prominently featured in bird guides from Audubon, Peterson, and the Smithsonian, among others, and has appeared in a variety of publications including *The New York Times*, *Audubon*, *Sunset*, and *Nature's Best Photography Magazine*. Additionally, his photography has been the subject of many regional and national television and radio pieces as well as several North American traveling exhibits.

An active public speaker, Paul presents multimedia owl and woodpecker programs at events throughout the United States and Canada every year. He serves as the director of major gifts for Conservation Northwest, a Seattle-based nonprofit dedicated to protecting, connecting, and restoring wildlands and wildlife from the coast of Washington to the Rockies of British Columbia.

More of his work can be seen at paulbannick.com, on Facebook at Paul Bannick Photography, and on Instagram @paulbannick.

A Note on the Photography

I strive to capture authentic moments of natural behavior in a manner that has the most minimal impact on the subject. This allows me to photograph nature as it is rather than as we wish it appeared. All of my images feature wild, unrestrained subjects. None of the images were taken with camera traps. I do not use Artificial Intelligence or any other tools to edit out distracting branches, birds, or other elements. I apply only global electronic modifications such as minimal adjustments to contrast, noise reduction, or saturation to match what I witnessed and do not create composite images.

The images in this book were captured with Canon and Sony Cameras with 600 mm f/4, 100–400 mm, and 24–70 mm lenses. Carbon tripods and gimbal heads were used for most images.

BRAIDED RIVER

BRAIDED RIVER, the conservation imprint of Mountaineers Books, combines photography and writing to bring a fresh perspective to key environmental issues facing western North America's wildest places. Our books reach beyond the printed page as we take these distinctive voices to a wider audience through lectures, exhibits, and multimedia events. Our goal is to build public support for wilderness preservation campaigns and inspire public action. This work is made possible through book sales and contributions made to Braided River, a 501(c)(3) nonprofit organization. Please visit BraidedRiver.org for more information on events, exhibits, speakers, and how to contribute to this work.

Braided River books may be purchased for corporate, educational, or other promotional sales. For special discounts and information, contact our sales department at 800.553.4453 or mbooks@mountaineersbooks.org.

THE MOUNTAINEERS, founded in 1906, is a nonprofit outdoor activity and conservation organization. Mountaineers Books, an independent nonprofit publisher, is a leader in outdoor education with publications that range from guidebooks to adventure stories, natural history to art and photography, and conservation to sustainable practice.

Mountaineers Books
1001 SW Klickitat Way, Suite 201
Seattle, WA 98134
800.553.4453
www.mountaineersbooks.org

Manufactured in China on FSC®-certified paper, using soy-based ink.

MIX
Paper | Supporting responsible forestry
FSC www.fsc.org
FSC® C008047

First edition, 2025

Braided River Executive Director: *Helen Cherullo*
Editor in Chief: *Kate Rogers*
Project Editor: *Janet Kimball*
Developmental Editor: *Linda Gunnarson*
Copyeditor: *Laura Whittemore*
Cover and Book Designer: *Cat Grishaver*
Cartographer: *Erin Greb*

Front cover: A Red-headed Woodpecker watches as its mate considers a potential nest cavity. Frontispiece: A Nuttall's Woodpecker demonstrates the species' trademark agility, allowing it to forage on small twigs.

Page 9: Two male Northern Red-shafted Flickers display, either in greeting or in a territorial encounter.

Page 10: The golden crown of a male American Three-toed Woodpecker could be mistaken for lichen on an old spruce snag.

Page 224: The parents of these Red-headed Sapsucker nestlings will often land near the nest as they encourage the nestlings to fledge.

Page 208: *Ivory-billed Woodpeckers* by John L. Ridgeway.

Page 209: *Imperial Woodpecker* by John James Audubon

Library of Congress Control Number: 2025935325

ISBN 978-1-68051-316-5

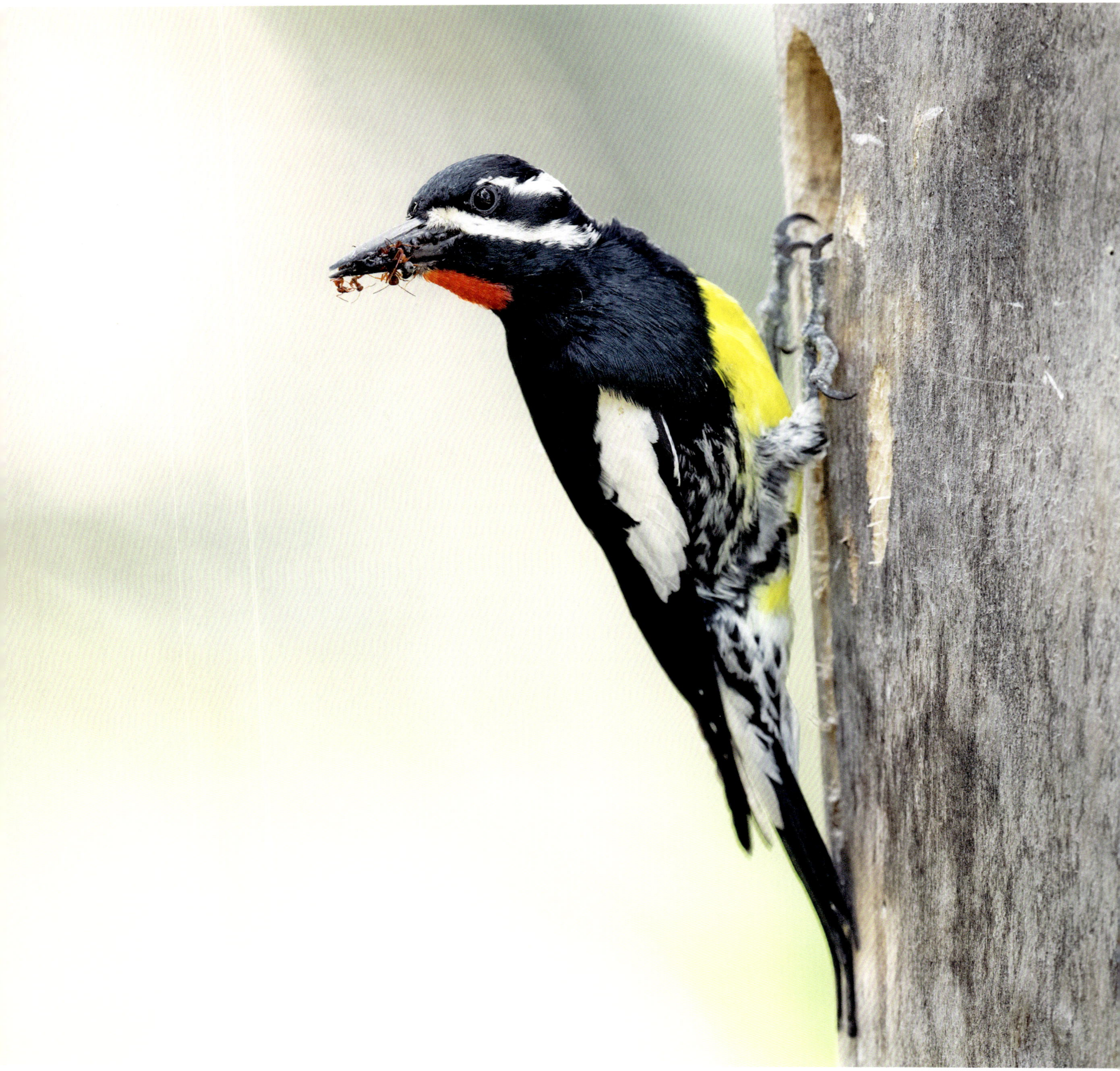

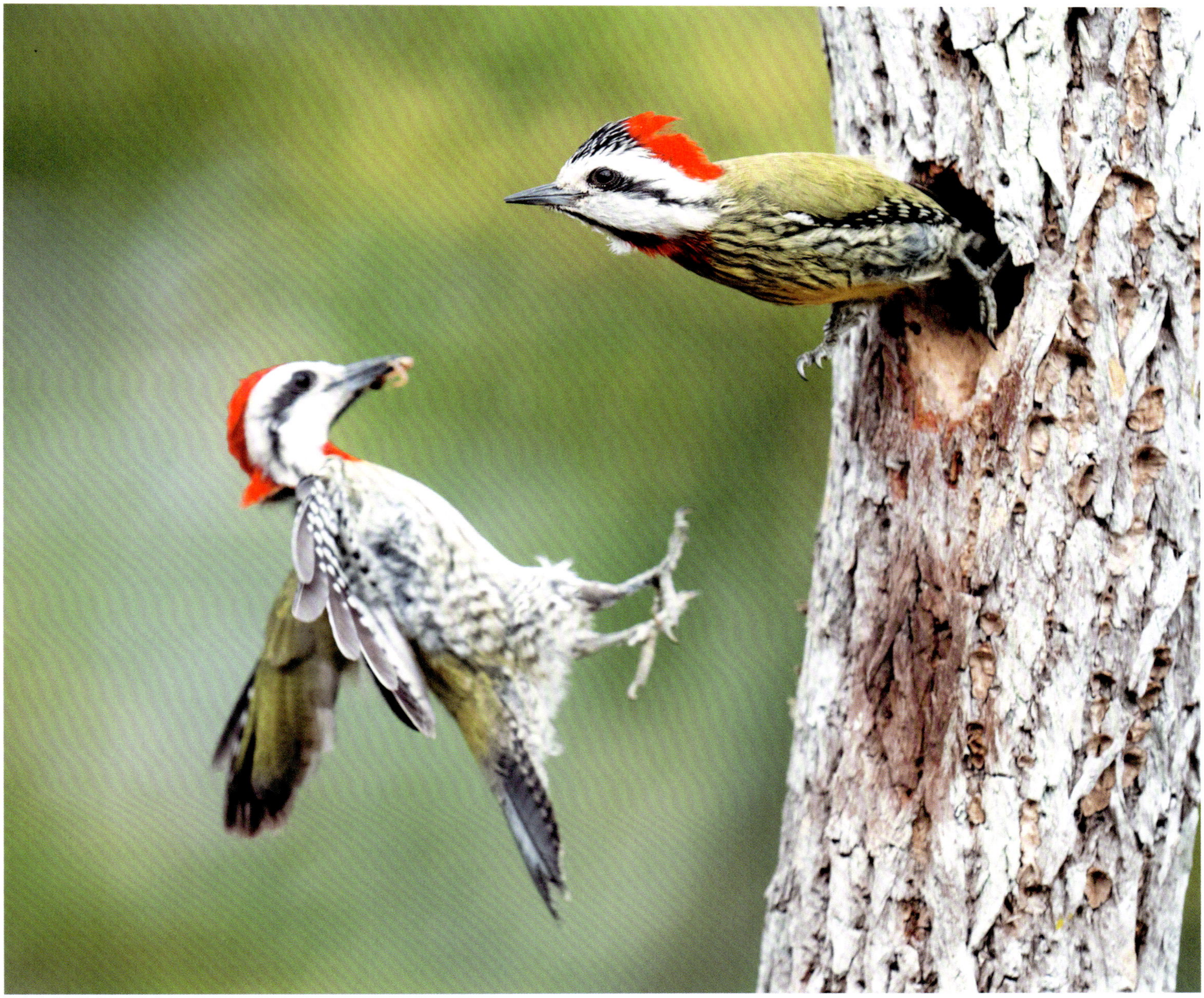

Opposite: Ants are among the most important food for woodpeckers. A male Williamson's Sapsucker arrives at his nest with an overflowing bill of them for his young.

Cuban Green Woodpeckers frequently bark an uncharacteristic series of harsh notes from a hidden location alerting their brooding mate of their intent to trade places in their nest. This, combined with their loud wing-beats as they approach, signal the mate to prepare to leave, minimizing the time the young or eggs are left unattended.

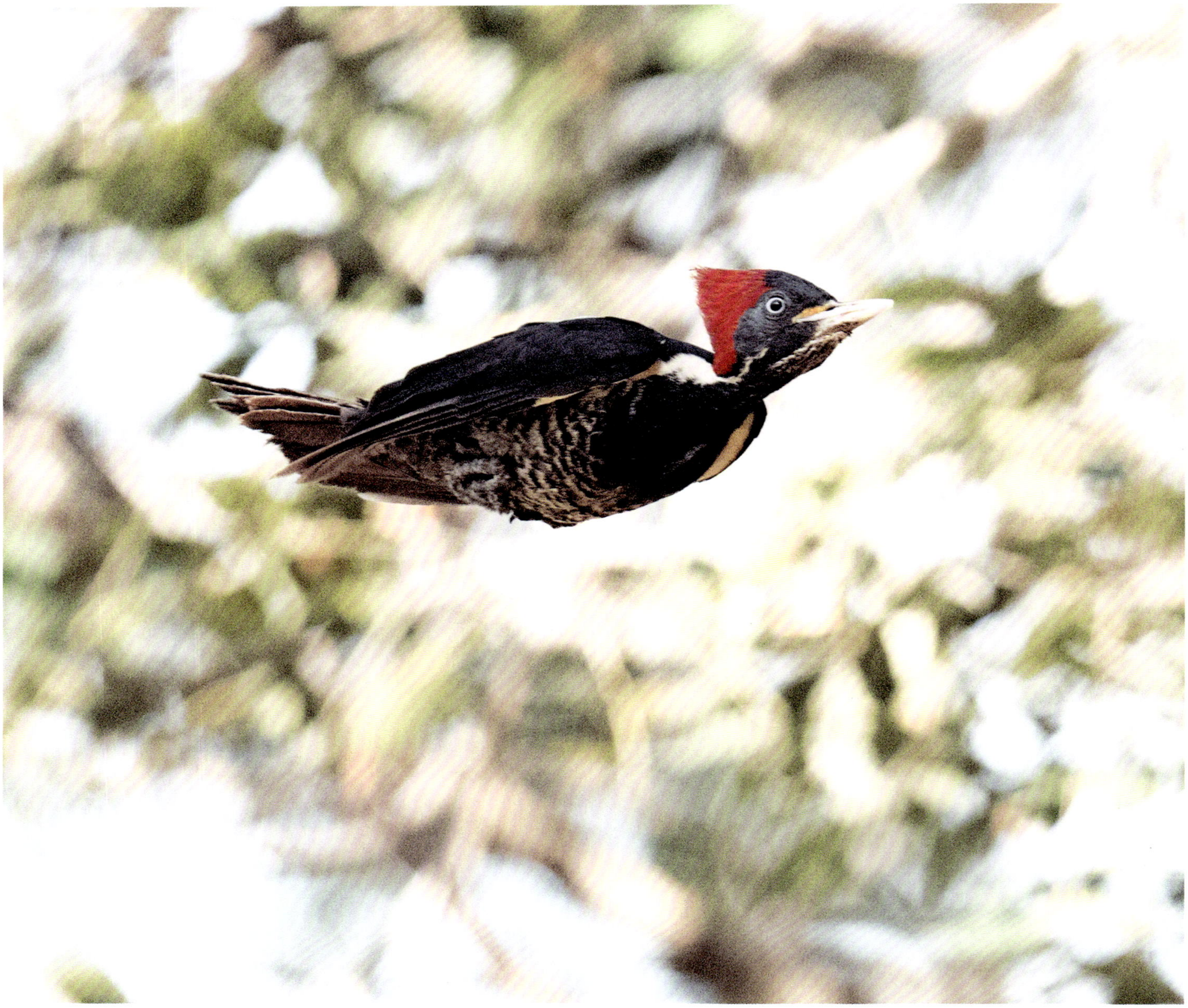

Above: A Linneated Woodpecker displays the flap-bounding flight style in which a woodpecker alternates between flapping for forward momentum and holding its wings against its body for speed, causing it to bounce through the air.

Right: Acorn Woodpeckers have a complex social structure and greet one another with wing spreading and calls, which in most other species relates to aggression or courtship.

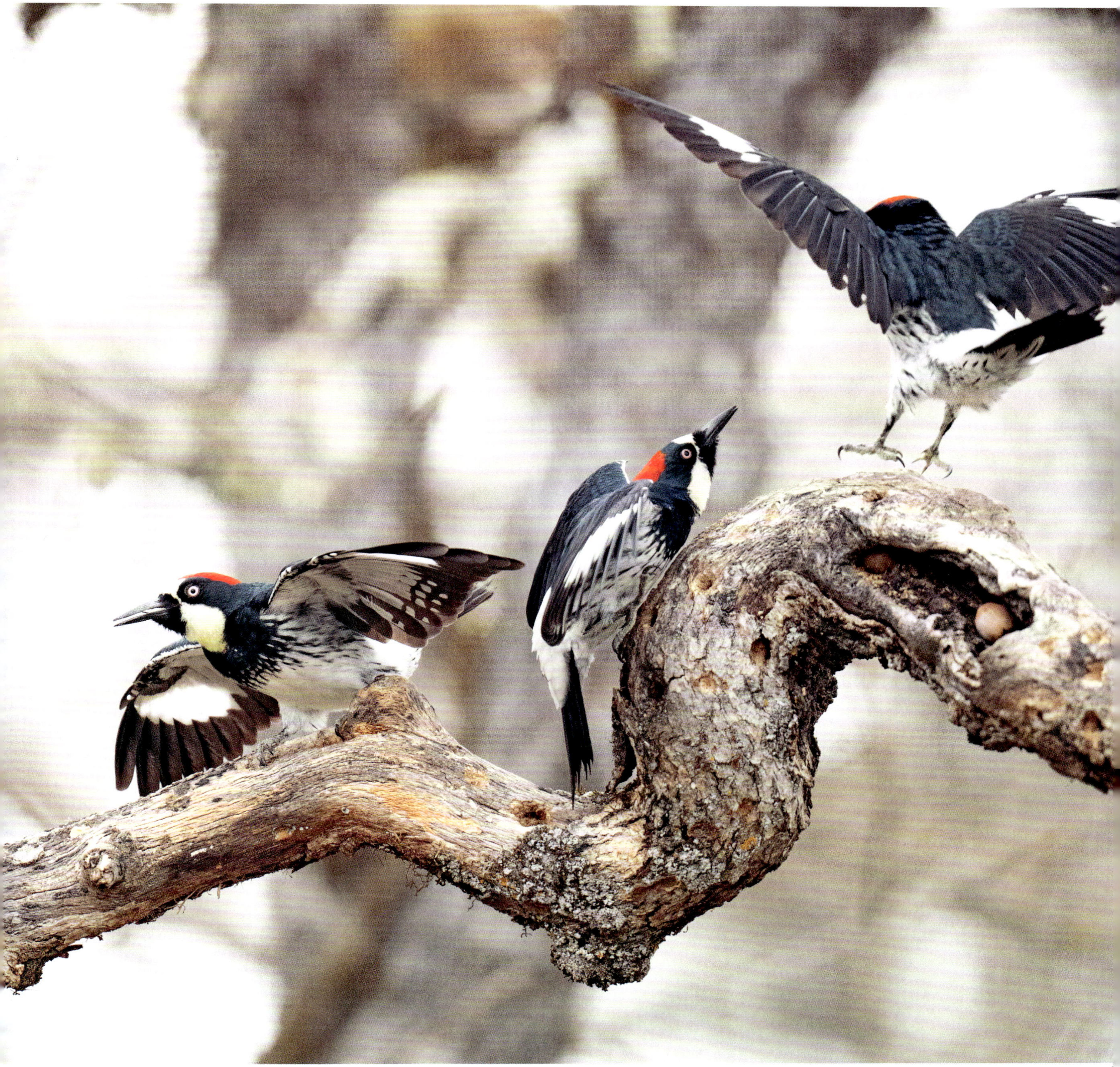

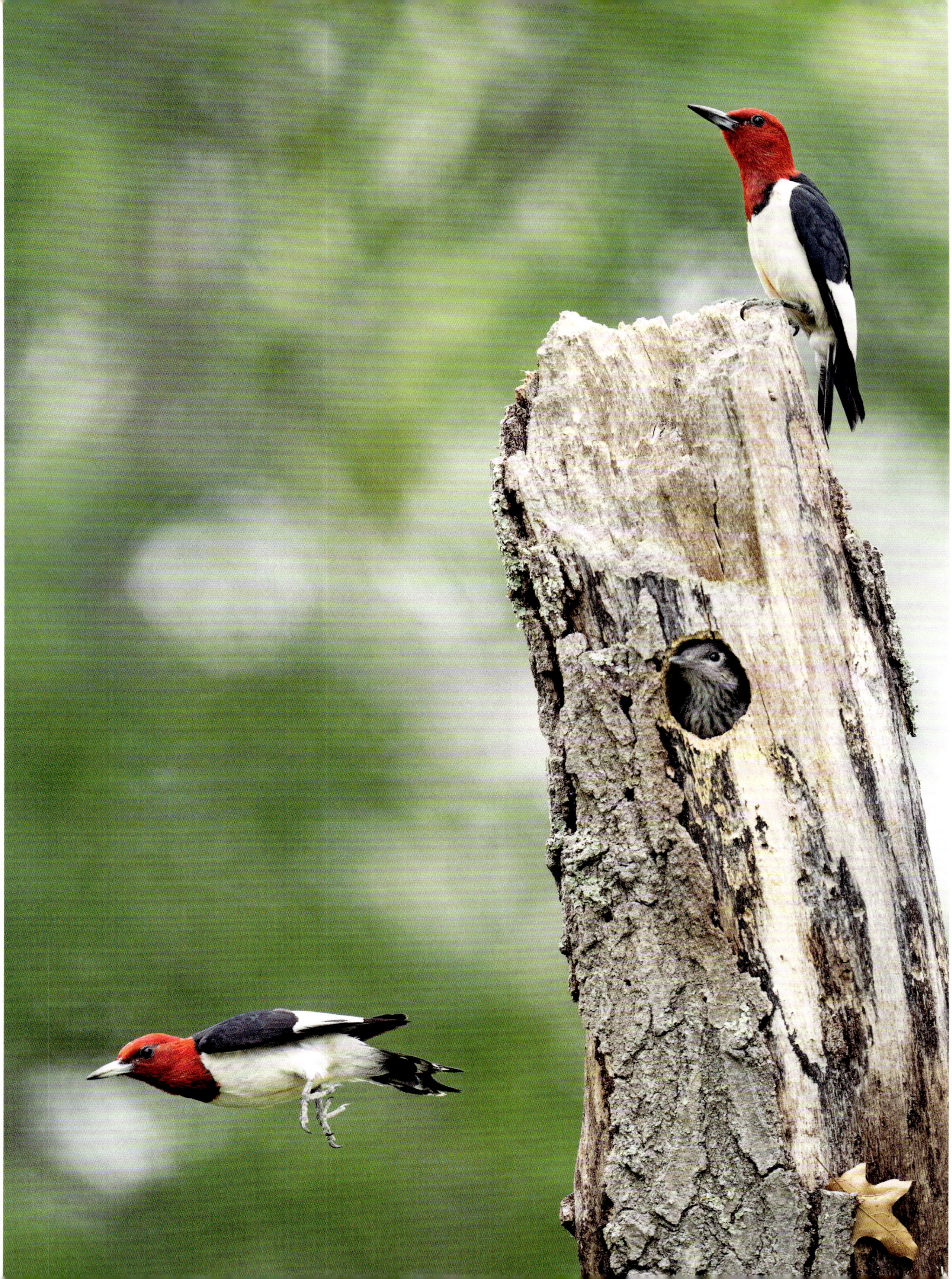